UPDATES IN ADVANCED LITHOGRAPHY

Edited by **Sumio Hosaka**

Updates in Advanced Lithography

http://dx.doi.org/10.5772/3348
Edited by Sumio Hosaka

Contributors

Hongbo Lan, Gregory Barbillon, Nima Arjmandi, Athanasios Milionis, Ilker Bayer, Despina Fragouli, Fernando Brandi, Athanassia Athanassiou, Gang Zhang, Khairudin Mohamed, Maan Alkaisi, Arnaud Spangenberg, Prem Prabhakaran, Fernand Wieder, Olivier Soppera, Jean-Pierre Malval, Patrice Baldeck, Nelly Hobeika, Fabrice Stehlin, Razvan Dabu, Michael Wang, Noritaka Kawasegi, Noboru Morita

Published by InTech

Janeza Trdine 9, 51000 Rijeka, Croatia

Copyright © 2013 InTech

Notice

Publishing Process Manager Sandra Bakic

Technical Editor InTech DTP team

Cover InTech Design team

Additional hard copies can be obtained from orders@intechopen.com

Updates in Advanced Lithography, Edited by Sumio Hosaka
p. cm.
ISBN 978-953-51-1175-7

For EU product safety concerns:
IN TECH d.o.o., Prolaz Marije Krucifikse Kozulić 3, 51000 Rijeka, Croatia,
info@intechopen.com or visit our website at intechopen.com.

Meet the editor

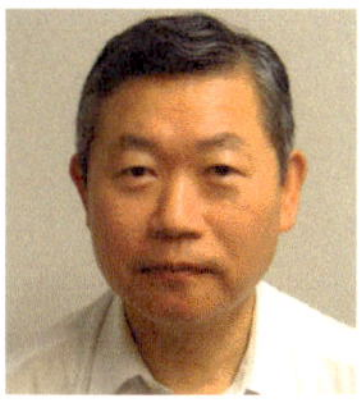

Sumio Hosaka was born in Japan in 1948. He received the Ph. D. degree in Engineering from Waseda University, Tokyo, in 1983. He joined the Hitachi Central Research Lab. (HCRL), Co. Ltd. Kokubunji, Tokyo in 1971, working in semiconductor lithography using electron, ion and light beams, nano-metrology using electron beam microscopy and scanning probe microscopy (SPM) techniques, in HCRL and terabyte huge storage using SPM in the Hitachi Advanced Research Lab. (HARL). After he joined the Hitachi Kenki Co., Tsuchiura, Ibaraki in 2000, working in-line atomic force microscope products based on his invention of first optical lever type atomic force microscope at 1987, he joined to department of electronic engineering, Gunma University, Kiryu, Gunma, in 2001. Today, he works in division of electronics and informatics, faculty of science and technology in Gunma University. Since 2001, he has been professor in the university, and is working on nano devices such as quantum conductance atomic switch (QCAS) and phase change memory, nano lithography and fabrication such as EB drawing, self-assembled nano dot arrays and patterned media in huge magnetic storage, and nano metrology based on atomic force controlling, near-field light, Raman spectroscopy and polarized light image in SPM applications.

Contents

Preface

Advanced lithography grows up to several fields such as nano-lithography, micro electro-mechanical system (MEMS) and nano-phonics, etc. Nano-lithography reaches to 20 nm size in advanced electron device. Consequently, we have to study and develop true single nano-meter size lithography. One of the solutions is to study a fusion of top down and bottom up technologies such as EB drawing and self-assembly with block copolymer. In MEMS and nano-photonics, 3 dimensional structures are needed to achieve some functions in the devices for the applications. Their formation are done by several methods such as colloid lithography, stereo-lithography, dry etching, sputtering, deposition, etc. This book covers a wide area regarding nano-lithography, nano structure and 3-dimensional structure, and introduces readers to the methods, methodology and its applications.

Prof. Sumio Hosaka
Graduate School of Engineering,
Gunma University,
Japan

Section 1

Lithography for 3D Structure and Nano Scale

Chapter 1

Colloidal Lithography

Ye Yu and Gang Zhang

Additional information is available at the end of the chapter

http://dx.doi.org/10.5772/56576

1. Introduction

The advent of nanoscience and nanotechnology has led to tremendous enthusiasm of researchers from different scientific disciplines such as physics, chemistry, and biology to engage with nanostructures with the intent of pursuing the innovative property derived from the nanometer dimension. In this context, fabrication of nanostructures accordingly becomes an increasing demand nowadays. Obviously, low-throughput and expensive maskless lithography is a less accessible choice for chemists, physicists, material scientists, and biologists. The success of extending mask-assisted lithography beyond microelectronics workshops is largely limited by the mask design and preparation. Recently a host of effort has been devoted to develop non-conventional lithographic techniques especially integrated with a bottom-up nanochemical procedure for surface patterning with low cost, flexible processing capability, and high throughput. However, most of the non-conventional lithographic techniques require an assistance of conventional lithographic techniques such as photolithography to design and make masks or masters. To develop ingenious, cheap, and non-lithographic ways to make masks or masters with high resolution (below 100 nm), a great deal of self-assembly nanostructures have been recruited for masking, including laterally structured Langmuir–Blodgett monolayers, liquid crystalline structures of surfactants, micro-phase separation structures of block copolymers, and self-assembly of proteins and nanoparticles.

Monodisperse colloidal particles with size ranging from tens of nanometers to tens of micrometers can be easily synthesized via wet chemistry ways such as emulsion polymerization and sol-gel synthesis. Due to the size and shape monodispersity, they can self-assemble into a two dimensional (2D) and three dimensional (3D) extended periodic array, usually referred to as colloidal crystal. Colloidal crystals are usually characterized by a brilliant iridescence arising from the Bragg reflection of light by their periodic structures. Despite the beauty, the iridescent color has recently inspired the explosive study of fabrication of 3D colloidal crystals or inverse opals – 3D inverted replication of the crystals – for pursuing a complete energy bandgap to manipulate electromagnetic waves, similar to that to do to electrons in semiconductors. Before being used as photonic materials, both the ordered arrays of solid particles and those of the interstices between the particles of colloidal crystals have already been used as masks or templates for surface patterning via for instance etching or deposition of materials. This bottom-up masking methodology has recently gained increasing attention for surface patterning due to the processing simplicity, the low cost, the flexibility of extending on various substrates with different surface chemistry and even curvatures, the ease of scaling down the feature size below 100 nm. In the present chapter, we refer to as various surface patterning processes based on use of colloidal crystals as masks as a whole as colloidal lithography (CL), overview the processing principles, and survey the recent advances.

2. Colloidal masks

The success of using colloidal crystals as masks for surface patterning is determined by the capability of directing self-assembly of colloidal particles and manipulating the crystal packing structures. Provided their size and shape are monodisperse, colloidal particles can be readily to self-assemble into long-range ordered arrays with a hexagonal packing, driven simply by entropic depletion and gravity. Up to date a variety of colloidal crystallization techniques – with and without the aid of templates – have successfully been developed to implement colloidal crystallization in a controlled fashion [1-3]. Due to enormous numbers of publications on colloidal crystallization and immense diversity of crystallization techniques reported thus far and especially by taking into account that colloidal lithography relies on masking of single layers or double layers of colloidal crystals, this section is centered mainly on techniques for 2D colloidal crystallization developed thus far.

2.1. Simple colloidal masks

2.1.1. Sedimentation

Sedimentation is a natural way for colloidal crystallization. In dispersion colloidal particles tend to settle out of the fluid under gravity and to accumulate and precipitate on a wall, which can be described by Stokes' law. This sedimentation process can be used to grow colloidal crystals with high quality, and the crystal thickness can be tuned by the particle concentration. However, the sedimentation time is always up to several hundreds of hours; time-consuming is a big drawback of this technique [4-6].

At the beginning of 1990's Nagayama's group has commenced a systematic study of sedimentation of colloidal particles in the presence of strongly attractive capillary forces [7]. With the help of optical microscopy and using a Teflon ring to confine the dispersions of colloidal particles, they have directly observed the particle sedimentation dynamics on a solid substrate. Their observations suggest a two-stage mechanism for 2D colloidal crystallization: 1) nucleation and 2) crystal growth (Fig. 1) [7]. Micheletto's group has fabricated 2D colloidal crystals

http://dx.doi.org/10.5772/56576

on a solid substrate via sedimentation by tilting the substrate about 9° and keeping the system temperature constantly using a Peltier cell [8].

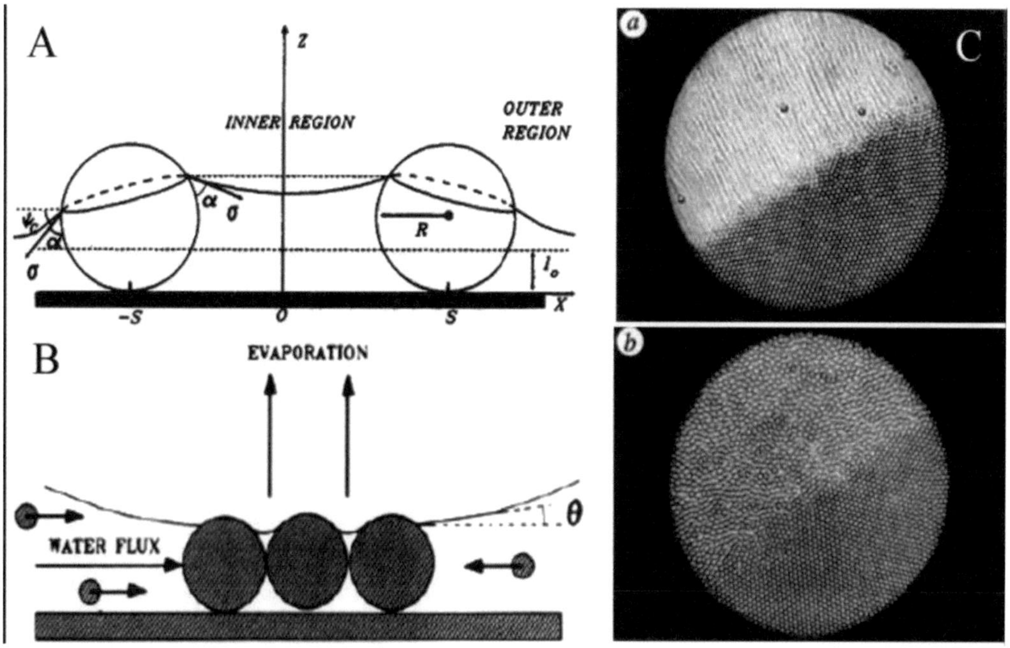

Figure 1. A) Two spheres partially immersed in a liquid layer on a horizontal solid substrate. The deformation of the liquid meniscus gives rise to interparticle attraction. (B) Convective flux toward the ordered phase due to the water evaporation from the menisci between the particles in the 2D array. (C) Photographs of 2D-crystal growth. Reprinted with permission [7].

2.1.2. *Vertical deposition*

When a supporting substrate is held vertically in a suspension of colloidal particles, moving the front of the suspension flow either by the solvent evaporation or by withdrawing the substrate out of the suspension can pin colloidal particles on the substrates – nucleation – and the convective transfer of the particles from the bulk phase to the drying front – crystallization (Fig. 2) [9]. The thickness of colloidal crystals obtained via vertical deposition is dependent on the ratio of the thickness of the liquid films remaining of supporting substrates to the diameter of the colloidal particles [9]. When the ratio is far larger than 1, 3D colloidal crystals are obtained with high quality; the crystal thickness can be tuned by the particle concentration [10]. When the ratio is comparable to or smaller than 1, 2D colloidal crystals can be obtained [9]. Vertical deposition may allow formation of large-area crack-free colloidal crystals provided the suspensions of colloidal particles wet well supporting substrates, there is no interaction between the particles and the substrates, the suspensions are sufficiently stable and the solvent evaporation is well controlled [9].

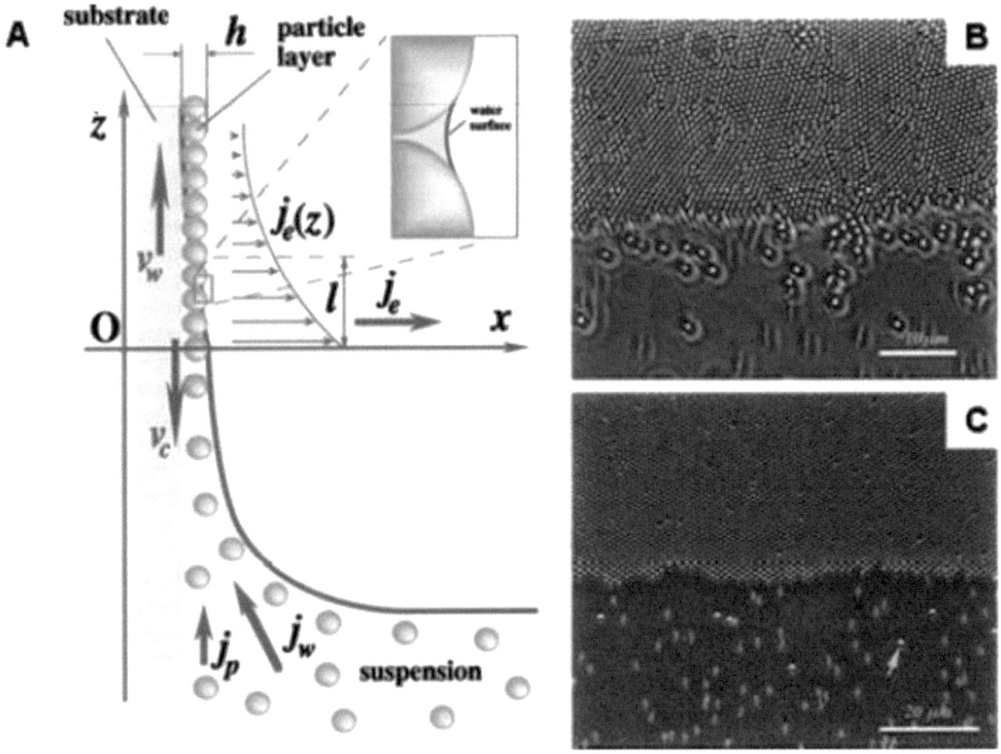

Figure 2. A) Sketch of the particle and water fluxes in the vicinity of monolayer particle arrays growing on a substrate plate that is being withdrawn from a suspension. The inset shows the menisci shape between neighbouring particles. (B and C) A part of the leading edge of a growing monolayer particle array. The upper-half of the photographs shows the formations of (B) differently oriented small domains of ordered 814-nm particles and (C) a single domain of ordered 953-nm particles. The lower-half shows particles dragged by the water flow toward the forming monolayer. Reprinted with permission [9].

Dip-coating is a fast and dip-coater assisted variant of vertical deposition [11]. Besides, a number of techniques have been developed to improve the efficiency and quality of colloidal crystallization via vertical deposition, such as variable-flow deposition [12], isothermal heating evaporation-induced self-assembly [13], two-substrate deposition [14], reduction of the humidity fluctuation [15], adjustment of the meniscus shape [16], temperature-induced convective flow [17] and vertical deposition with a tilted angle [18]. The maximal size of colloidal particles used for vertical deposition is limited by the particles sedimentation of colloidal particles, for instance 400-500 nm for silica particles and 1 μm for polystyrene particles. To compete with sedimentation, Kitaev and Ozin have used low pressure to accelerate the solvent evaporation, and successful growth of large-area 2D binary colloidal crystals with the diameter ratios of the large particles to the small ones in the range of 0.175 to 0.225 (Fig. 3) [19].

Vertical deposition has recently been extended to stepwise growth of 2D colloidal crystals with large and small colloidal particles on a substrate [20, 21]. In their procedure, the 2D colloidal

http://dx.doi.org/10.5772/56576

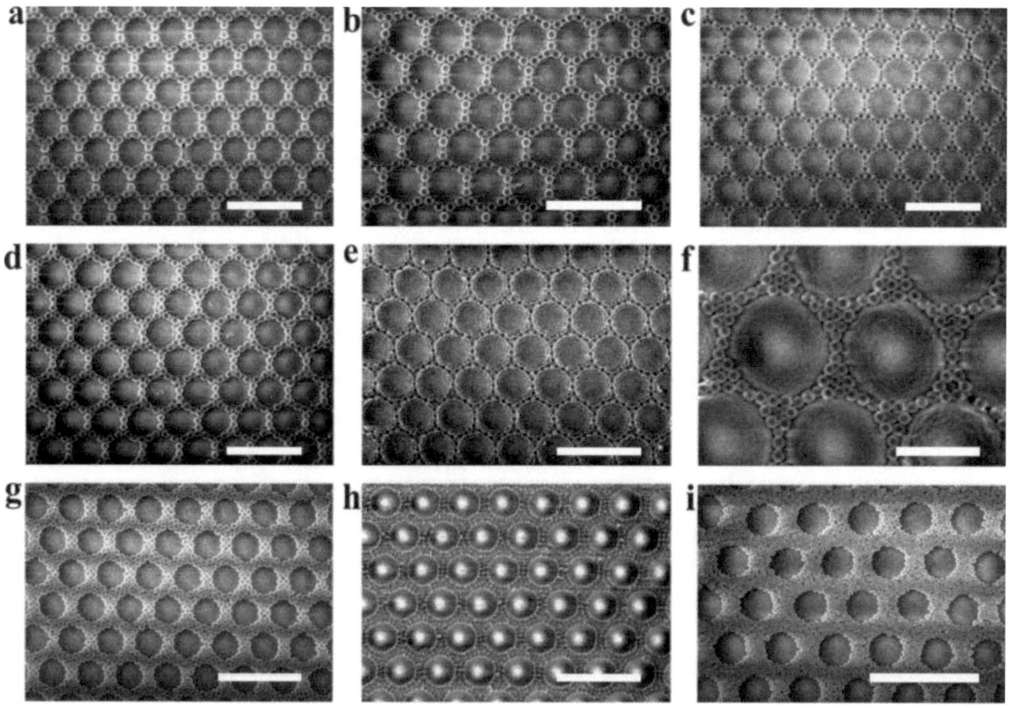

Figure 3. Library of surface micropatterns produced by accelerated evaporation co-assembly of binary dispersions of monodisperse microspheres with large size ratio and imaged with field emission scanning electron microscopy. Larger spheres of all binary dispersions were PS latex of size, d_L= 1.28 μm, while varying their volume fraction (φ_L), the volume fraction (φ_S), and size (d_S) of smaller spheres. Scale bars in (a-e) and (g-i) are 3 μm and that in (f) is 1 μm. Reprinted with permission [19].

crystal of the larger particles firstly formed on the substrate is used to template the growth of the 2D colloidal crystal of the smaller particles. By deliberately tuning the concentration of the small particle suspension, binary colloidal crystals with the stoichiometric ratios of large to small particle sizes of 1:2, 1:3, 1:4, or 1:5 have been constructed [20, 21].

2.1.3. Spin coating

Spin coating was the first technique for growth of 2D colloidal crystal masks for colloidal lithography due to the fact that it allows easy and quick formation of 2D crystals over large area [22]. The long range ordering degree of 2D colloidal crystals obtained via spin coating can be improved by increasing the wettability of the suspensions of colloidal particles on supporting substrates by for instance adding ethylene glycol into the suspension [23]. However, the spin coating is a process far more complicated than it appears and the underlying mechanism remains in debate. Rehg and Higgins have conducted a theoretical analysis of the physics governing spin coating of a colloidal particle suspension on a planar substrate [24]. Jiang and Mcfarland have succeeded in fabrication of wafer scale long-rang ordered and non-close-packed 2D and 3D colloidal crystals by spin coating of highly viscous triacrylate

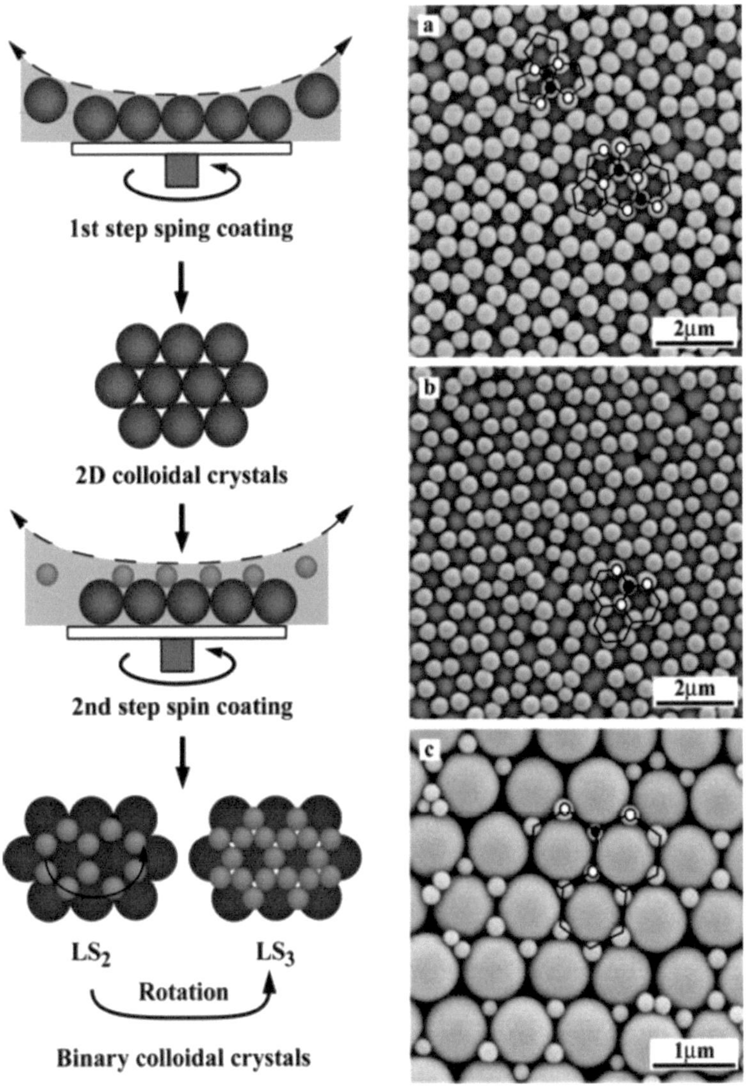

Figure 4. Left panel: Illustration of the procedure used to fabricate binary colloidal crystals by stepwise spin coating. Right panel: SEM micrographs of the binary colloidal crystals produced by stepwise spin coating at a spin speed of 3000 rpm, in which 519 nm (a), 442 nm (b), and 222 nm silica spheres (c) were confined within the interstices between hexagonal close packed 891 nm silica spheres. Reprinted with permission [27].

suspension of silica particles and subsequent polymerization of triacrylate, followed by partial removal of the polymer matrices [25, 26]. Wang and Möhwald have developed a stepwise spin coating protocol to consecutively deposit large and small colloidal particles into binary colloidal crystals, in which the interstitial arrays in the 2D colloidal crystal of the large particles are used to template the deposition of the small particles due to the spatial and depletion entrapment (Fig. 4) [27].

2.1.4. Colloidal crystallization at interface

Using the water/air interface as a platform for molecular self-assembly has been extensively studied. Langmuir-Blodgett (LB) technique has been proved as a powerful and versatile way to organize amphiphilic molecules (referring molecules that are hydrophobic on one end and hydrophilic on the other end) to macroscopic monolayer films at the water/air interface and transfer the films to solid substrates in a controlled manner [28]. It is also well studied but less recognized that in a biphasic system, e.g. water/oil, colloidal particles behave rather similar to amphiphilic molecules; they thermo- dynamically prefer to attach to the interface [29]. Due to this analogy, the water/air interface has been extended to support self-assembly of colloidal particles. Pieranski has conducted the first deliberate microscopic observation of 2D colloidal crystallization at the water/air interface and hypothesized the repulsive interaction between the dipoles of colloidal particles trapped at the interface due to the asymmetric charge distribution on the particle surface drives the particles to self-assemble into an ordered array (Fig. 5) [30]. Park *et al.* have developed heat-assisted interfacial colloidal crystallization, while the success of their technique relies on the convective flow generated during heating rather than the interface activity of colloidal particles [31]. Once 2D colloidal crystals are formed at the water/air interface, the LB technique has been used to transfer of them on different substrates [32-35]. Of significance is that the LB technique allows repetition of transfer of 2D colloidal crystals on a substrate into 3D colloidal crystals with precisely defined layer numbers [35].

Colloidal monolayers with high order and increased complexity beyond plain hexagonal packing geometries are useful for 2D templating of surface nanostructures and lithographic applications. Weiss and co-workers developed binary colloidal monolayers featuring a close-packed monolayer of large spheres with a superlattice of small particles in a single step using a Langmuir trough [36].

As compared with the water/air interface, the water/oil interface is a much better platform to trap colloidal particles due to the relatively low interfacial tension [29]. Thus, water/oil interfaces have been used for growth of 2D colloidal crystals [37, 38], while transfer of the resulting 2D colloidal crystals to solid substrates remains problematic. Besides water/air interfaces, air/water/air interfaces have also been utilized for colloidal crystallization. Velikov and coworkers have studied of colloidal crystallization in thinning foam films [39]. Using air/water/air interfaces for crystallization, Wang and co-workers have successfully obtained free-standing and crack-free colloidal crystal films with sizes over several square millimeters [40]. Instead of water/air interface, Zental and co-workers have used the interface between melted germanium and air for colloidal crystallization and obtained crack-free colloidal crystals [41].

2.2. Complex colloidal masks

2.2.1. Deformed colloidal masks

In general, polymers undergo a second-order phase transition from hard glassy state to soft rubbery state above a glass transition temperature (Tg) due to the free-volume change between the polymer chains. Therefore, annealing slightly above Tg can cause deformation of spherical

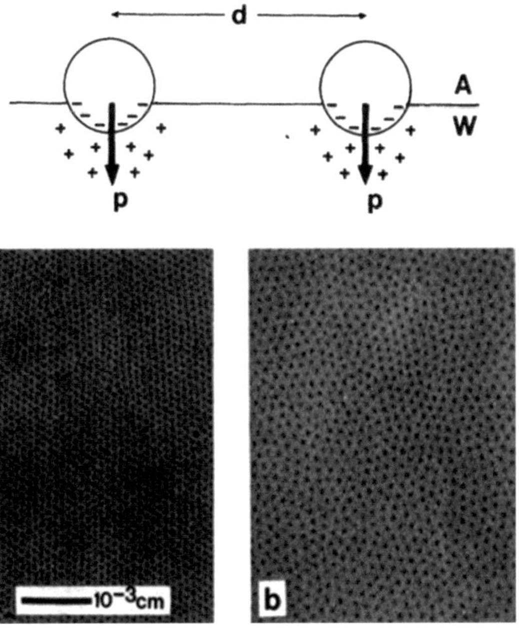

Figure 5. Upper panel: Schematic of the model of interaction of colloidal particles at the water (W)/air (A) interface. Lower panel: Photographs of polystyrene spheres (black dots) trapped at water/air interface. (a) Crystalline structure; (b) disordered structure. Reprinted with permission [30].

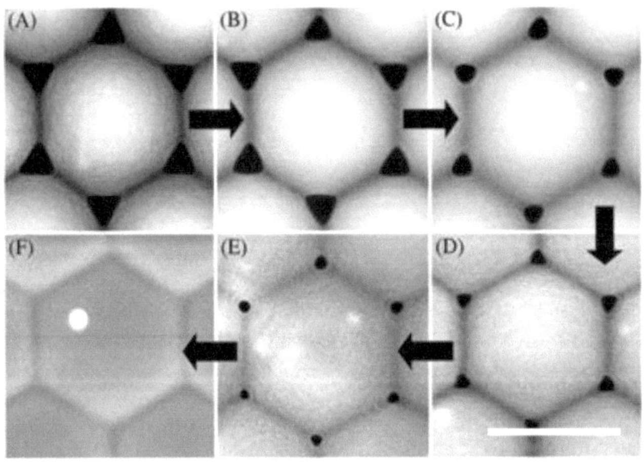

Figure 6. Precise control of the degree of annealing is achieved via adjustment of the number of microwave exposures: A 540 nm PS latex mask annealed in 25 mL of water/EtOH/acetone mixture by A) 1, B) 2, C) 4, D) 6, E) 7, and F) 10 microwave pulses. Scale bar: 500 nm. Reprinted with permission [43].

polymeric beads. It is demonstrated that microwave radiation can much more precisely control the deformation of spherical polymer particles by adjusting the microwave intensity than heating in oven [42].

Giersig and coworkers have recently developed a new annealing approach – using microwave pulse to heat polystyrene (PS) microspheres in a mixture of good and poor solvents for PS, which allows not only reduction of the sizes of the interstices of 2D PS colloidal crystals but also deformation of their geometry from triangular to rodlike, while preserving the interparticle spacing and packing order of the original crystals (Fig. 6) [43]. Recently, Yang *et al.* have demonstrated a photolithographic process to produce hierarchical arrays of nanopores or nanobowls with using colloidal crystals of photoresist particles [44]. In the case of inorganic particles, deformation is hard to achieve by thermal annealing. Polman and coworkers have successfully deformed silica@Au core-shell microspheres to oblate ellipsoids by using high energy ion irradiation due to the fact that the ion-induced deformation of the silica core is counteracted by the mechanical constraint of the gold shell [45]. Vossen and coworkers have recently reported that silica particles undergo anisotropic deformation under ion bombardment due to expansion in the plane perpendicular to the ion beam [46].

2.2.2. *Colloidal masks derived from modified colloidal particles*

Many specific colloidal masks have been made using methods mentioned above, usually utilizing one or two kinds of spherical colloidal particles as building blocks. Colloidal particles with anisotropic interactions are expected to enable a wide range of materials with novel optical and mechanical properties. While the self-assembly of spherical particles into periodic structures is relatively robust and well-characterized, the phase space describing the self-assembly of anisotropic particles is vast and has been only partially explored. It includes phases that are impossible for spherical particles to form, including gyroids, simple cubic lattices, and plastic crystals.

Eric R. Dufresne and co-workers demonstrated the use of an external electric field to align and assemble the dumbbells to make a birefringent suspension with structural color. In this way, dumbbells combine the structural color of photonic crystals with the field addressability of liquid crystals. In addition, if the solvent is removed in the presence of an electric field, the particles self-assemble into a novel, dense crystalline packing hundreds of particles thick, which was shown in Fig. 7 [47].

3. Colloidal lithography

3.1. Controllable etching

When a 2D colloidal crystal is formed on a solid substrate, the interstices between the solid particles can used as masks for reactive ions to create patterned bumps or pores on the substrate. In the beginning of 1980's Deckmann and Dunsmuir have pioneered the work of etching of a colloidal crystal into a textured surface using a reactive ion beam (RIE) [48]. Since then, reactive ion etching (RIE) has been widely used to interdependently reduce the particle sizes and thus widen the interstitial space in 2D colloidal crystal masks and eventually turn close-packing structures of the crystals to non-close packing one (vide infra). RIE in 3D colloidal crystals is an anisotropic process as the upper layers act as shadow masks for etching

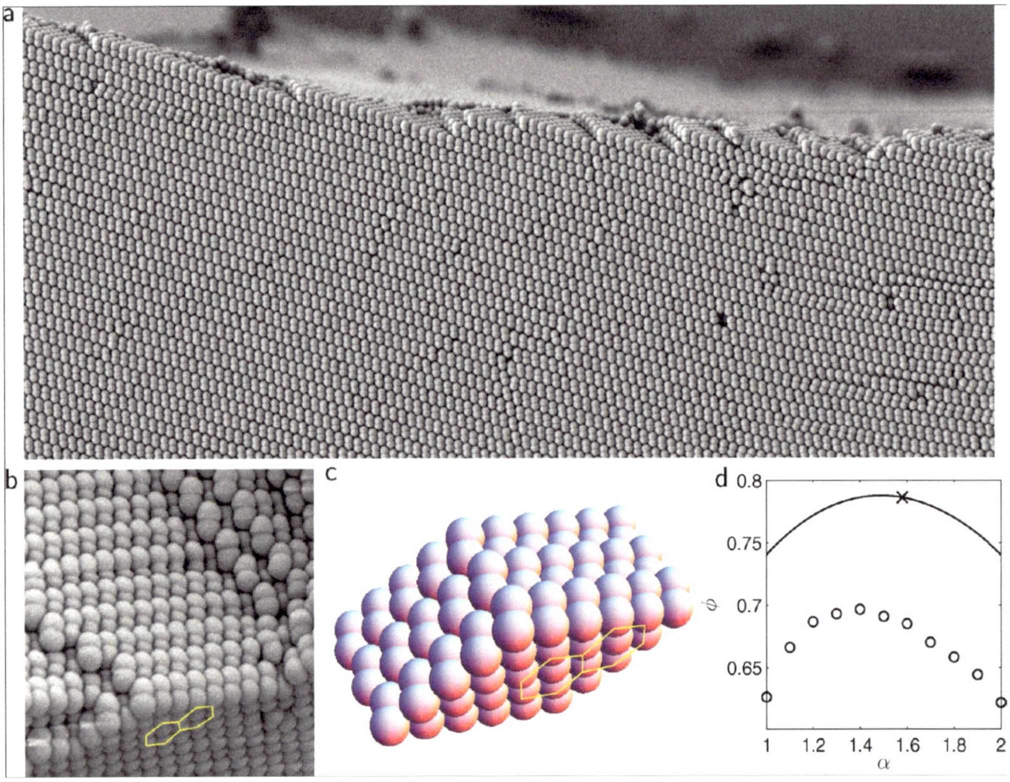

Figure 7. Crystal structure of suspension dried in an AC electric field. (a) SEM image of crystal formed by drying a suspension of dumbbells in the presence of an electric field. The field of view is 27 µm across. (b) SEM image highlighting crystal structure. Two adjoining hexagons formed by the dumbbell lobes are highlighted by the yellow hexagons. The field of view is 3.6 µm across. (c) Model of the crystal structure suggested by SEM images. Two adjoining hexagons are highlighted and correspond to the highlighted facet in (b). (d) Packing fraction versus aspect ratio for crystalline structures (line) and random, jammed packings (circles) generated from numerical simulations described in the Methods section. Reprinted with permission [47].

the lower layer particles. This anisotropic RIE can turn spherical particles to non-spherical particles, and the particle shapes and the hierarchical nanostructures obtained so strongly depends on the stacking sequence of the colloidal crystals, the crystal orientation relative to the substrate, the number of colloidal layers, and the RIE conditions (Fig. 8) [49]. Of most significance is that the anisotropic RIE paves a new way to machine the surfaces of colloidal particles. Such as nanopores arranged in threefold or fourfold symmetry, depending on the crystalline orientation of the original colloidal crystals, were machined on PS particles [50-52].

Forests of silicon pillars with diameters of sub-500 nm and an aspect ratio of up to 10 were have been fabricated by firstly conduct O_2 RIE to turn close-packed PS particle monolayers to non-close packed on and subsequently conduct a "Bosch" process to etch the supporting silicon wafers [53]. Sow *et al.* have demonstrated the characteristic features of a RIE silicon substrate using a PS colloidal crystal mask and produced a double dome structure by simultaneous etching of the mask and the regions beneath the particles [54]. As compared with convention-

http://dx.doi.org/10.5772/56576

ally used polymer masks such as photoresists removed by organic developers, colloidal masks can be removed easily by ultrasonication and thus cause less damage to nanostructured substrates obtained via RIE. Ordered arrays of polyacrylic acid domes have been fabricated by using 2D PS colloidal crystals as masks for O_2 RIE of the polymeric films; the removal of the PS masks has no damage to the surface chemistry and the structure of the resulting polymeric domes, thus enabling conjugation of proteins [55]. 2D PS colloidal crystals have been also used as masks for dry etching of SiO_2 slides to create periodic arrays of nanoplates, which can be transferred onto polymer films by imprinting [56]. Using colloidal crystals as masks for catalytic etching, Zhu *et al.* have fabricated large-scale periodic arrays of silicon nanowires and the diameters and heights of the nanowires and the center-to-center distances between the nanowires can be accurately controlled [57]. Using colloidal crystals as masks to create arrays of nanopores on supporting solid substrates via RIE, followed by consecutively deposition of gold films and removal of the colloidal masks, Ong *et al.* have fabricated 2D ordered arrays of gold nanoparticles nested in the nanopores of the templated substrate [58]. One potential extension of having gold nanoparticles confined in nanopores is to use them as catalysts for growth of nanowires of other materials such as ZnO nanowires.

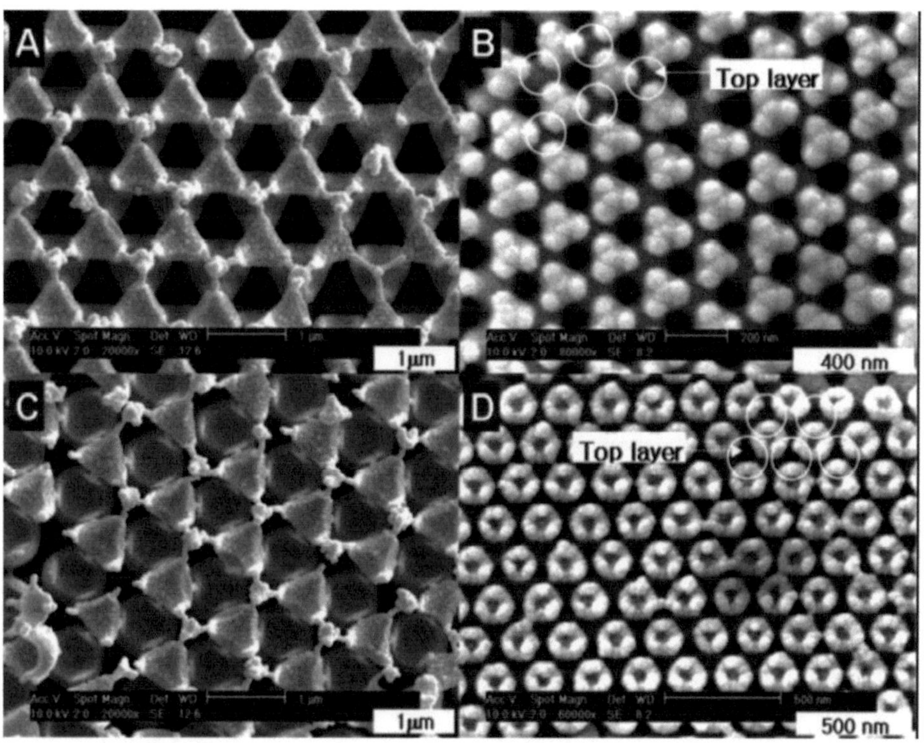

Figure 8. Modification of a mask using RIE for the fabrication of binary and ternary particle arrays with nonspherical building blocks. (a) and (b) Triangle arrays using binary and ternary colloidal spheres with an hcp arrangement. (c) and (d) Polygonal structures produced from colloidal layers with the (111) plane and the (100) plane of the *fcc* structure, respectively. Reprinted with permission [49].

3.2. Controllable deposition

3.2.1. Colloidal masks-assisted chemical deposition

Combining microcontact printing with colloidal crystal masking, Xia *et al.* have developed a simple method – edge spreading lithography (ESL) – to generate mesoscopic structures on substrates [59]. As the name suggests, ESL utilizes the edges of masks – the perimeters of the footprint of particles on substrates – to define the features of resultant structures. The ESL procedure begins with formation of 2D colloidal crystals of silica beads on the surfaces of gold or silver thin films [59]. As shown in Fig. 9(left panel), typically, a planar polydimethylsiloxane (PDMS) stamp bearing a thin film of the ethanol solution of an alkanethiol was placed on a 2D silica colloidal crystal.

The thiol molecules were released from the stamp to the silica particle during contact and subsequently transferred to the substrate along the surfaces of silica particles, leading to a self-assembled monolayer (SAM) circling the footprint of each silica particle. The area of the thiol SAM could expand laterally via reactive spreading as long as the thiols were continuously supplies. Upon removal of the stamp and lift-off of the beads, the ring pattern was developed by wet etching with aqueous Fe^{3+}/thiourea using the patterned SAM as a resist [59]. Of importance is that ESL allows generation of the concentric rings of different alkanethiol SAMs by successive printing different thiol inks, and the removal of silica particle templates and selective etching yield concentric gold rings and the width of the rings were determined by the printing period (Fig. 9, right panel) [60].

Shin *et al.* have developed another way to integrate colloidal masking and contact printing, referred to contact area lithography (CAL), to directly generate periodic surface chemical patterns at the sub-100 nm scale [61, 62]. Different from ESL, CAL relies on self-assembly of octadecyltrichlorosilane (OTS). After a 2D colloidal crystal of silica was formed on silicon wafer, the SAM of OTS was homogeneously grown both on silica particles and the supporting silicon wafer via a sol-gel process. The removal of silica particles left behind a periodically arranged array of the openings in the OTS SAM with the same symmetry as that of 2D colloidal crystals. The openings were subsequently used as masks for growth of the ordered arrays of nanoparticles of such as titania or for selectively etching of the ordered arrays of silica cavities on the silicon wafer. In the case of growth of titania, nucleation is rather site-selective due to the significant difference in the surface energy between the growing and surrounding surfaces [63].

3.2.2. Colloidal masks-assisted physical deposition

In 1981 Fischer and Zingsheim have used 2D colloidal crystals as masks for contact imaging with visible light [22]. A year later Deckman and Dunsmuir have demonstrated the feasibility of using 2D colloidal crystals as masks for both physical deposition of materials and in turn patterning the surfaces of supporting substrates [63]. They have coined the term "Natural Lithography" to describe this process as "naturally" assembled single layers of latex particles were used as masks rather than lithographic masks. Later on they have expanded the capability of "Natural Lithography" and especially developed the RIE

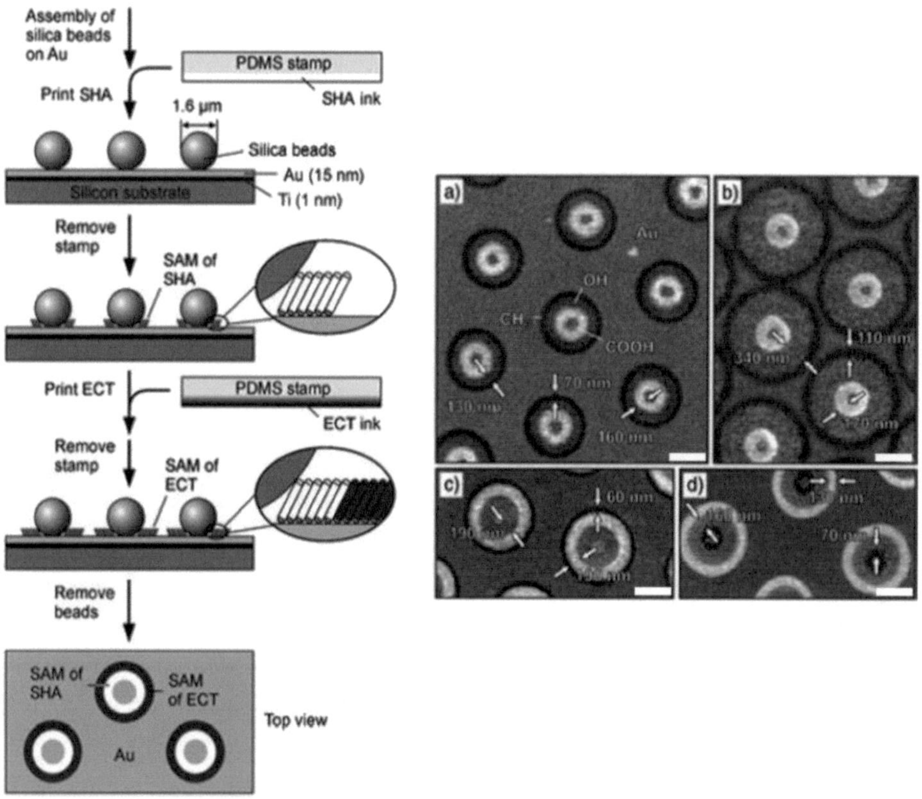

Figure 9. Left panel: Schematic illustration of the two-stepped ESL procedure used for side-by-side patterning of SHA and ECT monolayer rings on a gold substrate. Right panel. LFM images of concentric rings of carboxy- (bright), hydroxy-(gray), and methyl-terminated (dark) thiolate monolayers on gold. a) The rings were fabricated under the following conditions: 1 min for SHA, 1.5 min for HDDT, and 3 min for ECT. b) An increase in the printing times for HDDT and ECT to 3.5 and 4 min, respectively, resulted in wider rings for these two monolayers. c, d) The position of each monolayer in the concentric structure could be varied by changing the printing order. The pattern in (c) was generated by printing HDDT for 1.5 min, followed by printing of SHA and ECT for 3 min each. The sample shown in (d) was prepared by printing both ECT and HDDT for 1 min, and SHA for 2 min. All scale bars correspond to 500 nm. Reprinted with permission [60].

process for increasing the structural complexity of 2D colloidal crystal masks [48]. Since then the group of Van Duyne has devoted numerous efforts to develop patterning techniques using colloidal crystals as masks for metallic vapor deposition [23, 64-66]. In the context of nanoscience, they changed the name "Natural Lithography" to "Nanosphere Lithography" (NSL). Most important is that they have intensively investigated the plasmon resonance properties of metallic patterns obtained via NSL and their correlation with the feature morphology with the intent of developing high sensitive biosensors based on surface enhanced Raman spectroscopy (SERS) [67].

In a NSL procedure, a 2D colloidal crystal is used as masks for material deposition. The materials for physical deposition can be freely chosen without any limitations; commonly used are various metals such as gold and silver. The projection of the interstices between ordered

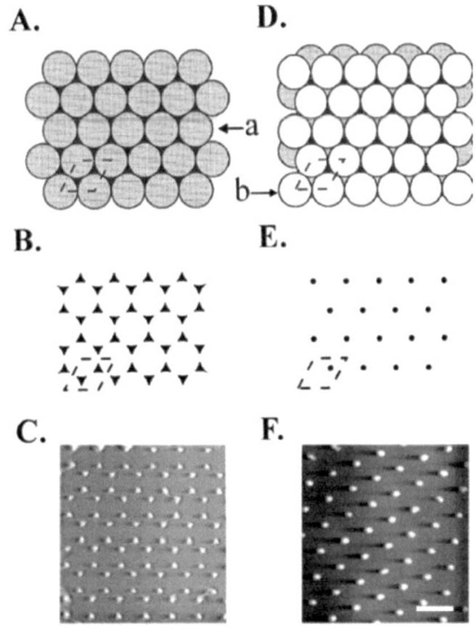

Figure 10. Schematic diagrams of single-layer (SL) and double-layer (DL) nanosphere masks and the corresponding periodic particle array (PPA) surfaces. (A) *a*(111) SL mask, dotted line represents the unit cell, *a* refers to the first layer nanosphere; (B) SL PPA, 2 particles per unit cell; (C) 1.7× 1.7 μm constant height AFM image of a SL PPA with M = Ag, S = mica, D = 264 nm, d_m = 22 nm, r_d = 0.2 nm s^{-1}. (D) *a*(111)*p*(1× 1)-*b* DL mask, dotted line represents the unit cell, *b*refers to the second layer nanosphere; (E) DL PPA, 1 particle per unit cell; (F) 2.0× 2.0 μm constant height AFM image of a DL PPA with M = Ag, S = mica, D = 264 nm, d_m = 22 nm, r_d = 0.2 nm s^{-1}. Scale bar: 300 nm. Reprinted with permission [23].

close-packed particles defines the shape of the nanodots deposited on substrates; the dots usually show a quasi-triangular shape and are arranged in a space group P6mm array due to the hexagonal packing of the colloidal crystal mask (Fig. 10a-c). Van Duyne *et al.* have extended colloidal crystal masking from single layer of hexagonally close packed particles to double layers [23]. Since the overlapping of the interstices between the upper and lower layers leads to a hexagonal array of quasi-hexagonal projection on a substrate, using double layer colloidal crystals as masks yields a hexagonal array of quasi-hexagonal nanodots (Fig. 10d-e).

In a general NSL procedure, the substrate to be patterned is positioned normal to the direction of material deposition. The in-plane shape of the nanodots and the spacing of the nearest-neighboring dots derived from NSL are dictated by the projection of the interstices of single or double layers of colloidal crystals on substrates. They can be tuned by varying the projection geometry of the interstices on substrates by titling the masks with respect to the incidence of the vapor beam for instance. This has inspired development of angle-resolved NSL (AR-NSL), pioneered by the group of Van Duyne [66].

In a AR-NSL process, the incidence angle of the propagation vector of the material deposition beam with respect to the normal direction of the colloidal mask (θ) and/or the azimuth angle of the propagation vector with respect to the nearest neighboring particles in the colloidal masks (ϕ) – the mask registry with respect to the vector of the material deposition beam – have been employed to reduce the size of the nanodots obtained and, at the same time, elongate their triangular shape (Fig. 11). By rotating substrates, Giersig and coworkers have recently found that AR-NSL can generate much more complicate metallic nanostructures and they referred to this process as shadow NSL [43, 68, 69]. Zhang and Wang have recently demonstrated the feasibility of consecutively depositing two different metals, such as gold and silver, at two different incidence angles, to construct ordered binary arrays of gold and silver nanoparticles [70].Due to the rotation of the colloidal mask, shadow NSL relies in a process resolved by the azimuth angle (ϕ) of the incidence deposition beam rather than the incidence angle (θ).

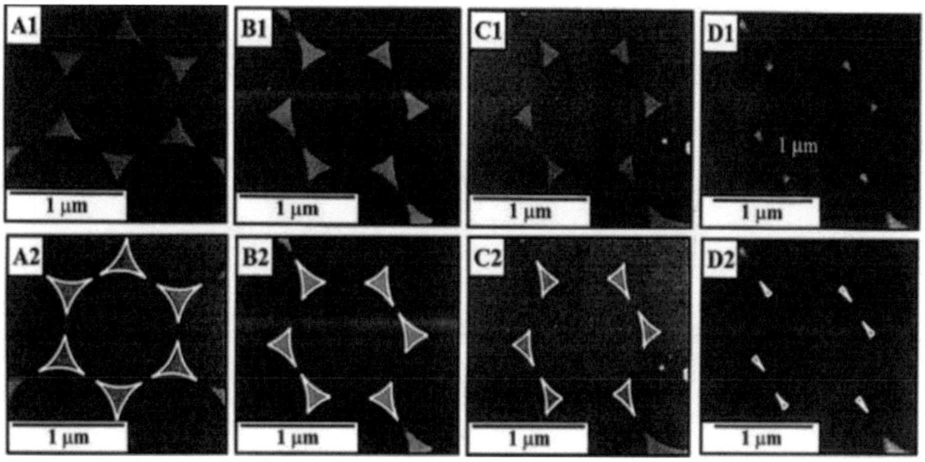

Figure 11. Field emission SEM images of AR NSL fabricated - gold nanodot arrays and images with simulated geometry superimposed, respectively. (A1, A2) θ = 108 °, ϕ = 288 °, (B1, B2) θ = 208 °, ϕ = 28 °, (C1, C2) θ = 268 °, ϕ = 168 °, and (D1, D2) θ = 408 °, ϕ = 28 °. All samples are Cr deposited onto Si (111) substrates. Images were collected at 40k magnification. θ is the incidence angle and ϕ the azimuth angle. Reprinted with permission [66].

The elegant extension of AR-NSL is to stepwise conduct physical vapor deposition of identical or different materials at the different incidence angles. The group of Van Duyne has succeeded in growth of surface patterning features composed of two triangular nanodots either overlapped or separated by two deposition steps at θ = 0º and θ > 0º, respectively [65]. Giersig *et al.* have also developed a stepwise shadow NSL protocol to deposit different materials at different incidence angles when the colloidal masks were rotating and they have succeeded in encapsulation of the metallic structures to prevent them from oxidation [69].

Prior to physical vapor deposition, colloidal crystal masks can undergo RIE to reduce the sizes of the particles and widen the interstitial spaces, thus increasing the dimension of triangular nanodots obtained via NSL. Increasing the RIE time can turn close-packed colloidal crystal masks to non-close packed ones, which leads to thin films with hexagonally arranged pores [71, 72]. Wang *et al.* have recently integrated AR-NSL with the use of RIE-modified colloidal crystals as masks to diversify the structural complexity of the patterning feature derived from NSL from triangular (or deformed) nanodots to nanorods and nanowires [73]. Laterally arranging different nanowires into a periodic array with a defined alignment is hard to implement by otherwise means, either conventional lithographic techniques or self-assembly techniques.

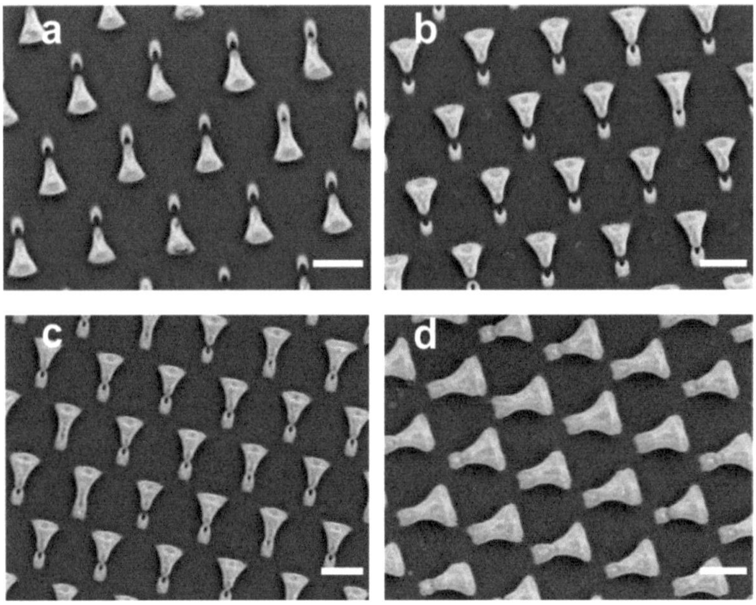

Figure 12. SEM pictures of hexagonally arranged Au nano-shuttlecocks obtained by using bilayers of hexagonal close-packed 925 nm PS spheres, etched by O_2-plasma for 10 (a), 20 (b), 25 (c), and 30 min (d), as masks for Au vapor deposition. The incidence angle of Au vapor flow was set as 15°. The scale bars are 500 nm. Reprinted with permission [74].

Wang *et al.* have recently extended the RIE process for modification of double layers of colloidal crystals for AR-NSL [74]. Using O_2 plasma etched bilayers of hexagonally packed particles as masks for gold deposition, highly ordered binary arrays of gold nanoparticles with varied shapes, for instance, with a shuttlecock-like shape composed of a small crescent-shaped nanoparticle and a big fan-shaped one, have been fabricated (Fig. 12). As compared to that of the corresponding bulk materials, the melting point of nanoparticles is much lower and especially more sensitive to the surface tension. Since the large curvature causes a high surface tension, the annealing of non-round nanoparticles may give rise to a retraction of their apexes, eventually generating a round shape [75]. Wang *et al.* have successfully transformed the shape of Au nanoparticles obtained from crescent-like or fan-shaped to round with a rather narrow distribution in terms of size and shape [74].

Dmitriev *et al.* have extended colloidal crystal masking from the use for material deposition to that for controlled etching and developed an interesting variant of NSL – hole-mask colloidal lithography (HCL) [76]. Different from conventional NSL, the essential new feature of HCL is that the substrate and the colloidal crystal mask are interspaced by a sacrificial layer. After physical vapor deposition, the removal of the colloidal mask leads to a thin film mask with nanoholes, which is known as "hole-mask". The hole-mask is subsequently used for vapor deposition and/or etching steps to further define a patterning feature on the substrate. HCL includes a number of the advantages of NSL, such as large area coverage, high fabrication speed, independent control over feature size and spacing, and processing simplicity. It can be applied to a wide range of materials, including Au, Ag, Pd, Pt, SiO_2 etc.

Various colloidal spheres, organic and inorganic, can be produced that are exceedingly monodisperse in terms of size and shape. Nevertheless, their surfaces still remain chemically homogeneous or heterogeneous. Controlling the surface properties of colloidal particles is one of the oldest and, at the same time, the most vital topics in colloid science and physical chemistry. Patchy particles, i.e., particles with more than one patch or patches that are less than 50% of the total particle surface, should present the next generation of particles for assembly [77-79]. However, patterning the surface of colloidal particles with sizes of micrometers or submicrometers is a formidable challenge due to lack of the proper mask.

When 2D colloidal crystals are used as masks for physical vapor deposition, it is expected that only the upper surfaces of the colloidal particles, exposing directly to the vapor beam, will be coated with new materials, which leads to two spatially well-separated halves on the colloidal particles, coated and non-coated, with two distinct surface chemical functionalities [80, 81]. Such particles are usually referred to as Janus particles. By embedding a monomer of close-packed colloidal particles in a photoresist layer, Bao *et al.* have succeeded in tuning the surface areas of the colloidal particles exposing to the vapor beam during material deposition via etching the photoresist layer with O_2 plasma, thus leading to a good control of the domain sizes deposited on the particles [82]. When a monolayer of close-packed colloidal particles is constructed at the water/air interface or the wax/liquid interfaces, selective modification can be implemented in either of the two phases, leading to Janus particles [83, 84].

Wang *et al.* have pioneered the study of using the upper single layers of colloidal crystals as masks for the lower layer particles during physical vapor deposition [85]. Of importance is that the methodology reported by Wang *et al.* – using colloidal crystals for self-masking – is independent of the curvature and chemical composition of the surfaces (Fig. 13). By using O_2 plasma to etch colloidal crystal templates, mainly the top layer, and conducting physical vapor deposition at the non-zero incidence angle, Wang *et al.* have also demonstrated that the size and shape of the patterns obtained on the second layer particles show a pronounced dependence on the plasma etching time and the incidence angle [86]. Pawar and Kretzschmar have recently extended the concept of using colloidal crystal for self-masking for glancing angle deposition [87].

Wang *et al.* have recently employed the upper double layers as masks for patterning the particles in the third layers via physical deposition [88]. As a consequence, they have succeeded in stereo-decoration of colloidal particles with two, three, four, or five nanodots. The number of dots per sphere is dependent on the crystalline structure of the colloidal crystal masks, the plasma etching time, and the incidence angle. The nanodots decorated on particles are arranged in a linear, trigonal, tetrahedral, or right-pyramidal fashion, which provides the nanoscale analogues of sp, sp2, sp3 sphybridized atomic orbitals of carbon. The Au nanodots obtained on microspheres, therefore, can be recruited as the bonding site to dictate the integration of the spheres, thus paving a new way of colloidal self-assembly – colloidal valent chemistry of spheres [89] – to create hierarchical and complicated "supraparticles" [77].

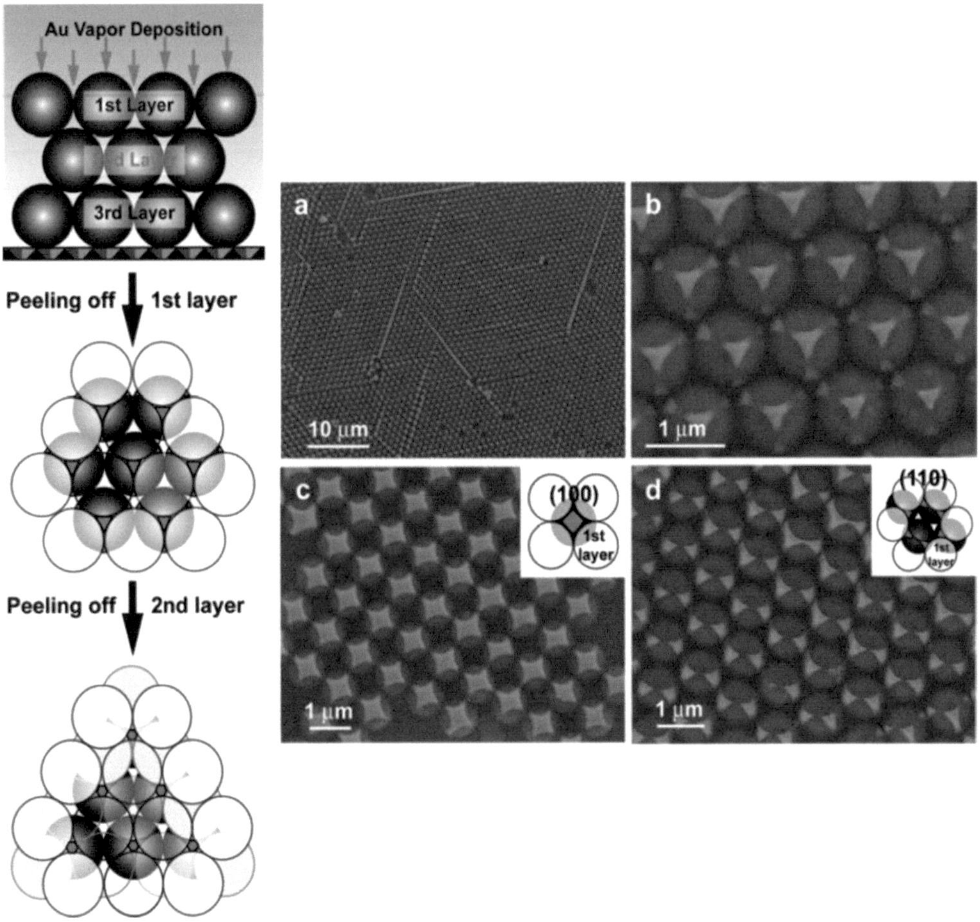

Figure 13. Left panel: Schematic illustration of the procedure to create colloidal spheres with Au-patterned surfaces by the combination of Au vapor deposition and using the top mono- or bilayers of colloidal crystals with (111) facets parallel to the substrates as masks. Right panel: Low (a) and high magnification scanning electron microscope (SEM) picture of 925 nm polystyrene spheres with Au patterned surfaces generated by templating the top monolayers of colloidal crystals with (111) (b) (100) (c) and (110) (d) facets parallel to the substrates. Reprinted with permission [85]

3.2.3. Extension of colloidal lithography

One extension of NSL is to use the surface patterns obtained as templates to grow nanostructures of a variety of materials via bottom-up self-assembly. Mulvaney's group has grown monolayer and multilayer films of semiconductor quantum dots on surface patterns derived from NSL, leading to nanostructured luminescent thin films [90, 91]. Valsesia *et al.* have used ordered arrays of polyacrylic acid domes derived via NSL to selectively couple with bovine serum albumin [55]. Using NSL-derived surface patterns as templates to grow proteins, Sutherland *et al* have found that the surface topography enhances the binding selectivity of fibrinogens to platelets [92].

http://dx.doi.org/10.5772/56576

The second extension is to use NSL-derived surface patterns as etching masks to create surface topography. Chen *et al.* have fabricated silicon nanopillar arrays with diameters as small as 40 nm and aspect ratios as high as seven [93]. The size and shape of the nanopillars can be controlled by the size and shape of the sputtered aluminum masks, which are again determined by the feature size of the colloidal mask and the number of the colloidal layers. Nanopillars with different shapes can also be fabricated by adjusting the RIE conditions such as the gas species, bias voltage, and exposure duration for an aluminum mask with a given shape. As-prepared nanopillar arrays can be utilized for imprinting a layer of PMMA above its glass transition temperature [94]. Similarly, Weekes *et al.* have fabricated ordered arrays of cobalt nanodots for patterned magnetic media [95]. By introducing intermediary layers of SiO_2 between colloidal crystal masks and substrates, this etching strategy can be applied to a wide range of materials without much concern for the surface hydrophilicity of the targeted substrates. Using the similar protocol, large-area ordered arrays of 512 nm pitch hole, with vertical and smooth sidewalls, has been successfully formed on GaAs substrates [96].

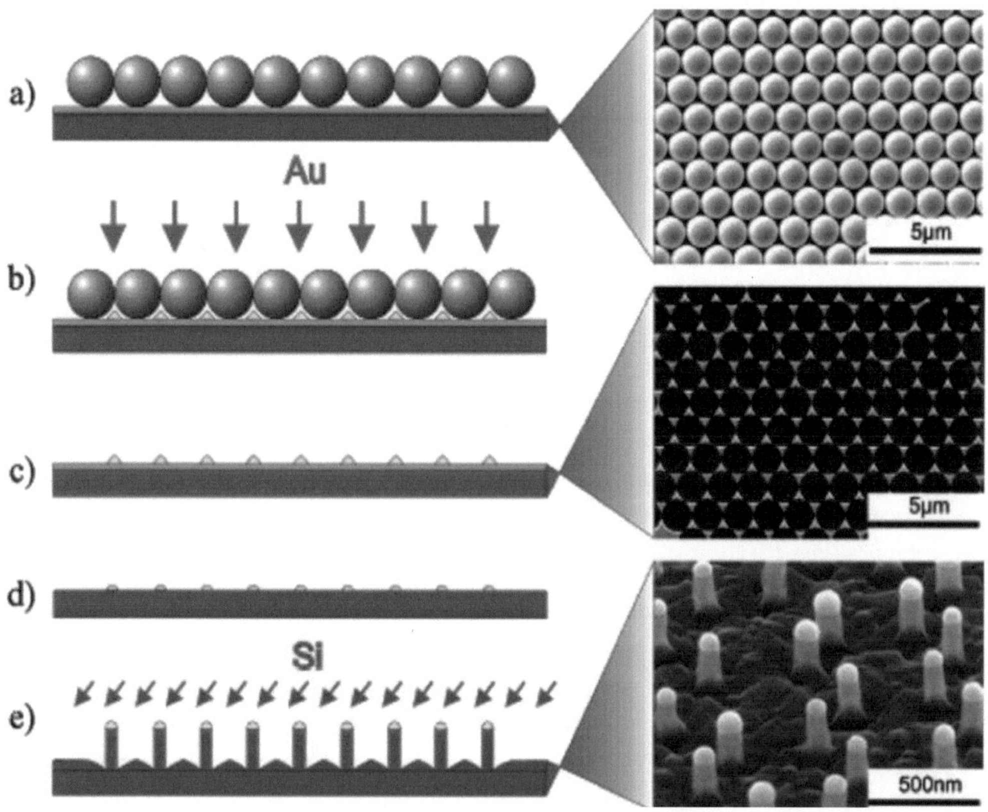

Figure 14. Steps of Si nanowire fabrication. NSL: (a) deposition of a mask of polystyrene particles (b) deposition of gold by thermal evaporation, (c) removal of the spheres, (d) thermal annealing and cleaning step to remove the oxide layer, and (e) Si deposition and growth of nanowires by MBE. (Right) corresponding SEM micrographs of wafers at different steps. Reprinted with permission [100].

The third extension is to use NSL-derived surface patterns to template or catalyze the growth of other functional materials. Zhou *et al.* have successfully used the ordered arrays of gold nanodots derived from NSL as seeds to highly aligned single-walled carbon nanotubes laid on quartz and sapphire substrates [97]. This method has great potential to produce carbon nanotube arrays with simultaneous control over the nanotube orientation, position, density, diameter, and even chirality, which may work as building blocks for future nanoelectronics and ultra-high-speed electronics [98]. Wang *et al.* have used gold nanodot arrays as seeds for hexagonally arranged arrays of zinc oxide nanorods aligned perpendicular to the substrates [99]. Similarly Fuhrmann *et al.* have obtained ordered arrays of Si nanorods by using the gold nanodots as seeds for molecular beam epitaxy (Fig. 14) [100]. Similarly, discretely ordered arrays of organic light-emitting nanodiode (OLED) have been fabricated based on NSL-derived surface patterns [101].

4. Applications

4.1. Optical properties

Surface patterns derived from CL, especially NSL, are usually made up of metals such as gold and silver. Noble metal nanostructure arrays have pronounced surface plasmon resonance, which results from incident electromagnetic radiation exciting coherent oscillations of conduction electrons near a metal-dielectric interface [102]. Giessen and co-workers introduced an angle-controlled colloidal lithography as a fast and low-cost fabrication technique for large-area periodic plasmonic oligomers with complex shape and tunable geometry parameters, and investigated the optical properties and found highly modulated plasmon modes in oligomers with triangular building blocks. Fundamental modes, higher-order modes, as well as Fano resonances due to coupling between bright and dark modes within the same complex structure are present, depending on polarization and structure geometry. This process is well-suited for mass fabrication of novel large-area plasmonic sensing devices and nanoantennas (Fig. 15) [103].

One of the straightforward technical applications of CL is to use as highly sensitive biosensors relied on the localized-surface plasmon resonance (LSPR) of metallic nanostructures [67]. The LSPR of metallic nanostructures composed of gold rings [104] and disks [105], obtained via NSL has been studied. It is found that the LSPR can be tuned by varying either the diameter of the disks at a constant disk height or the ring thickness. The shape-dependent red shift originates from the electromagnetic coupling between the inner and outer ring surfaces, which leads to energy shifts and splitting of degenerate modes [106]. NSL has been also used to create nanocaps and nanocups; their LSPR behavior has been studied [107]. Lee *et al.* have generated gold crescent moon structures with a sub-10 nm sharp edge via NSL, which exhibit a very strong SERS [108].

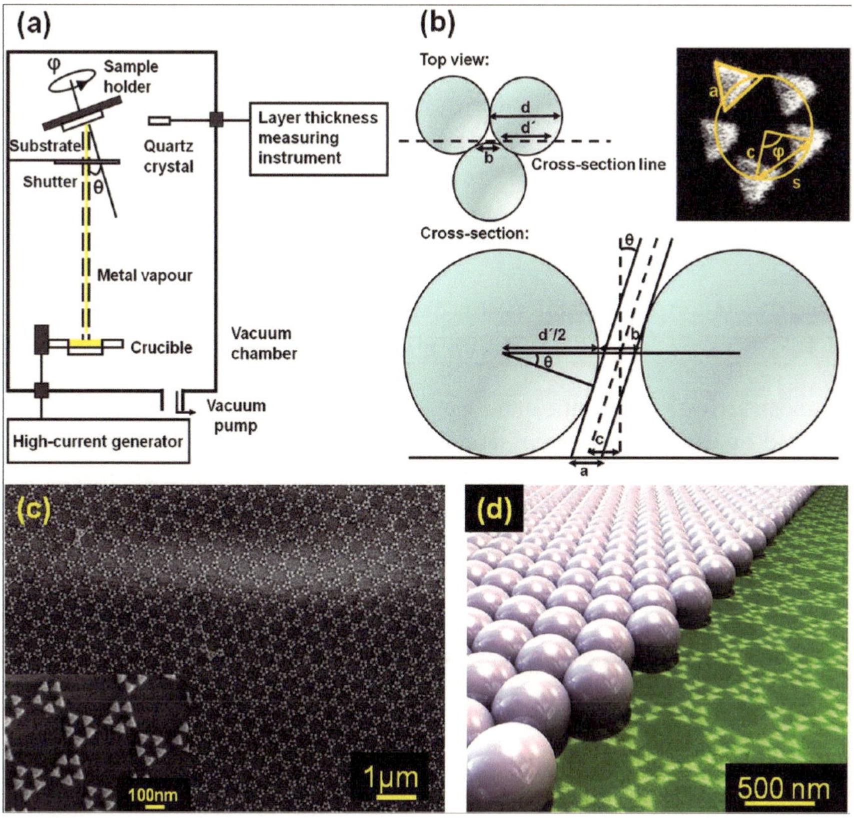

Figure 15. Fabrication of oligomers with triangular building blocks. (a) Schematic diagram of the evaporation setup. (b)Simplified geometrical model for symmetric pentamer fabrication. Top left: top view. Bottom: side view at cross section line as indicated in top view. Top right: geometrical parameters. (c) Scanning electron microscopy (SEM) images of the large are asymmetric pentamers. The gap size is as small as 20 nm. (d) Artistic view of our fabrication scheme, using a real SEM image of a symmetric pentamer. The transmittance spectrum of the sample using a large-area optical beam diameter. Reprinted by permission of [103].

Light trapping across a wide band of frequencies is important for applications such as solar cells and photo detectors. Yao Y. and Yao J. *et al.* demonstrated a new approach based on colloidal lithography to light management by forming whispering-gallery resonant modes inside a spherical nanoshell structure. A broadband absorption enhancement across a large range of incident angles was observed. The absorption of a single layer of 50-nm-thick spherical nanoshells is equivalent to a 1-μm-thick planar nc-Si film. This light-trapping structure could enable the manufacturing of high-throughput ultra-thin film absorbers in a variety of material systems that demand shorter deposition time, less material usage and transferability to flexible substrates. [109]

4.2. Wettability

The wettability of solid surfaces is a significant property depending on both chemical compositions and the surface structure. A great number of ordered arrays generated through simple or modified colloidal lithography could induce different wettabilities. Recently, Koshizaki and co-workers fabricated vertically ordered Co_3O_4 hierarchical nanorod arrays using pulsed laser deposition (PLD) onto colloidal crystal masks followed by an annealing process, and the as-prepared Co_3O_4 nanorod arrays demonstrated stable superhydrophilicity without UV irradiation even after half a year owing to the improved roughness of the hierarchical structure and the abundant OH− groups induced by the PLD and annealing processes.

4.3. Other applications

Besides the exploitation of CL and the patterns obtained thereof in LSPR-assisted sensing, the magnetic properties of CL-derived nanostructures gain increasing attention. In general, nanoscale magnetic materials often exhibit superparamagnetic behavior. Moreover, an ordered nanostructure of magnetic materials is required for investigation of the mesoscopic effects induced by the confinement of magnetic materials in nanoscale domains [111]. Since magnetic properties are strongly dependent on the domain size and the distance between domains, Weekes *et al.* have created ordered arrays of isolated magnetic nanodots via NSL [95]. The coercivity and switching width of the isolated nanodot arrays are enhanced as compared to those of continuous magnetic films. Well-organized arrays of magnetic nanorings over a large area have been prepared via NSL and they show a stable vortex state due to the absence of a destabilizing vortex core, which should hold promise in vertical magnetic random access memories [112, 113]. Albrecht *et al.* have found that Co/Pd multilayers on a colloid surface exhibited a pronounced magnetic anisotropy [114].

What's more, a core question in materials science is how to encode non-trivial organized structures within simple building blocks. A recent report from this laboratory described methods for functionalizing latex spheres to make them hydrophobic at their poles, leading to the directed self-assembly of a kagome lattice pattern in which each sphere was coordinated with four neighbors, two at each pole. Granick and co-workers developed methods for functionalizing micrometer-sized colloidal spheres with three or more zones of chemical functionality through colloidal lithography, literally combining double-sided angle-resolved physical deposition and controllable chemical etching. These synthesis methods allowed targeting of various lattice structures whose bonding between neighboring particles in liquid suspension was visualized in situ by optical microscopy [115,116].

5. Summary

The recent development of CL, especially the integration of etching the colloidal mask, altering the incidence angle, and stepwise and regularly changing the mask registry, leads to a powerful nanochemical patterning tool with low cost in capital and operation, high throughput, and ease to be adopted on various planar and curved surfaces and even on microparticles. Different from conventional mask-assisted lithographic processes in which the mask design and production usually remain a challenge for scaling down the feature size and diversifying the feature shape, CL embodies a simple way for masking – self-assembly of monodisperse microspheres on a targeted substrate. The feature size can easily shrink below 100 nm by reducing the diameter of the microspheres used according to the simple correlation between the interstice size and the sphere diameter. The feature shape can be easily diversified by the crystalline structure of a colloidal crystal mask, the time of anisotropic etching of the mask, the incidence angle of vapor beam and the mask registry (the azimuth angle of vapor beam). Currently, CL allows fabrication of very complicated 2D and 3D nanostructured features, such as multiplex nanostructures with a clear-cut lateral and vertical heterogeneity. A number of new nanostructures are hard to be implemented, or cannot be in some cases, by conventional lithographic techniques. As such, CL provides a nanochemical and complementary tool of conventional and fully top-down lithographic techniques, and thus holds immense promise in surface patterning.

However, CL is still in a very early stage of development. Despite the great progress in colloidal crystallization it still remains a formidable challenge to create a defect-free single crystal with a defined crystalline face. The presence of defects dramatically reduces the patterning precision of CL. For instance, the random orientation of polycrystalline crystalline domains in a colloidal mask is a disaster for collimating the mask registry. In this aspect, template-assisted epitaxy for colloidal crystallization is promising as it allows growth of colloidal crystals with defined packing structure and orientation. Since a patterned substrate is necessitated for the colloidal epitaxy, its applicability for patterning is limited. How to transfer a colloidal crystal derived from this colloidal epitaxy onto different substrates without deterioration of the crystal quality should be an ensuing task for CL. Besides, fabrication of large area monolayer of periodically close-packed microspheres with sizes smaller than 100 nm remains highly challenging, which brings a technical problem to reduce the feature size below 10 nm via CL. In a CL patterning process, furthermore, the feature size and the interspace size between the features cannot be separately manipulate, as they both are directly proportional with the sphere size in a colloidal mask, which largely limit the patterning capability of CL.

Author details

Ye Yu and Gang Zhang*

*Address all correspondence to: gang@jlu.edu.cn

State Key Lab of Supramolecular Structure and Materials, College of Chemistry, Jilin University, Changchun, China

References

[1] D. Wang, H. Möhwald, Template-directed colloidal self-assembly – the route to 'top-down' nanochemical engineering, *J. Mater. Chem.* 2004, *14*, 459-468.

[2] Y. Xia, B. Gates, Y. Yin, Y. Lu, Monodisperse colloidal particles: old materials with new applications, *Adv. Mater.* 2000, *12*, 693-713.

[3] M. E. Leunissen, C. G. Christova, A-P, Hynninen, C. P. Royall, A. I. Campbell, A. Imhof, M. Dijkstra, R. van Roij, A. van Blaaderen, Ionic colloidal crystals of oppositely charged particles, *Nature* 2005, *437*, 235-240.

[4] R. Mayoral, J. Requena, J. S. Moya, C. López, A. Cintas, H. Miguez, F. Meseguer, L. Vazquez, M. Holgado, A. Blanco, 3D long-range ordering in and SiO_2 submicrometer-sphere sintered superstructure, *Adv. Mater.* 1997, *9*, 257-260.

[5] H. Miguez, F. Meseguer, C. López, A. Mifsud, J. S. Moya, L. Vazquez, Evidence of FCC crystallization of SiO_2 nanospheres, *Langmuir* 1997, *13*, 6009-6011.

[6] H. Migurez, C. López, F. Meseguer, A. Blanco, L. Vazquez, R. Mayoral, M. Ocafia, V. Fornes, A. Mifsud, Photonic crystal properties of packed submicrometric SiO_2, *Appl. Phys. Lett.* 1997, *71*, 1148-1150.

[7] N. D. Denkov, O. D. Velev, P. A. Kralchevsky, I. B. Invanov, H. Yoshimura, K. Nagayama, Mechanism of formation of 2D crystals from latex-particles on substrates, *Langmiur*1992, *8*, 3183-3190.

[8] R. Micheletto, H. Fukuda, M. Ohtsu, A simple method for the production of a 2D ordered array of small latex-particles, *Langmuir*, 1995, *11*, 3333-3336.

[9] A. S. Dimitrov, K. Nagayama, Continuous convective assembling of fine particles into 2D arrays on solid surfaces, *Langmuir* 1996, *12*, 1303-1311.

[10] P. Jiang, J. F. Bertone, K. S. Hwang, V. L. Colvin, Single-crystal colloidal multilayers of controlled thickness, *Chem. Mater.* 1999, *11*, 2132-2140.

[11] Z. –Z., Gu, A. Fujishima, O. Sato, Fabrication of high-quality opal films with controllable thickness, *Chem. Mater.* 2002, *14*, 760-765.

[12] Z. Zhou, X. S. Zhao, Flow-controlled vertical deposition method for the fabrication of photonic crystals, *Langmuir* 2004, *20*, 1524-1526.

[13] S. Wong, V. Kitaev, G. A. Ozin, Colloidal crystals films: advances in universality and perfection, *J. Am. Chem. Soc.* 2003, *125*, 15589-15598.

[14] X. Chen, Z. Chen, N. Fu, G. Lu, B. Yang, Versatile nanopatterned surfaces generated via 3D colloidal crystals, *Adv. Mater.* 2003, *15*, 1413-1417.

[15] Y. W. Chung, I. C. Leu, J. H. Lee, M.H. Hon, Influence of humidity on the fabrication of high-quality colloidal crystals via a capillary-enhanced process, *Langmuir* 2006, *22*, 6454-6460.

[16] M. H. Kim, S. H. Im, O. O. Park, Rapid fabrication of two- and three-dimensional colloidal crystal films via confined convective assembly, *Adv. Funct. Mater.* 2005, *15*, 1329-1335.

[17] Z. Cheng, W. B. Russel, P. M. Chaikin, Controlled growth of hard-sphere colloidal crystals, *Nature* 1999, *401*, 893-895.

[18] S. H. Im, M. H. Kim, O. O. Park, Thickness control of colloidal crystals with a substrate dipped at a tilted angle into a colloidal suspension, *Chem. Mater*. 2003, *15*, 1797-1802.

[19] V. Kitaev, G. A. Ozin, Self-assembled surface patterns of binary colloidal crystals, *Adv. Mater*. 2003, *15*, 75-78.

[20] K. P. Velikov, C. G. Christova, R. P. A. Dullens, A. van Blaaderen, Layer-by-layer growth of binary colloidal crystals, *Science* 2002, *296*, 106-109.

[21] M. H. Kim, S. H. Im, O. O. Park, Fabrication and structural analysis of binary colloidal crystals with 2D superlattices, *Adv. Mater*. 2005, *17*, 2501-2505.

[22] U. C. Fischer, H. P. Zingsheim, Submicroscopic pattern replication with visible light, *J. Vac. Sci. Technol.* 1981, *19*, 881-885.

[23] J. C. Hulteen, R. P. van Duyne, Nanosphere lithography-a materials general fabrication process for periodic particle array surfaces, *J. Vac. Sci. Technol. A*. 1995, *13*, 1553-1558.

[24] T. J. Rehg, B. G. Higgins, Spin coating of colloidal suspensions, *AIChE*1992, *38*, 489-501.

[25] P. Jiang, M. J. McFarland, Large-scale fabrication of wafer-size colloidal crystals, macroporous polymers and nanocomposites by spin-coating, *J. Am. Chem. Soc.* 2004, *126*, 13778-13786.

[26] P. Jiang, M. J. McFarland, Wafer-scale periodic nanoholes arrays templated from 2D non-close-packed colloidal crystals, *J. Am. Chem. Soc.* 2005, *127*, 3710-3711.

[27] D. Wang, H. Möhwald, Rapid fabrication of binary colloidal crystals by stepwise spin-coating. *Adv. Mater.* 2004, *16*, 244-247.

[28] A. Ulman, An introduction to ultrathin organic films: from Langmuir-Blodgett to self-assembly, *Academic Press*, Boston, 1991.

[29] B. P. Binks, Particles as surfactants-similarities and differences. *Current Opinion in Colloid & Interface Science* 2002, *7*, 21-41.

[30] P. Pieranski, Two dimensional interfacial colloidal crystals. *Phys. Rev. Lett.* 1980, *45*, 569-572.

[31] S. H. Im, Y. T. Lim, D. J. Suh, O. O. Park, 3D self-assembly of colloids at a water-air interface: A novel technique for the fabrication of photonic bandgap crystals. *Adv. Mater.* 2002, *14*, 1367-1369.

[32] M. Kondo, K. Shinozaki, I. Bergström, N. Mizutani, Preparation of colloidal monolayers of alkoxylated silica particles at the air-liquid interface. *Langmuir* 1995, *11*, 394-397.

[33] K. -U. Fulda, B. Tieke, Langmuir films of monodisperse 0.5μm spherical polymer particles with a hydrophobic core and a hydrophilic shell. *Adv. Mater.* 1994, *6*, 288-290.

[34] B. van Duffel, R. H. A. Ras, F. C. De Schryver, R. A. Schoonheydt, Langmuir-Blodgett depostion and optical diffraction of two dimensional opal. *J. Mater. Chem.* 2001, *11*, 3333-3336.

[35] S. Reculusa, S. Ravaine, Synthesis of colloidal crystals of controllable thickness through the Langmuir-Blodgett technique. *Chem. Mater.* 2003, *15*, 598-605.

[36] N. Vogel, L. de Viguerie, U. Jonas, C. K. Weiss, K. Landfester, Wafer-scale fabrication of ordered binary colloidal monolayers with adjustable stoichiometries. *Adv. Funct. Mater.*2011, *21*, 3064-3073.

[37] L. M. Goldenberg, J. Wagner, J. Stumpe, B-R. Paulke, E. Grnitz, Simple method for the preparation of colloidal particle monolayers at the water/alkane interface. *Langmuir* 2002, *18*, 5627-5629.

[38] S. Reynaert, P. Moldenaers, J. Vermant, Control over colloidal aggregation in monolayers of latex particles at the oil-water interface. *Langmuir* 2006, *22*, 4936-4945.

[39] K. P. Velikov, F. Durst. O. D. Velev, Direct observation of the dynamics of latex particles confined inside thinning water-air films. *Langmuir* 1998, *14*, 1148-1155.

[40] Z-Z Gu, D. Wang, H. Möhwald, Self-assembly of microspheres at the air/water/air interface into free-standing colloidal crystal films. *Soft Matter* 2007, *3*, 68-70.

[41] B. Griesebock, M. Egen, R. Zental, Large photonic films by crystallization on fluid substrates. *Chem. Mater.* 2002, *14*, 4023-4025.

[42] P. Hanarp, M. Kall, D. S. Sutherland, Optical properties of short range ordered arrays of nanometer gold disks prepared by colloidal lithography. *J. Phys. Chem. B* 2003, *107*, 5768-5772.

[43] A. Kosiorek, W. Kandulski, H. Glaczynska, M. Giersig, Fabrication of nanoscale rings, dots, and rods by combining shadow nanosphere lithography and annealed polystyrene nanosphere masks. *Small* 2005, *1*, 439-444.

[44] J. H. Moon, W.-S. Kim, J.-W. Ha, S. G. Jang, S.-M. Yang, J. K. Park, Colloidal lithography with crosslinkable particles: Fabrication of hierarchical nanopore arrays, *Chem. Commun.* 2005, *32*, 4107-4109.

[45] J. J. Peninkhof, C. Graf, T. van Dillen, A. M. Vredenberg, A. van Blaaderen, A. Polman, Angle-dependent extinction of anisotropic silica/Au core/shell colloids made via ion irradiation. *Adv. Mater.* 2005, *17*, 1484-1488.

[46] D. L. J. Vossen, D. Fific, J. Penninkhof, T. Van Dillen, A. Polman, A. Van Blaaderen, Combined optical tweezers/ion beam technique to tune colloidal masks for nanolithography, *Nano Lett.* 2005, *5*, 1175-1179.

[47] J. D. Forster, J. Park, M. Mittal, H. Noh, C. F. Schreck, C. S. O'Hern, H. Cao, E. M. Furst, E. R. Dufresne, Assembly of Optical-Scale Dumbbells into Dense Photonic Crystals, ACS Nano 2011, 5, 6695-6700.

[48] H. W. Deckmann, J. H. Dunsmuir, Applications of surface textures produced with natural lithography. *J. Vac. Sci. Technol. B* 1983, *1*, 1109-1112.

[49] D.-G. Choi, H. K. Yu, S. G. Jang, S.-M. Yang, Colloidal lithographic nanopatterning via reactive ion etching. *J. Am. Chem. Soc.* 2004, *126*, 7019-7025.

[50] S.-M. Yang, S. G. Jang, D.-G. Choi, S. Kim, H. K. Yu, Nanomachining by colloidal lithography. *Small* 2006, *2*, 458-475.

[51] D.-G. Choi, S. G. Jang, S. Kim, E. Lee, C.-S. Han, S.-M. Yang, Multifaceted and nanobored particle arrays sculpted using colloidal lithography. *Adv. Funct. Mater.* 2006, *16*, 33-40.

[52] Y. Zheng, Y. Wang, S. Wang, C. H. A. Huan, Fabrication of nonspherical colloidal particles via reactive ion etching of surface-patterned colloidal crystals. *Colloid Surf. A* 2006, *277*, 27-36.

[53] C. L. Cheung, R. J. Nikolić, C. E. Reinhardt, T. F. Wang, Fabrication of nanopillars by nanosphere lithography. *Nanotechnology* 2006, *17*, 1339-1343.

[54] B. J.-Y. Tan, C.-H. Sow, K.-Y. Lim, F.-C. Cheong, G.-L. Chong, A. T.-S. Wee, C.-K. Ong, Fabrication of a two-dimensional periodic non-close-packed array of polystyrene particles. *J. Phys. Chem. B* 2004, *108*, 18575-18579.

[55] A. Valsesia, P. Colpo, M. M. Silvan, T. Meziani, G. Ceccone, F. Rossi, Fabrication of nanostructured polymeric surfaces for biosensing devices. *Nano Lett.* 2004, *4*, 1047-1050.

[56] B. Wang, W. Zhao, A. Chen, S.-J. Chua, Formation of nanoimprintingmould through use of nanosphere lithography. *J. Crystal Growth* 2006, *288*, 200-204.

[57] Z. Huang, H. Fang, J. Zhu, Fabrication of silicon nanowire arrays with controlled diameter, length, and density. *Adv. Mater.* 2007, *19*, 744-748.

[58] B. J. Y. Tan, C. H. Sow, T. S. Koh, K. C. Chin, A. T. S. Wee, C. K. Ong, Fabrication of size-tunable gold nanoparticles array with nanosphere lithography, reactive ion etching, and thermal annealing. *J. Phys. Chem. B* 2005, *109*, 11100-11109.

[59] J. M. McLellan, M. Geissler, Y. Xia, Edge spreading lithography and its application to the fabrication of mesoscopic gold and silver rings. *J. Am. Chem. Soc.* 2004, *126*, 10830-10831.

[60] M. Geissler, J. M. McLellan, J. Chen, Y. Xia, Side-by-side patterning of multiple alkanethiolate monolayers on gold by edge-spreading lithography. *Angew. Chem. Int. Ed.* 2005, *44*, 3596-3600.

[61] C. Bae, H. Shin, J. Moon, M. M. Sung, Contact area lithography (CAL): A new approach to direct formation of nanometric chemical patterns. *Chem. Mater.* 2006, *18*, 1085-1088.

[62] C. Bae, J. Moon, H. Shin, J. Kim, M. M. Sung, Fabrication of monodisperse asymmetric colloidal clusters by using contact area lithography (CAL). *J. Am. Chem. Soc.*2007, *129*, 14232-14239.

[63] H. W. Deckmann, J. H. Dunsmuir, Natural lithography. *Appl. Phys. Lett.* 1982, *41*, 377-379.

[64] J. C. Hulteen, D. A. Treichel, M. T. Smith, M. L. Duval, T. R. Jensen, R. P. Van Duyene, Nanosphere lithography: size-tunable silver nanoparticle and surface cluster arrays. *J. Phys. Chem. B* 1999, *103*, 3854-3863.

[65] C. L. Haynes, R. P. Van Duyne, Nanosphere lithography: A versatile nanofabrication tool for studies of size-dependent nanoparticle optics. *J. Phys. Chem. B* 2001, *105*, 5599-5611.

[66] C. L. Haynes, A. D. McFarland, M. T. Smith, J. C. Hulteen, R. P. Van Duyne, Angle-resolved nanosphere lithography: Manipulation of nanoparticle size, shape, and interparticle spacing. *J. Phys. Chem. B* 2002, *106*, 1898-1902.

[67] A. Willets, R. P. Van Duyne, Localized surface plasmon resonance spectroscopy and sension. *Annu. Rev. Phys. Chem.* 2007, *58*, 267-297.

[68] M. Giersig, M. Hilgendorff, Magnetic nanoparticle superstructures. *Eur. J. Inorg. Chem.* 2005, 3571-3583.

[69] A. Kosiorek, W. Kandulski, P. Chudzinski, K. Kempa, M. Giersig, Shadow nanosphere lithography: Simulation and experiment. *Nano Lett.* 2004, *4*, 1359-1363.

[70] G. Zhang, D. Wang, Fabrication of heterogeneous binary arrays of nanoparticles via colloidal lithography. *J. Am. Chem. Soc.* 2008, *130*, 5616-5617.

[71] S. M. Weekes, F. Y. Ogrin, Torque studies of large-area Co arrays fabricated by etched nanosphere lithography. *J. Appl. Phys.* 2005, *97*, 10J503.

[72] D.-G. Choi, S. Kim, S. G. Jang, S.-M. Yang, J.-R. Jeong, S.-C. Shin, Nanopatterned magnetic metal via colloidal lithography with reactive ion etching. *Chem. Mater.* 2004, *16*, 4208-4211.

[73] G. Zhang, D. Wang, H. Möhwald, Fabrication of multiplex quasi-three-dimensional grids of one-dimensional nanostructures via stepwise colloidal lithography. *Nano Lett.* 2007, *7*, 3410-3413.

[74] G. Zhang, D. Wang, H. Möhwald, Ordered binary arrays of Au nanoparticles derived from colloidal lithography. *Nano Lett.* 2007, *7*, 127-132.

[75] A. Habenicht, M. Olapinski, F. Burmeister, P. Leiderer, J. Boneberg, Jumping nanodroplets. *Science,* 2005, *309,* 2043-2045.

[76] H. Fredriksson, Y. Alaverdyan, A. Dmitriev, C. Langhammer, D. S. Sutherland, M. Zäch, B. Kasemo, Hole-mask colloidal lithography. *Adv. Mater.* 2007, *19,* 4297-4302.

[77] E. W. Edwards, D. Wang, H. Möhwald, Hierarchical organization of colloidal particles: from colloidal crystallization to supraparticle chemistry. *Macromol. Chem. Phys.* 2007, *208,* 439-445.

[78] H. Zhang, E. W. Edwards, D. Wang, H. Möhwald, Directing the self-assembly of nanocrystals beyond colloidal crystallization. *Phys. Chem. Chem. Phys.* 2006, *8,* 3288-3299.

[79] S. C. Glotzer, M. J. Solomon, Anisotropy of building blocks and their assembly into complex structures. *Nature Mater.* 2007, *6,* 557-562.

[80] K. Fujimoto, K. Nakahama, M. Shidara, H. Kawaguchi, Preparation of unsymmetrical microspheres at the interfaces, *Langmuir,* 1999, *15,* 4630-4635.

[81] Y. Lu, H. Xiong, X. Jiang, Y. Xia, M. Prentiss, G. M. Whitesides, Asymmetric dimers can be formed by dewetting half-shells of gold deposited on the surfaces of spherical oxide colloids. *J. Am. Chem. Soc.* 2003, *125,* 12724-12725.

[82] Z. Bao, L. Chen, M. Weldon, E. Chandross, O. Cherniavskaya, Y. Dai, J. Tok, Toward controllable self-assembly of microstructures: Selective functionalization and fabrication of patterned spheres. *Chem. Mater.* 2002, *14,* 24-26.

[83] A. Perro, S. Reculusa, S. Ravaine, E. Bourgeat-Lami, E. Duguet, Design and synthesis of Janus micro- and nanoparticles. *J. Mater. Chem.* 2005, *15,* 3745-3760.

[84] L. Hong, A. Cacciuto, E. Luijten, S. Granick, Clusters of charged Janus spheres, *Nano Lett.* 2006, *6,* 2510-2514.

[85] G. Zhang, D. Wang, H. Möhwald, Patterning microsphere surfaces by templating colloidal crystals. *Nano Lett.* 2005, *5,* 143-146.

[86] G. Zhang, D. Wang, H. Möhwald, Nanoembossment of Au patterns on microspheres. *Chem. Mater.* 2006. *18,* 3985-3992.

[87] A. B. Pawar, I. Kretzschmar, Patchy particles by glancing angle deposition. *Langmuir* 2008, *24,* 355-358.

[88] G. Zhang, D. Wang, H. Möhwald, Decoration of microspheres with gold nanodots—giving colloidal spheres valences. *Angew. Chem. Int. Ed.* 2005, *44,* 7767-7770.

[89] D. R. Nelson, Toward a tetravalent chemistry of colloids. *Nano Lett.* 2002, *2,* 1125-1129.

[90] J. Pacifico, D. Gómez, P. Mulvaney, A simple route to tunable two-dimensional arrays of quantum dots. *Adv. Mater.* 2005, *17*, 415-418.

[91] J. Pacifico, J. Jasieniak, D. E. Gómez, P. Mulvaney, Tunable D-3 arrays of quantum dots: Synthesis and luminescence properties. *Small* 2006, *2*, 199-203.

[92] D. S. Sutherland, M. Broberg, H. Nygren, B. Kasemo, Influence of nanoscale surface topography and chemistry on the functional behaviour of an adsorbed model macromolecule. *Macromol. Biosci.* 2001, *1*, 270-273.

[93] C.-W. Kuo, J.-Y. Shiu, P. Chen, Size- and shape-controlled fabrication of large-area periodic nanopillar arrays. *Chem. Mater.* 2003, *15*, 2917-2920.

[94] C.-W. Kuo, J.-Y. Shiu, Y.-H. Cho, P. Chen, Fabrication of large-area periodic nanopillar arrays for nanoimprint lithography using polymer colloid masks. *Adv. Mater.* 2003, *15*, 1065-1068.

[95] S. M. Weekes, F. Y. Ogrin, W. A. Murray, Fabrication of large-area ferromagnetic arrays using etched nanosphere lithography. *Langmuir* 2004, *20*, 11208-11212.

[96] S. Han, Z. Hao, J. Wang, Y. Luo, Controllable two-dimensional photonic crystal patterns fabricated by nanosphere lithography. *J. Vac. Sci. Technol. B* 2005, *23*, 1585-1588.

[97] K. Ryu, A. Badmaev, L. Gomez, F. Ishikawa, B. Lei, C. Zhou, Synthesis of aligned single-walled nanotubes using catalysts defined by nanosphere lithography. *J. Am. Chem. Soc.* 2007, *129*, 10104-10105.

[98] K. H. Park, S. Lee, K. H. Koh, R. Lacerda, K. B. K. Teo, W. I. Milne, Advanced nanosphere lithography for the areal-density variation of periodic arrays of vertically aligned carbon nanofibers. *J. Appl. Phys.* 2005, *97*, 024311.

[99] X. Wang, C. J. Summers, Z. L. Wang, Large-scale hexagonal-patterned growth of aligned ZnOnanorods for nano-optoelectronics and nanosensor arrays. *Nano Lett.* 2004, *4*, 423-426.

[100] B. Fuhrmann, H. S. Leipner, H.-R. Höche, L. Schubert, P. Werner, U. Gösele, Ordered arrays of silicon nanowires produced by nanosphere lithography and molecular beam epitaxy. *Nano Lett.* 2005, *5*, 2524-2537.

[101] J. G. C. Veinot, H. Yan, S. M. Smith, J. Cui, Q. Huang, T. J. Marks, Fabrication and properties of organic light-emitting "nanodiode" arrays. *Nano Lett.* 2002, *2*, 333-335.

[102] J. M. Yao, A. P. Le, S. K. Gray, J. S. Moore, J. A. Rogers, R. G.L Nuzzo, Functional Nanostructured Plasmonic Materials. *Adv. Mater.*2010, *22*, 1102-1110.

[103] Large-area high-quality plasmonic oligomers fabricated y angle-controlled coilloidal nanolithography. *ACS Nano*2011, *5*, 9009-9016.

[104] J. Aizpurua, P. Hanarp, D. S. Sutherland, M. Kall, G. W. Bryant, F. J. Garcia de Abajo, Optical properties of gold nanorings. *Phys. Rev. Lett.* 2003, *90*, 057401.

[105] P. Hanarp, M. Kall, D. S. Sutherland, Optical properties of short range ordered arrays of nanometer gold disks prepared by colloidal lithography. *J. Phys. Chem. B* 2003, *107*, 5768-5772.

[106] B. Lamprecht, G. Schider, R. T. Lechner, H. Ditlbacher, J. R. Krenn, A. Leitner, F. R Aussenegg, Metal nanoparticle gratings: Influence of dipolar particle interaction on the plasmon resonance. *Phys. Rev. Lett.* 2000, *84*, 4721-4724.

[107] J. Liu, A. I. Maaroof, L. Wieczorek, M. B. Cortie, Fabrication of hollow metal "nanocaps" and their red-shifted optical absorption spectra. *Adv. Mater.* 2005, *17*, 1276-1281.

[108] Y. Lu, G. L. Liu, J. Kim, Y. X. Mejia, L. P. Lee, Nanophotonic crescent moon structures with sharp edge for ultrasensitive biomolecular detection by local electromagnetic field enhancement effect, *Nano Lett.* 2005, *5*, 119-124.

[109] Y. Yao, J. Yao, V. K. Narasimhan, Z. Ruan, C. Xie, S. Fan, Y. Cui, Broadband light management using low-Q whispering gallery modes in spherical nanoshells. Nat. Commun. 3:664.

[110] L. Li, Y. Li, S. Gao, N. Koshizaki, Ordered Co_3O_4 hierarchical nanorod arrays: tunable superhydrophilicity without UV irradiation and transition to superhydrophobicity. J. Mater. Chem. 2009, 19, 8366-8371.

[111] A. Moser, K. Takano, D. T. Margulis, M. Albrecht, Y. Sonobe, Y. Ikeda, S. Sun, E. E. Fullerton, Magnetic recording: advancing into the future. *J. Phys. D* 2002, *35*, R157-167.

[112] J. Zhu, Y. Zheng, G. A. Prinz, Ultrahigh density vertical magnetoresistive random access memory. *J. Appl. Phys.* 2000, *87*, 6668-6673.

[113] F. Zhu, D. Fan, X. Zhu, J.-G. Zhu, R. C. Cammarata, C.-L. Chien, Ultrahigh-density arrays of ferromagnetic nanorings on macroscopic areas. *Adv. Mater.* 2004, *16*, 2155-2159.

[114] M. Albrecht, G. Hu, I. L. Guhr, T. C. Ulbrich, J. Boneberg, P. Leiderer, G. Schatz, Magnetic multilayers on nanospheres. *Nature Mater.* 2005, *4*, 203-206.

[115] Q. Chen, E. Diesel, J. K. Whitmer, S. C. Bae, E. Luijten, S. Granick, Triblock colloids for directed self-assembly. *J. Am. Chem. Soc.* 2011, *133*, 7725-7727.

[116] Q. Chen, S. C. Bae, S. Granick, Directed self-assembly of a colloidal kagome lattice. *Nature* 2011, *469*, 381-384.

Chapter 2

Recent Advances in Two-Photon Stereolithography

Arnaud Spangenberg, Nelly Hobeika,
Fabrice Stehlin, Jean-Pierre Malval, Fernand Wieder,
Prem Prabhakaran, Patrice Baldeck and
Olivier Soppera

Additional information is available at the end of the chapter

http://dx.doi.org/10.5772/56165

1. Introduction

Recent developments in nanoscience and nanotechnology were strongly supported by significant advances in nanofabrication. The growing demand for the fabrication of nano-structured materials has become increasingly important because of the ever-decreasing dimensions of various devices, including those used in electronics, optics, photonics, biology, electrochemistry, and electromechanics (Henzie et al., 2004; Fan et al., 2006). In particular, a societal revolution is expected with the miniaturization of mechanical, chemical or biological systems known as microlectromechanical systems (MEMS) (Lee et al. 2012), or micrototal analysis systems (μTAS) (Reyes et al. 2002, Dittrich et al. 2006, West et al. 2008).

Among all fabrication processes, photolithography has been strongly developed since few decades to fulfil to the needs of the microelectronics industry. Researches in this area were essentially motivated by finding ways to provide new solutions to pursue the trend towards a constant decrease of the size of the transistors as stated in the "Moore's law" (Moore, 1965). To reach these objectives, lithographic fabrication methods have been widely diversified leading to DUV lithography (Ridaoui et al., 2010), X-ray lithography (Im et al., 2009) and e-beam lithography (Gonsalves et al. 2009) to quote a few. Although some of these techniques exhibit resolution of less than 10 nm, these methods are inherently 2D. Unlike conventional microelectronics components, many MEMS or μTAS devices (motors, pumps, valves…) require 3D fabrication capability to insure the same function as the corresponding macroscopic device. Thus, lithographic fabrication of 3D microstructures has emerged and has been divided in two categories depending if they give access to restricted or arbitrary 3D pattern fabrication. On the first hand, specific structures as periodic patterns have been made using self-assembly

(Shevchenko et al. 2002), layer-by-layer assembly (Kovacs et al. 1998), soft lithography (Quake et al. 2000), and holographic photopolymerization (Campbell et al. 2000). However by these techniques, no free-moving or complex microstructures have been achieved. On the other hand, arbitrary 3D patterns have been realized by using the so-called direct write technologies which gathers ink-based writing (Lewis et al. 2004), microstereolithography (Maruo et al. 2002) and two-photon stereolithography (TPS) (Kawata et al. 2001; Maruo et al. 1997). Though examples of submicrometer resolution have been demonstrated for ink-based writing (Lewis et al. 2004) and microstereolithography (Maruo et al. 2002), these techniques are mainly used for micro or macrofabrication.

In this context, two-photon stereolithography which is an advanced version of microstereolithography appears of high interest since it offers intrinsically sub-100 nm resolution. Additionally to its unique ability of writing arbitrary structures with sub-100 nm features without use of any mask, TPS is also an attractive fabrication process due to the versatility of materials used including polymers, biopolymers, ceramics, metals, and hybrid materials.

The aim of this chapter is to review some recent works about two-photon stereolithography and its applications. In the first part, a brief introduction to TPS and the fundamental concepts will allow illustrating its interest and its current development. The second part will be dedicated to the most relevant materials developed for TPS regarding to the applications targeted. Furthermore, some typical applications where 3 dimensionalities play a crucial role will be highlighted. Finally, the last part will describe the recent advances in TPS both from the writing speed and the resolution (Li et al. 2009) in order to compete with other nanofabrication techniques. As the result, the contribution of this chapter is to propose a comprehensive overview of fundamental issues in TPS as well as its current and future promising potential.

2. Two-photon stereolithography

2.1. Introduction

One of the first attempts to fabricate 3D structures arised from IBM in 1969 (Cerrina, 1997). By combining electrodeposition and X-ray lithography, high-aspect ratio metal structures were obtained. Further works on X-ray lithography gave rise to the well-known process LIGA in the early 1980s (Becker et al. 1984). Despite the maturity of the technique and demonstration of some 3D complex structures, their application has not been widespread due to the availability of synchrotron radiation sources and X-ray masks.

Historically speaking, TPS began with the 3D microfabrication process using photopolymers developed by Kodama in 1981 (Kodama, 1981). Further developments lead to the birth of stereolithography, then microstereolithography to achieve resolution down to 1 μm. Even if in some cases, submicrometer resolution has been demonstrated (Maruo et al. 2002), it is still challenging to obtain microstructures with nano or submicron features due to the layer-by-layer nature of this technique. To overcome this drawback, Wu et al. (Wu et al. 1992) proposed the two-photon lithography concept which is based on the nonlinear optical process of two photon absorption (TPA). This work was directly inspired of the first demonstration of localized excitation in two-photon fluorescence microscopy by Denk et al. two years before (Denk et al. 1990). However, the feasibility of TPS[1] as a real 3D fabrication technique was illustrated by Maruo et al. in 1997 through the fabrication of a 7 μm-diameter and 50 μm-long spiral coil (Maruo et al. 1997). Finally, in 2001, micromachines and microbull with feature sizes close to the diffraction limit were realized (Kawata et al. 2001). By using non-linear effects of TPS, subdiffraction-limit spatial resolution of 120 nanometers has been successfully achieved.

2.2. Two-photon absorption

Contrary to conventional stereolithography techniques where polymerization is induced by absorption of a single photon, TPS is based on two photon absorption (TPA) process. TPA and more generally multiphoton absorption (MPA) process have been first predicted in 1931 by Marie Goeppert-Mayer (Goeppert-Mayer, 1931) and then verified experimentally thirty years later (Kaiser et al. 1961), thanks to the advent of laser. Finally, two-photon photopolymerization was experimentally reported for the first time in 1965 as the first example of multiphoton excitation-induced photochemical reactions (Pao et al. 1965). However, it's only with the commercialization of tunable solide ultrashort pulse laser like Ti:Sapphire laser in the 1990s that application of TPA is widespread in various domains like biology imaging (two photon fluorescence microscopy) or microfabrication (TPS).

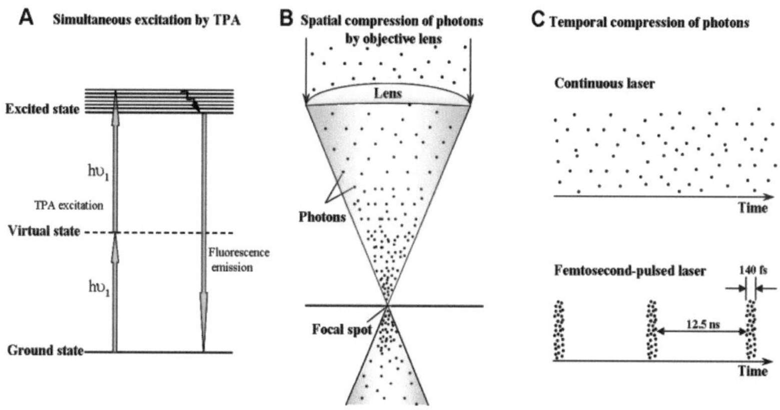

Figure 1. A. Mechanism of TPA when simultaneous excitation occurs. B and C. Illustration of two methods for increasing the probability that TPA occurs: density of photon is increased by B. spatial compression using objectives with high numerical aperture, C. temporal compression using ultrafast lasers.

Two different mechanisms have been described for TPA: the sequential excitation and the simultaneous two-photon excitation. In the frame of TPS, only the second one is involved (Figure 1A). In this case, a virtual intermediate state is created by interaction of the material with the first absorbed photon. In order to reach the first real excited state, a second photon

1 TPS is also called two-photon polymerization (TPP), multiphoton absorption polymerization (MAP), 3-dimensional Direct Laser Writing (3D DLW) or 3-dimensional lithography.

has to be absorbed during the short lifetime (around 10^{-15} s) of this virtual state. To increase the probability of such a non-linear absorption process, high density of photons is requested. Consequently, in main applications (including TPS) where TPA is involved, objectives with high numerical aperture (NA) and ultra short pulse laser are employed for increasing spatial and temporal density of photons, respectively (Figure 1B and C). The main interest of TPA compared to single photon absorption is that excitation is localized within the focal volume of a laser beam. Consequently, it gives access to 3D microfabrication since the polymerization threshold is not reached out of the focal volume. Typically, volume less than 1 μm^3 can be addressed. In parallel to the technical developments for TPA, molecular engineering has been strongly developed to design molecules or molecular architectures with large TPA cross section. An exhaustive review on this point can be found in reference (He et al. 2008) and few typical examples will be given in the next section.

2.3. Experimental set-up for TPS

The typical TPS setup is composed of three main parts: (i) the excitation source, (ii) the computer-aided design (CAD) system and (iii) the scan method. The excitation source with high intensity is important to favor TPA process. Even if Ti:Sapphire laser operating at 800 nm are often used, Baldeck and coworkers have demonstrated that TPS could be performed successfully by using a cheap Nd-YAG microlaser operating at 532 nm (Wang et al. 2002). The CAD system has to be chosen carefully since trajectories can influence the writing time and more important the mechanical resistance of the final structure. Finally, the scan method will have a crucial impact of the throughput of the writing process. The first possibility is to use Galvano mirrors for horizontal scanning coupled to piezoelectronic stage for vertical scanning which presents the advantage to scan with high speed. However, the total horizontal range accessible by this optical system is limited to few ten of micrometers due to spherical aberration when using objectives with high numerical aperture. The most popular solution consists to scan in x, y and z direction by using a piezoelectric stage. While the scan speed is low compared to previous option, few hundred of micrometer can be scanned.

As depicted in Figure 2, a typical set-up is composed of a mode-locked Ti:Sapphire laser as excitation source which presents duration pulse of ten hundred of femtosecond at 800 nm, repetition rate around 80 MHz and average output power of 1-3 W. The intensity of the laser is controlled by an optical or mechanical shutter. Before the introduction of the beam into the microscope, it is expanded so as to overfill the back aperture of the objective. By tightly focusing the pulsed laser beam (ns to fs pulses) into a multi-photon absorbing material, it is possible to trigger a photoreaction (e.g. photopolymerization) inside a volume below the dimension of the voxel. Complex structures (such as in Figures 3, 4, 6 and 8) can then be generated by moving in the laser focus in the 3 dimensions inside the monomer substrate. Usually, samples are placed on a 3D piezoelectric stage, and then move above the fixed laser beam by CAD. Upon the irradiation, only areas exposed at the focal point are polymerized further than the polymerization threshold, leading to the desired structures after washing away the unsolidified photoresist. Finally, the back reflection is collected by an additional port and send to a CCD to monitor the fabrication in real time.

http://dx.doi.org/10.5772/56165

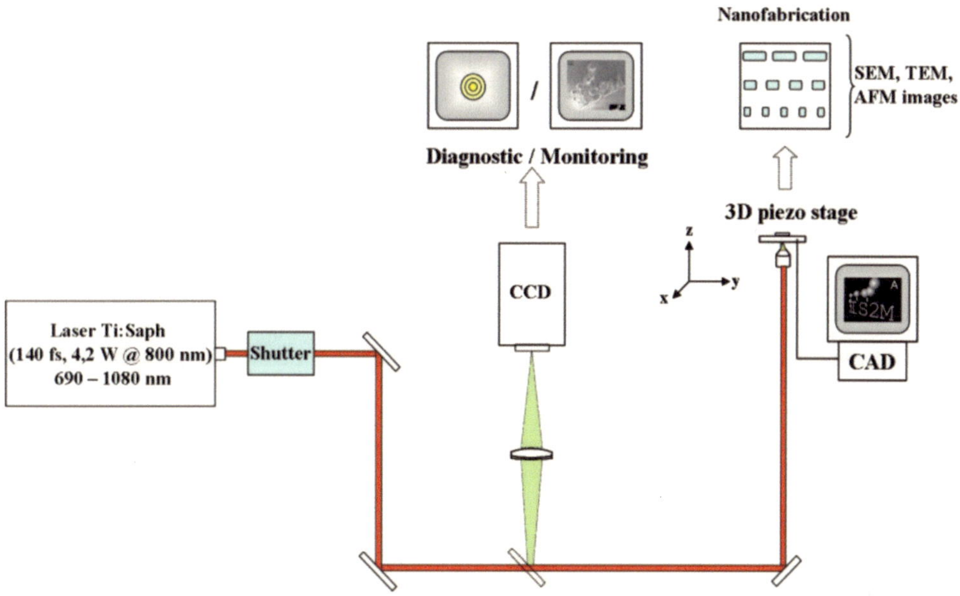

Figure 2. Schematic typical experimental TPS set-up.

3. Materials for two-photon stereolithography

Basically, the formulation destined to TPS is composed of 2 main components: a photoinitiator system and a monomer. The two-photon photoinitiator is a species, or a combination of chemicals able to absorb efficiently two photons to generate excited states from which reactive species can be created. One of the most important parameter is the two-photon absorption cross section that directly characterizes the capacity of the photoinitiator to absorb two photons. The reactive species (radicals or ions) must be able to initiate polymerization of the monomers that constitute the building blocks of the final material. After initiation, propagation and termination reactions take place as observed in the sequences of classical polymerization scheme.

In principle, any one-photon photopolymerizable system can be adapted to TPS, provided that a suitable TPS photoinitiator can be added to the monomer system. Most of the published works deal with free radical photopolymers. The main reason is relative to a wider availability of free radical photoinitiators.

Additionally to the photoinitiator and monomers, other chemicals can be added in TPS systems like inhibitors (to control the polymerization threshold and thus spatial extend of polymerization - some examples will be given in the next section), and additives to bring specific properties (fluorophores, metal nanoparticles, quantum dots, etc...). The choice of the monomer is in relation with the final application whereas the choice of the photoinitiator integrates the irradiation wavelength and nature of the monomer.

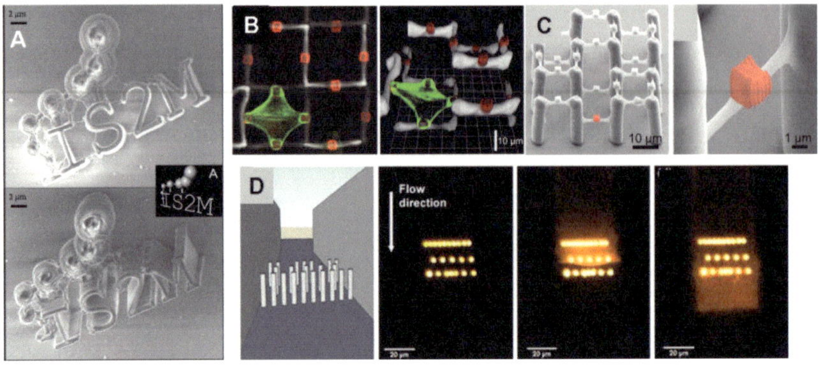

Figure 3. structures realized by TPS with A. polyacrylate derivatives (λ_{exc}: 532 nm; average power : 20 μW; writing speed: 45 μm.s^{-1} ; Malval et al. 2011), B. and C. sol-gel materials for biocompatible 3D scaffold (λ_{exc}: 808 nm; average power : 20 mW; writing speed: 200 μm.s^{-1} Klein et al. 2011), D. trypsine derivative for biocatalysis degradation (λ_{exc}: 532 nm; average power : 1 mW; writing speed: nc, Iosin et al. 2012). Reproduced from the respective references.

As illustrated in Figure 3, different examples of 3D structures have been realized with various materials. Interest of such structure will be discussed in each corresponding material sub-section. The next paragraphs are aimed at giving some examples of systems and associated applications.

3.1. Free-radical photoresists

Among different monomers available, acrylate monomers have been the most widely used for TPS. The reason for this success is a wide range of commercially available acrylate monomers with tailored properties: chain length, number of reactive function, viscosity, polarity to quote a few. Moreover, acrylate monomers exhibit a high reactivity and good mechanical properties that allow complex 3D structures being created.

In parallel, a wide choice of free-radical TPP photoinitiators has been developed. Highly efficient two-photon absorbing systems such as 4,4'-dialkylamino trans-stilbene (Cumpston et al. 1999) and other bis-donor bis(styryl)benzene or bis(phenyl)polyene (Lu et al. 2004; Zhang et al. 2005; Rumi et al. 2000) were employed for two-photon initiated free radical polymerization. Other conjugated photoinitiators as fluorenes (Belfield et al. 2000; Martineau et al. 2002, Jin et al. 2008) and ketocoumarins (Li et al. 2007) derivatives also demonstrated remarkable TPS properties. Finally, another promising strategy is the direct photogeneration of highly reactive radicals such as α-aminoalkyl ones.

In particular, Malval et al. demonstrated an elegant strategy to improve the efficiency of thioxanthone-based systems (Malval et al. 2011, Figure 3A). New hybrid anthracene-thioxanthone system assembled into a chevron-shaped molecular architecture was proposed. A strong increase in the two-photon absorption cross section by more than a factor of 30 as compared to thioxanthone was observed. As a consequence, anthracene-thioxanthone constitutes a suitable two-photon initiating chromophore with a much higher efficiency as thioxanthone used as reference. At λ_{exc} = 710 nm for instance, the two-photon polymerization threshold of anthracene-thioxanthone was shown to be five times lower with respect to that of thioxanthone.

Additional examples of polyacrylate structures realized by TPS can be found in figure 4 section 4.1.

3.2. Cationic photoresist

Cationic photoinitiated polymerization of epoxides, vinyl ethers and methylenedioxolanes has received increasing attention, owing in large part to the oxygen insensitivity of the cationic process (Belfield et al. 1997a and 1997b). Moreover, cationic photoresist appears as an interesting choice from application point of view since UV negative tone photoresists have demonstrated their interest for microelectronics, optics, microfluidic or MEMS.

However, the difficulty to design efficient TPA photoacid generators has limited the development of TPA cationic photoresists. For this reasons, many efforts have been devoted to increase the sensitivity of such systems. First approach was based on sensitizers such as coumarin (Li et al. 2001), phenothiazine (Billone et al. 2009), or thioxanthone (Steidl et al. 2009) associated to a commercial PAG such as onium salts. Second approach relies on a molecular association of the acid generator functionality into the structure of the two-photon active chromophore. In the latest case, the reactivity of the PAG is no longer limited by diffusion and thus a significant improvement of the photopolymerization efficiency was demonstrated (Zhou et al. 2002; Yanez et al. 2009; Xia et al. 2012).

Among other application, the epoxy-based photoresists are extremely interesting when complex structures with high aspect ratio are needed. Indeed, thanks to their good mechanical properties, they have been successfully used for application in microfluidic (Maruo et al. 2006) or MEMS (Bückmann et al. 2012).

3.3. Advanced functional materials

Despite their advantages, polymers have intrinsic limitations for some applications. For instance, their mechanical properties at high temperature or in contact with solvents degrade rapidly. They also present low refractive index that limits their use in optical applications. Their toxicity may prevent them from use in contact with living organisms. For these reasons, alternative strategies have been developed to combine the advantages of 3D structuration by TPA and functional materials.

The sol-gel route is interesting in the frame of micro-nano-fabrication since it allows the fabrication of inorganic or hybrid organic-inorganic materials at relatively low temperature. The first strategy followed for combining lithography and sol-gel materials consisted in developing hybrid precursors that can undergo both sol-gel hydrolysis-condensation reaction and photoinduced crosslinking (Blanc et al. 1999; Soppera et al. 2001). These materials, also called Ormorcer® or Ormorsil® have been adapted to TPA by use of suitable photoinitiators and interesting applications in the frame of optics (Ovsianikov et al. 2008) or biology (Klein et al. 2011, see Figure 3B, C) were demonstrated. These materials were mostly used in optics since the refractive index of the material can be tuned by adding metal alcoxides. However, in these hybrid materials, the proportion of organic moieties in the crosslinked material is still important, so many efforts have been dedicated to the formulation of fully inorganic materials (Passinger et al. 2007).

Another important class of inorganic materials is metals. Metals nanoparticles, nanostructures or thin layers are indeed very interesting for electrical connections in devices and also for their plasmonic properties. Recent works have been reported on the fabrication of 3D metallic structures by combining TPS and silver evaporation (Rill et al. 2008). However, by this process, full metal coverage is challenging and induces a supplementary step. Therefore, several groups have developed more direct strategies based only on TPS. For instance, Prasad et al. have fabricated submicrometric plasmonic structures which exhibit interesting conductive properties (Shukla et al. 2011). More recently, Spangenberg et al. also demonstrated that a silver complexe can be used as TPA photoinitiating system and as precursors for nanoparticles fabrications, leading in a single step to a polymer-metal nanoparticles nanocomposite (Spangenberg et al. 2012). Such routes open the doors towards microstructures with conductive properties or magnetic properties that can be useful for MEMS actuation.

One other very important and growing field of applications for TPA materials is relative to biological applications. TPA microstructuring has been extensively used the last years to propose micro and nanostructured surfaces with tailored chemical composition to be used as model substrates to investigate the development of biofilms. The unique advantage of TPS is to propose real 3D structures that mimic with more accuracy the local environment of cells or bacteria than planar surfaces. In this context, polymer matrixes have been widely used but also, sol-gel materials were proposed since they are biocompatbile materials that can be used as inert topological matrixes, as illustrated in Figure 3B and 3C (Klein et al. 2011). Additionally, biological materials like trypsin or collagen precursors were developed to propose a direct writing route for 3D biocompatible structures. Besides the interest of allowing a direct writing of complex structures, an advantage of TPA microfabrication is to be adapted for integration of microstructures in closed environment. For example, Iosin et al. demonstrated the possibility of integrating trypsin's micropillars in a microfluidic system (Figure 3D). Trypsin is an enzyme used for catalyze the degradation of specific peptide. Interestingly, by following the variation of fluorescence intensity resulting from the peptide clivage, the authors have shown that trypsin structures kept its enzyme catalysis activity.

Although numerous materials have been designed to fulfill the requirements of various applications, there is still an important demand for optimizing current systems. Besides, with the emergence of STED–like lithography (for STimulated Emission Depletion, see next section), designs of new photosystems will be crucial for the development of such technique.

4. Current challenges in TPS

As shown through several examples, TPS is a powerful and attractive technique for present and futures applications. Due to the need of a well-controlled 3D nanofabrication technique, several commercial set-ups have emerged on the market since 5 years. However TPS suffers from two main drawbacks for a more largely widespread in other scientific area or in industries. The first roadblock concerns the low-throughput of the process. Indeed, TPS is based on a serial process (i.e. point-by-point writing) which is a serious problem when mass production is needed. Moreover, compared to low-throughput techniques like e-beam lithography, resolution achievable by TPS is still 1 or 2 order of magnitude lower.

In this section, we will discuss about different approaches to address these specific points and highlight some recent developments which answer mostly to these drawbacks and promise a brilliant future to TPS.

4.1. Resolution

Because of the rapid improvement of TPS resolution in the past decade, a special attention has to be care on the way to define and measure it. In the most of works, "resolution" corresponds to the lateral and/or axial features size of single voxel or single line. Different methods such as ascending scan method (Sun et al. 2002) or suspending bridge method (DeVoe et al. 2003) have been proposed in order to improve the accuracy of the measurements. However, these two methods suffer from drawbacks which are the difficulty to avoid the truncation effect and the unknown influence of material's shrinkage. Therefore, though less information is provided, most of the groups define resolution as the width of a single line on the surface.

Nevertheless, with the emergence of several STED-like lithographies, more precise definition of resolution has become necessary in order to compare their abilities. Although extension of the famous Abbe's criterion introduced in conventional or two-photon absorption microscopy can be extended to TPS to describe the optical limitation of the lithography system, ultimate resolution for a given optical lithography system has to be determined by considering the role of the photopolymer. Indeed, in the frame of the writing of two close lines, due to consumption of photoinitiators and diffusion of various species (photoinitiator, scavengers), the writing process of the second line can be strongly affected. This effect is sometimes referred as the resin's memory. Thus strong dependence with respect to the initial concentrations of photoinitiator as well as the viscosity of the matrix is expected. Up to date, no mathematical model includes all the parameters. Therefore, as suggested by Fischer et al. (Fischer et al. 2012), a better solution for determine both axial and lateral resolution would be the fabrication of a 3D periodic unit as a crystal photonic for a given photopolymerizable system. It has to be mentioned that typical ratio between axial over lateral resolution in TPS is ranging from 2 to 5 depending of optical conditions. In the next part, one has to keep in mind that the resolution or feature size is given for both an optical and chemical system.

4.1.1. Overview of the different strategies

Since the microbull with 120 nm features size realized by Kawata and coworkers (Kawata et al. 2001), various approaches have been attempted to improve the resolution of TPS (Figure 4).The first approach which is still used nowadays relies on the design of high-efficiency photoinitiators. By this method, linewidths of 80 nm have been measured (Xing et al. 2007).

Another approach based on the use of a shorter wavelength has allowed writing of 3D structure with 60 nm feature size (Haske et al. 2007). Indeed, as dictated by the extended to TPS Abbe's criterion, lateral resolution is proportional to the wavelength. However, the wavelength can not be reduced indefinitely since the material may absorb linearly at shorter wavelength and consequently lead to the lost of the intrinsic resolution of TPS. Finally, more recent and impressive feature size was obtained by an enhanced version of TPS inspired by STED microscopy (Li et al. 2009). The principle of this technique will be described in detail in the next section. With 800 nm excitation wavelength, voxel of 40 nm height have been achieved, that represents $\lambda/20$. This spectacular result has to be compared with the voxel of 600 nm height obtained by using conventional TPS where excitation wavelength is set at 800 nm which corresponds to $\lambda/1.33$. Even if no experimental evidence has been shown for lateral resolution by this technique, $\lambda/20$ is also clearly achievable. Further insight of this new technique is addressed in the next subsection.

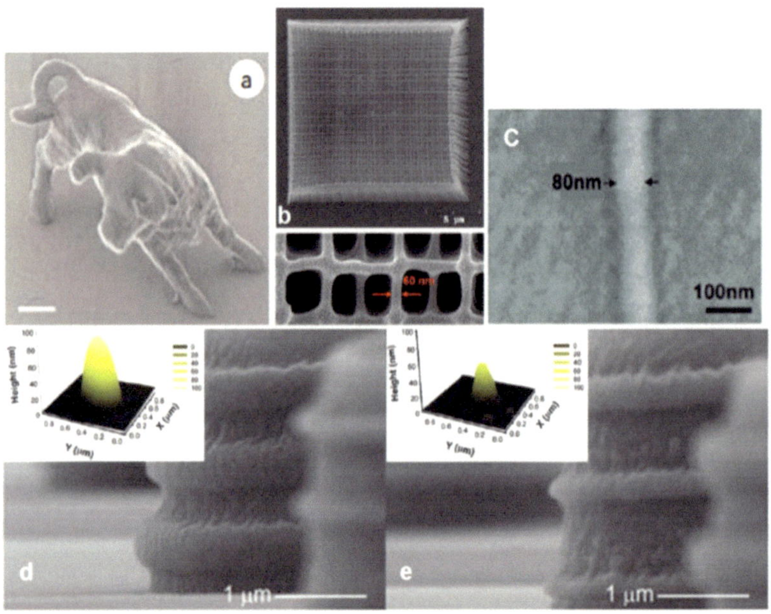

Figure 4. Improving spatial resolution of two-photon microfabrication by different strategies during the past decade. a) the famous microbull exhibiting 120 nm features size due to intrinsic properties of TPS (λ = 780 nm, λ/6.5; Kawata et al. 2001), b) Photonic crystal with 60 nm features size due to the use of shorter wavelength (λ = 532 nm, λ/8; Haske et al. 2007), c) 80 nm linewidth by using a high efficient photoinitiator (λ = 800 nm, λ/10; Xing et al. 2007), d) and e) microtower fabricated by using a conventional TPS and a STED-like TPS, respectively. In d and e insets, AFM measurements of voxel size are shown. e) inset is the current smallest axial resolution (40 nm) for a voxel thanks to a STED-like method (λ = 800 nm, λ/20; Li et al. 2009). Reproduced from the respective references.

4.1.2. STED-like lithography

In the recent past, the diffraction resolution barrier of fluorescence light microscopy has been radically overcome by stimulated emission depletion (STED) microscopy. Since its theoretically (Hell et al. 1994) and experimentally (Klar et al. 1999) birth, STED delivers nowadays routinely images of biological samples with a resolution down to 10-20 nm. Due to its great achievements in life-science, STED and more globally super resolution microscopy have been recognized as the "method of the year" in 2008 in Nature Methods. Finally, world record lateral resolution down to 5.6 nm using visible light has been reported by Hell's group (Rittweger et al. 2009). In STED, a first short laser pulse is used to bring fluorescent molecules in their excited state. In order to de-excite these chromophores through stimulated emission, a second laser pulse (usually at longer wavelength to avoid one photon absorption) has to occur after vibrational relaxation of the excited electronic state but before fluorescence occurs i.e. few ps to few ns later than the first laser pulse. The efficiency of the deactivation strongly depends on the intensity and the wavelength of the depletion pulse, as well as the time delay of depletion pulse versus the excitation pulse. The precise localization of fluorescence arises from the spatial phase shaping of the depletion beam. The latter causes de-excitation to occur everywhere except in a region at the center of the original focal volume. The idea to translate these groundbreaking concepts to optical lithography has been evoked in 2003 (Hell et al. 2003), but first demonstration applied to TPS has been published only 6 years later (Li et al. 2009). Nowadays, in the frame of STED-like optical lithography, 3 different depletion mechanisms have been reported in the literature. In all cases, two laser beams are used, one for excitation and a second one for deactivation as illustrated in Figure 5. Whereas the excitation beam allows the formation of species (i.e radicals) which initiate the polymerization, the phase shaped deactivation beam allow photophysically or photochemically inhibition of the reticulation around the central excited zone. Depending of the phase mask used, the voxel can be reduced along the axial direction (bottle-beam shape, see Figure 5B) or along the lateral direction (donut shape).

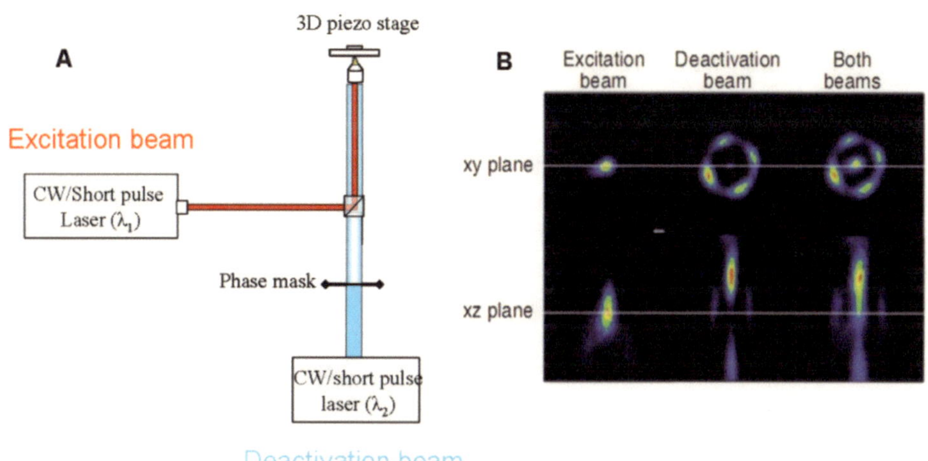

Figure 5. A. Schematic experimental set-up for STED-like optical lithographies. B. False-color, multiphoton-absorption–induced luminescence images of the cubes of the PSFs of the excitation beam, the phase shaped deactivation beam, and both beams together. Adapted from reference (Li et al. 2009).

4.1.2.1. Two-color photoinitiation/inhibition lithography

Among the three STED-like optical lithography methods, the so called two-color photoinitiation / inhibition lithography (2PII) is the only one based on a *photochemical* deactivation of photoinitiator (Scott et al. 2009, McLeod et al. 2010). In this case, by using continuous laser, the excitation is performed by a single photon absorption process at 473 nm and deactivation occurs also by a linear process at a distinct wavelength (364 nm). Upon deactivation beam, two weakly reactive radicals are produced which can interact with the initiating radical to stop the polymerization. By using a donut mode for shaping the deactivation beam, reduction of the lateral extent of single voxel until 65 nm is reached. More recently, with a similar set-up, another group has claimed having designed a more efficiency photopolymerizable system which has been illustrated by the fabrication of 40 nm dots (Cao et al. 2011). While the idea to use inexpensive continuous laser sources is attractive, this process has been demonstrated only for fabrication of 2 dimensional structures. Although manipulation of the photoinhibiting wavelength into a "bottle beam" profile would induce confinement along the third axis, when focusing deeper into the photoresist volume, both excitation and deactivation might be attenuated by the linear absorption of the material. Besides, consumption of photoinitiator and/or photoinhibitor along the pathway of the beam could lead to a time and space dependence of concentration of photoactivable species during the 3D fabrication. 2PII based on two-photon absorption process for both excitation and deactivation could avoid this problem, but up to now, no such experiment has been reported in the literature.

4.1.2.2. RAPID lithography

The first attempt to translate the spectacular optical resolution from STED microscopy to lithography has to be attributed to Fourkas's group (Li et al. 2009). Contrary to 2PII, deactivation is based on a *photophysical* process. On preliminary experiments, Fourkas and coworkers have used the conventional STED configuration. In this case, the excitation beam (800 nm) was overlaid by a second beam red-shifted with respect to the excitation beam wavelength. Then, in order to enhance the efficiency of stimulated emission, the depletion pulse was stretched to duration of 50 ps. to guarantee enough intensity of the depletion beam. Deactivation i.e. inhibition of the polymerization was observed for a wide range of wavelength (760 to 840 nm). However, whereas tuning the pulse delay from 0 to 13 ns between the two beams should affect the efficiency of the inhibition, no significant effect had been observed. Consequently the deactivation was not assigned to stimulated emission like in STED, but the authors have ascribed this effect to the depletion of an intermediate state with longer lifetime. Therefore they have decided to name this method RAPID for Resolution Augmentation through Photo-Induced Deactivation. Further experiments are under investigation to determine the nature of the intermediate specie.

Thanks to the longer lifetime of the intermediate, a second configuration has been successfully used: depletion effect has been performed with continuous laser which cancel the need to control the delay between excitation (800 nm) and depletion (800 nm) beams. By using the later configuration and a bottle beam profile for the depletion beam, voxels of 40 nm height have been achieved. In the frame of this study, the depletion effect is so sensitive that the excitation beam can induce itself the deactivation. Interestingly, for a RAPID compatible photoinitiator, the linewidth increases at faster scan speed. In the opposite, in conventional photoinitiator, faster scan and consequently weaker exposure dose yields to decrease of the linewidth. This opposite effect for RAPID photoinitiator is explained by the depletion effect done by the excitation beam. Indeed, slow down the scan speed allow to excited photoinitiator to be deactivate. Smartly, the authors have taken benefit of this unexpected dependence towards scan speed to propose system insensitive to abrupt change of trajectory (Figure 6). In this case, photopolymerizable system combines both conventional and RAPID photoinitiators. Finally, according to conclusions of Wegener and coworkers (Fischer et al. 2012), because the depletion time-constant is between 15 and 350 ms in case of RAPID lithography, writing speed is comprised between 30 μm.s^{-1} to 150 μm.s^{-1}. While this speed corresponds to typical speeds used in academic field, this slow speed might be an obstacle for its use in industry (see section 4.2).

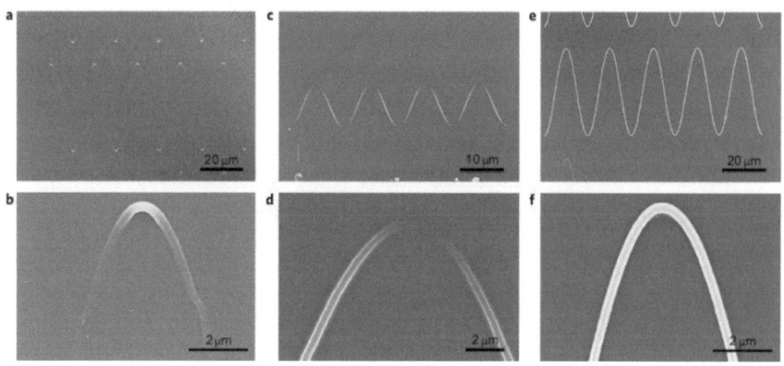

Figure 6. A, B Large and close view of fabrication of sinusoidal structures with a conventional photoinitiator, respectively. C, D Large and close view of fabrication of sinusoidal structures with a RAPID photoinitiator, respectively. E, F Large and close view of fabrication of sinusoidal structures with a mixture of conventional and RAPID photoinitiators, respectively. Reproduced from reference (Stocker et al. 2010).

4.1.2.3. STED lithography

For efficient STED, molecules have to present large oscillator strength between the ground state S_0 and the first excited sate S_1 to favor later depletion. Because of their use in fluorescence microscopy, such type of molecules has also to exhibit strong fluorescence quantum yield. But this relatively long lifetime excited state (usually few ns) allows the depletion to take place.

In contrary, common photoinitiator exhibit low oscillator strength and are designed to present efficient intersystem crossing (ISC) yielding to reactive species which can give rise to radical and further to polymerization. Moreover, excited state lifetime of photoinitiator is usually found to be around 100 ps which would result in the use of high power pulses shorter than 100 ps to induce depletion. Unfortunately, with this large pulse energy, depletion could be competed by multiphoton absorption leading to undesired polymerization. Compare to previous STED-like lithography technique, STED lithography requires the use of two distinct-wavelength short pulse lasers for both excitation and deactivation. In addition, pulse delay between the two beams has to be controlled carefully. While this configuration seems more constraining than 2PII and RAPID, higher scan speed (around 5 $m.s^{-1}$) is expected (Fischer et al. 2012).

STED lithography experiment has been attempted with isopropylthioxantone (ITX) as a photoinitiator. But further experiments such as pump-probe experiment have shown that the STED mechanism was not the main depletion pathway (Wolf et al. 2011) as claimed in previous work (Fischer et al. 2010). Based on the pump-probe experiment in ethanol (Wolf et al. 2011), a better suitable candidate appears to be the dye (7-diethylamino-3-thenoylcoumarin) (DETC) since stimulated emission was clearly demonstrated.

However, because the S_1 lifetime of a molecule usually depends on the solvent, a detailed and adapted pump probe study of DETC in the monomer has been realized (Fischer et al; 2012). While it has been shown that stimulated emission was not the only possible pathway, it was the first clear evidence of the possibility to perform true STED lithography. Fast and slow components of depletion were observed to exhibit opposite wavelength dependencies which indicate the existence of two distinct depletion mechanisms (Figure 7A). The fast component was ascribed to stimulated emission depletion, since its spectral dependence fits nicely the spectrum of the stimulated-emission (SE) cross-section. The slow component was not assigned in the frame of this study and further studies have to be accomplished to unravel this point. It has to be noted that at longer wavelength the relative strength of the fast component is weak regarding those of slow component.

Interestingly, STED lithography pump-probe experiment with the same photopolymerizable system, but with a depletion wavelength set at 642 nm has been performed by Harke and coworkers (Harke et al. 2012). In this experiment, pulse delay experiment has been realized, but no evidence of STED has been observed. Nevertheless, the unique component of depletion effect presents a timescale in the same range as typical triplet lifetime. The authors assume that the depleted excited state is not the singlet state S_1 as in STED, but the triplet state T_1. Besides, one possible and well-known pathway to deplete the triplet state proposed by the authors is the reverse intersystem crossing (ReISC).

Until now, only 3 pump-probe experiments have been performed by two distinct groups. Even if these studies lead to different observations and conclusions, it can certainly be explained by the use of different experimental conditions (depletion wavelength, pulse delay, excitation wavelength). Thus results illustrate the need to improve the knowledge in STED-like lithography process to define requirements list for efficient STED lithography photoinitiators.

The recent progresses of STED-like lithography have allowed new very promising applications in photonics. 3D polarization-independent carpet cloak for visible light have been fabricated for the first time which demonstrate the unique ability of TPS as 3D fabrication method (Fischer et al. 2011). In this case, 3D photonic crystal exhibits distance between two lines of 375 nm and 175 nm in axial and lateral directions, respectively. This has to be compared with the 510 and 210 nm values found in the frame of conventional TPS for axial and lateral directions, respectively. Whereas noticeable improvement has been shown for axial resolution,

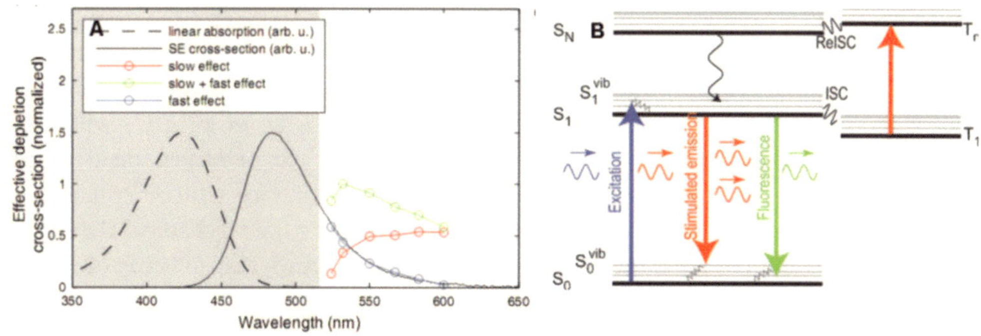

Figure 7. A. Spectral sensitivity of the different processes for 10 mW depletion power. Due to pronounced single photon absorption, depletion is not possible in grey area. B. schematic illustration of the different pathway involved in the depletion of DETC. Reproduced from references (Fischer et al. 2012 and Harke et al. 2012).

the gain in lateral resolution is less remarkable. This is explained by the use of a bottleneck beam shaping for the depletion beam. While combination of bottleneck and donut phase masks could be used to shape the beam and so to improve simultaneously lateral and axial resolution, it may be interested for specific applications to use only bottleneck beam since it can induce a more spherical voxel (ratio of 2.1 in this example).

Owing to its current and unique fabrication ability and its potential ability regarding to high-throughput (5 $m.s^{-1}$ scan have been predicted for pure STED lithography), an exponential increase of works on this becoming hot topic is expected in the near future.

To conclude about these STED-like section, it has to be mentioned that because of the relative novelty of STED-like optical lithographies (since 2009) and the fact that until now only 5 groups in the world have shown their skills to design such type of experiment, new insights are expected to appear rapidly in the near future. Interdisciplinary research has to be lead in order to propose a STED lithographic set-up with the dedicated optimized materials for few tens of nanometers in three dimensions. This will give birth of the first 3D arbitrary nanofabrication technique.

4.1.3. Diffusion-assisted TPS

An alternative method to improve the resolution is to add a quencher in the photopolymerizable system. In presence of quencher, the photoinduced radical can be quenched which consequently prevents polymerization. By this way, it has been shown that the radical diffusion can be controlled resulting in the confinement of the polymerization region (Tanaka et al. 2005). However, in this reported work, the concentration of quenching molecules has to be much larger than those of radical produced in order to result in an effective deactivation. Therefore, Lu and coworkers designed a novel photoinitiator with a radical quenching moiety (Lu et al. 2011). In this case, an intramolecular radical deactivation can occur leading to a more efficiently control of radical diffusion than in the case of an intermolecular one. As a result, finer features can be formed. However, by these methods no sub-diffraction gaps between two lines have been demonstrated, and only small effects on the feature size have been observed.

More recently, Sakellari and coworkers proposed another route to control the extent of the polymerization region (Sakellari et al. 2012). From their point of view, since a nondiffusing quencher results only in an increase of the polymerization threshold, they proposed to add a mobile quencher. Contrary to other works (Tanaka et al. 2005, Lu et al. 2011) where the quencher plays its inhibitor role by interacting with the photoinitiator or the generated radical, the quencher used in this work is an amine-based monomer. It interacts with other monomer or become part of the polymer backbone without compromising the mechanical stability of the structure. Last but not least, the amine functions allow a future metallization or further chemical functionalization. By this method, fabrication of woodpile structures with 400 nm intralayer period has been achieved for the first time with a *single beam* (Figure 8). Without the amine-based monomer, the authors have already shown in previous works that the minimum intralayer period achievable for an equivalent crystal photonic was around 900 nm (Sun et al. 2010). Moreover, this 400 nm intralayer period obtain by diffusion method has to be compare with the best result obtained by STED-like lithography, i.e. 375 nm intralayer period for the same type of photonic crystal (Fischer et al. 2011). Interestingly, while a single beam is used in the frame of this diffusion assisted high resolution TPS (as named by their authors), comparable resolution are obtained in both cases. While this method is easier to implement in laboratory compare to STED-like lithography, it has to be mentioned that in order to get such impressive resolution, the scan speed is intrinsically low to allow diffusion of the quencher into the scanned area. Typical scan speed in this study is around 20 $\mu m.s^{-1}$. In the last section, solutions to overcome this speed limitation will be addressed.

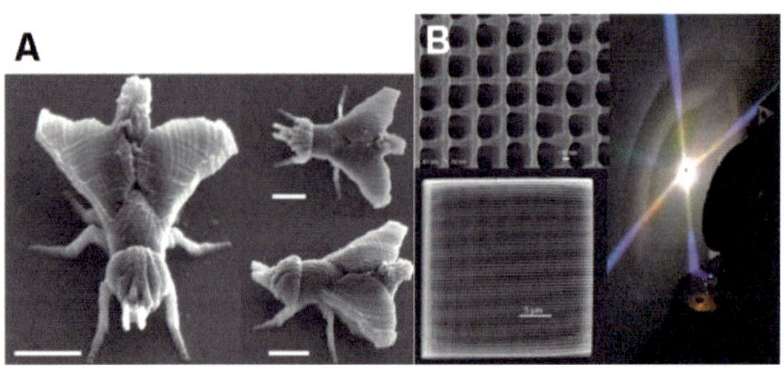

Figure 8. A. Microstructures realized by intramolecular quenching method. Scale bars are 5 µm. B. SEM images of photonic crystal fabricated by diffusion assisted high resolution TPS and diffraction pattern generated by the photonic device. Reproduced from references (λ_{exc}: 780 nm; average power : 7.70 mW; writing speed: 66 $\mu m.s^{-1}$, Lu et al 2011) and (λ_{exc}: 800 nm; average power : 20 mW; writing speed: 20 $\mu m.s^{-1}$ Sakellari et al. 2012), respectively.

4.1.4. Other methods

Recently, 40 nm feature size has been obtained by combining chemical and optical approaches (Emons et al. 2012). The measurement of feature size has been performed by suspending bridge method. From a chemical point of view, the authors have demonstrated that the addition of a crosslinker (pentacrylate derivative) allow a resolution enhancement of a factor 2 (150 nm feature size versus 82.5 nm with a 50 fs pulse laser). As expected, the addition of crosslinker should play a positive effect on the resolution since it allow to the suspended line to be maintained during the development step. In the other hand, an additional resolution enhancement has been achieved by using shorter laser pulse: the use of 8 fs instead of 50 fs allows improving feature size of the line from 150 nm to 90 nm for photoresin without cross linker.

In addition to the above optical and chemical tricks, further improvement of resolution can be achieved by other minor technical development like high hybrid optics diffractive (Burmeiter et al. 2012).

To conclude, recent technical developments of TPS open the doors to strong improvement of the resolution. Even if the diffraction limit has been beaten both in lateral and axial direction thanks to different methods such as the STED-like lithography or the diffusion-assisted high resolution TPS, effort research has to be focus on new photopolymerizable system to benefit completely of the intrinsic resolution achievable by the different techniques. For STED-like lithography, optimization of the photopolymerizable system should lead to feature size around 10 nm. Finally, when comparing the different technique, a particular attention has to be taken into account concerning the maximum scan speed for future use in industry. This will be the object of the next section.

4.2. Recent advances in throughput: From prototyping towards production of semi-series

Despite the possibilities to fabricate 3D objects with sub-100 nm features in a single step, TPS use is as far as we know limited to scientific community. Indeed, owing to the point-by-point writing process, TPS appears as an extremely slow technique for mass production in industry. Typical writing speed range in academic research goes from few $\mu m.s^{-1}$ to few $mm.s^{-1}$ which has to be compared with the few $m.s^{-1}$ used in industry for different laser process (ablation, laser control, rapid prototyping by inspired-stereolithography methods,…). Until 2003, as for resolution, the research effort for increasing the throughput of TPS was mainly focused to the synthesis of high-efficiency photoinitiators. Nevertheless, in the past decade several research groups have proposed various strategies to break down this technological bolt.

4.2.1. Influence of the scanning mode

As an attempt to solve this problem, Sun et al. demonstrated the impact of the laser beam trajectory over the manufacturing time by significantly increasing the fabrication efficiency of 90% when using CSM (contour scanning method, also called vector mode) mode rather than RSM (raster scanning method) mode. As shown in Figure 9, in the raster mode, all voxels which constitute the volume whose contains the microstructure are scanned. In case of CSM mode, only the voxels defining the surface of the microstructure are scanned. As a result, it took 3 hours or 13 minutes to manufacture the micro-bull using RSM and CSM mode respectively (Sun, 2003). Further information on the role played by trajectories can be found laser and photonics review (Park et al., 2009).

For applications where the objects have to be completely filled such as microlens, an additional UV exposure step is done. However cracks in the structure might occur to leading to a dramatic decrease of desired performance.

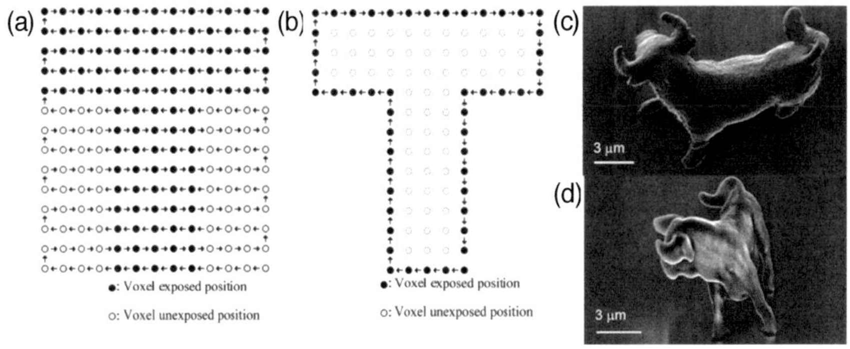

Figure 9. Two fabrication strategies based on scanning mode: a) Raster mode, b) Vector mode (or contour mode), c) and d) SEM images of a microbull structures using RSM and CSM respectively. Reproduced from reference Sun 2003.

4.2.2. Replication

To address the low-throughput of the method, LaFratta et al. have proposed to use TPS in tandem with soft lithography technique known as microtransfer molding (Xia et al. 1998). By this way, high-fidelity molds of structures with extremely high aspect ratios and large overhangs have been realized (LaFratta et al. 2004, Figure 10). Besides, in the frame of this work, more than ten replicas have been made from a single master without any significant deterioration of the resulting structures. Even if this technique have been applied to more complex structures such as arches or coils (LaFratta et al. 2006), a range of geometries or objects such as closed traps or micropumps are still impossible or currently too challenging to replicate using microtransfer molding. Finally, from our knowledge, no improvement or additional example of this technique has been recently reported in the literature, underlying the difficulties to separate the molder to the master without damage. Recent advances in soft lithography might facilitate the delivering step.

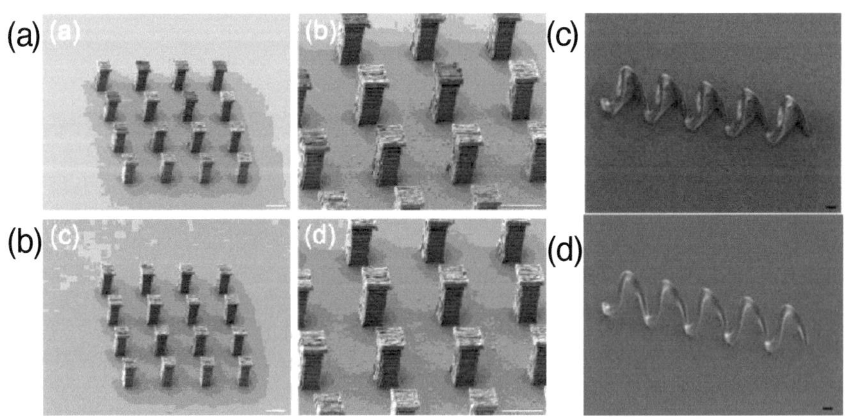

Figure 10. a) and b) SEM images of master and replica array of towers respectively, c) and d) SEM images of master and replica of coil respectively. Scale bars are 10 µm. Reproduced from reference LaFratta 2004 and 2006.

4.2.3. Multi-focal TPS

Another solution for boosting fabrication speed while avoiding geometric limitations associated with molding is the use of multi-focal strategy. This technical innovation has been first demonstrated by Kawata and coworkers in 2005 (Kato, 2005). By combining TPS with a microlens array, more than one hundred identical and individual 3D objects have been written simultaneously resulting in a two-order increase in the fabrication yield compared to single-beam TPS (Figure 11). In 2006, Kawata et al. succeed to write in parallel more than 700 hundred identical structures (Formanek, 2006) illustrating the high potential for large scale production.

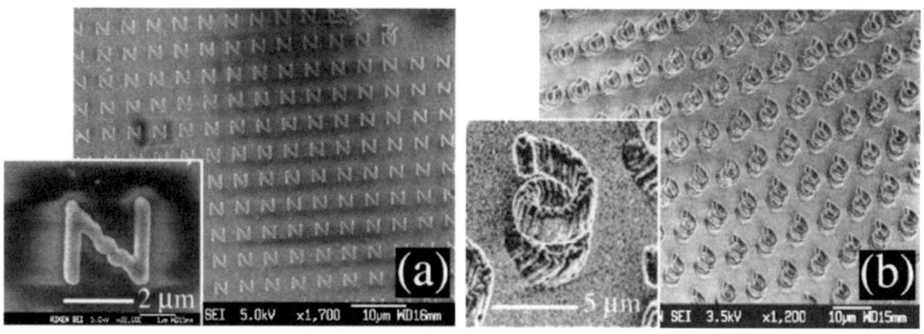

Figure 11. a) and b) examples of two- and three-dimensional fabrication by mean of microlens array.

Recently, Ritschdorff et al. proposed a more general approach of multi-focal TPS to allow parallel but independently writing of different objects (Ritschdorff et al. 2012). Indeed, until this work, advanced TPS based on multi-focal strategy was dedicated to the creation of identical replicas or to the fabrication of a single structure with many identical sub-units. In the frame of this recent work, a proof-of-concept has been illustrated with the construction of biocompatible networks by using two independent sub-beams. In order to control each beam separately, the main beam is directed through a dynamic mask (typically a spatial light modulator). To extend this appealing strategy to numerous parallel and independently sub-beam, high-power lasers are required. Additionally to the increase of the fabrication's speed, this work opens the doors to numerous and more flexible applications. For example, one could imagine making unlimited modifications into microfluidic systems, but more interestingly, generation of pattern with different exposure time will result to display gradients in chemical functionality, mechanical functionality or porosity which play a key role in tissue engineering.

However the promising potential of multi-focal TPS for speed up fabrication is quite obvious, two points have to be taken in consideration to use it as a tool in laboratory or industries. The first point concerns the use of an expensive amplified femtosecond laser in order to provide enough energy after each lens or dynamic mask. In addition, the laser beam intensity distribution has also to be perfectly controlled to deliver the same amount of energy for each lens and to fabricate uniform structures. It has to be mentioned that though this point may be a brake for scientific community, this is clearly not the case for its use in industry. A more serious issue with parallel fabrication is the precise control of alignment of the hundred forming laser beams with respect to the plan of the substrate. A tilt of less than 1° of the substrate will result in the fabrication of inhomogeneous structures which is unacceptable from a metrology point of view in industries.

4.2.4. Currently speed of TPS process

In the literature, usual process speeds of several 100 $\mu m.s^{-1}$ are reported with sub-100 nm resolution. More rarely, $mm.s^{-1}$ can be reached while keeping a submicrometric resolution. Until recently, the fastest demonstration of microstructures with micrometric resolution has been realized by Fourkas's group by using a very sensitive photoinitiator (Kumi et al. 2010). In the frame of this work, speed of 1 $cm.s^{-1}$ was reported.

Since march 2012, a 300-micrometer long model of a Formula 1 race car has been fabricated by TPS in only four minutes while keeping micrometric resolution. Thus spectacular result means that a process speed of 5 $m.s^{-1}$ is involved, which is the same order of other laser process used in industry. A video of the construction can be found on the website of the Vienna University (Vienna, 2012). Unfortunately, certainly due to economic interests, little information can be freely accessed. According to their website, the increase in speed results from efforts from a chemical and mechanical point of view.

5. Conclusion and perspectives

After the pioneers works on TPS (Maruo, 1997), the research efforts were mainly focused on the synthesis of high efficient photoinitiators and materials in order to respectively speed up the writing process and to improve the mechanical, optical or chemical properties of the resulting 3D objects. Thanks to both the versatility of photopolymerizable systems and to the possibility to incorporate additional materials into the structures, TPS has attracted considerable attention over the past decade leading to enough mature technology. Indeed, despite the novelty of TPS, this is now daily used for broad range of applications such as creation of 3D components for microfluidic systems, tissue scaffolding, optical components, and so on. Besides, only 10 years after the first instance of 3D microstructures created by TPS, several companies have developed commercial 3D microfabrication set-ups which have supported the widespread of the technique in various research fields.

Recently, the rapid technical development of TPS provided much better structural resolution and high-throughput. The combination of all these improvements in a single commercial set-up will certainly boost the use of TPS in industry. From resolution point of view, the Abbe diffraction limit in optical lithography has been overcome by using and/or adapting a concept called STED originally from optical microscopy. The latter is already commercialized since 2005 by several well-known optical microscopy companies and is well expanded in life sciences. Interestingly, even if both in optical microscopy and lithography the diffraction limit has been beaten thanks to the STED principle, record for lateral spatial resolution in optical lithography (175 nm) is still far away from thus in optical microscopy (5.6 nm). In order to obtain comparable resolution, further investigations are required to enhance the comprehension of the photophysical and photochemical mechanism underlying the STED lithography.

In particularly, a better understanding will give a list of criteria for novel photoinitiators devoted to this promising technique.

Concerning the throughput of the technique, speed of 5 $m.s^{-1}$ has been recently announced by an European consortium (march 2012). For comparison, this speed is quite close to those used in conventional process in microelectronics industry such as control or ablation process (10 to 50 $m.s^{-1}$). Such promising advances should allow overcome limit for mass production and consequently should reinforce the highly potential of TPS for industry.

To conclude, this technology opens up new perspectives in a wide range of applications such as rapid prototyping of micro- and nanofluidics, small-scale production of microoptics components, or 3D frameworks for cell biology. Finally, owing to its currently fast expansion and to the versatile science involved in all the chain, TPS appears as a fantastic and so appealing field of research for the next decades.

Acknowledgements

Agence Nationale pour la Recherche (ANR - Projects 2-PAGmicrofab ANR-BLAN-0815-03, NANOQUENCHING and NIR-OPTICS), CNRS and Région Alsace are gratefully acknowledged for financial supports.

Author details

Arnaud Spangenberg[1*], Nelly Hobeika[1], Fabrice Stehlin[1], Jean-Pierre Malval[1], Fernand Wieder[1], Prem Prabhakaran[2], Patrice Baldeck[2] and Olivier Soppera[1]

*Address all correspondence to: arnaud.spangenberg@uha.fr

1 Institut de Science des Matériaux de Mulhouse, Mulhouse, France

2 Laboratoire de Spectrométrie Physique, Université Joseph-Fourier, Saint Martin d'Hères, France

References

[1] Amato, L, Gu, Y, Bellini, N, Eaton, S. M, Cerullo, G, & Osellame, R. (2012). Integrated three-dimensional filter separates nanoscale from microscale elements in a microfluidic chip. *Lab on a Chip*, , 12, 1135-1142.

[2] Becker, E. W, Ehrfeld, W, & Muenchmeyer, D. (1984). Accuracy of X-ray lithography using synchrotron radiation for the fabrication of technical separation nozzle ele-

ments. Inst. Kernverfahrenstech., Kernforschungszent. Karlsruhe G.m.b.H., Karlsruhe, Fed.Rep.Ger., , 92.

[3] Belfield, K. D, & Abdelrazzaq, F. B. (1997). a. Novel photoinitiated cationic copolymerizations of 4-methylene-2-phenyl-1,3-dioxolane. *Journal of Polymer Science Part A: Polymer Chemistry*, , 35, 2207-2219.

[4] Belfield, K. D, & Abdelrazzaq, F. B. (1997). b. Photoinitiated Cationic Crosslinking of 4-Methylene-2-phenyl-1,3-dioxolane with 2,2'-(1,4-Phenylene)bis-4-methylene-1,3-dioxolane. *Macromolecules*, , 30, 6985-6988.

[5] Belfield, K. D, Schafer, K. J, Liu, Y, Liu, J, Ren, X, & Stryland, E. W. V. (2000). Multiphoton-absorbing organic materials for microfabrication, emerging optical applications and non-destructive three-dimensional imaging. *Journal of Physical Organic Chemistry*, , 13, 837-849.

[6] Billone, P. S, Park, J. M, Blackwell, J. M, Bristol, R, & Scaiano, J. C. (2009). Two-Photon Acid Generation in Thin Polymer Films. Photoinduced Electron Transfer As a Promising Tool for Subwavelength Lithography. *Chemistry of Materials*, , 22, 15-17.

[7] Blanc, D, Pelissier, S, Saravanamuttu, K, Najafi, S. I, & Andrews, M. P. (1999). Self-processing of surface relief gratings in photosensitive hybrid sol-gel glasses. *Advanced Materials*, , 11, 1508-1511.

[8] Bückmann, T, Stenger, N, Kadic, M, Kaschke, J, Frölich, A, Kennerknecht, T, Eberl, C, Thiel, M, & Wegener, M. (2012). Tailored 3D Mechanical Metamaterials Made by Dip-in Direct-Laser-Writing Optical Lithography. *Advanced Materials*, , 24, 2710-2714.

[9] Burmeister, F, Zeitner, U. D, Nolte, S, & Tünnermann, A. (2012). High numerical aperture hybrid optics for two-photon polymerization. *Optics Express*, , 20, 7994-8005.

[10] Campbell, M, Sharp, D. N, Harrison, M. T, Denning, R. G, & Turberfield, A. J. (2000). Fabrication of photonic crystals for the visible spectrum by holographic lithography. *Nature* , 404, 53-56.

[11] Cao, Y, Gan, Z, Jia, B, Evans, R, & Gu, M. (2011). High-photosensitive resin for super-resolution direct-laser-writing based on photoinhibited polymerization. *Optics Express*, , 19, 19486-19494.

[12] Cerrina, F. (1997). Application of X-rays to Nanolithography. *Proc. IEEE*, , 84, 644-651.

[13] Cumpston, B. H, Ananthavel, S. P, Barlow, S, Dyer, D. L, Ehrlich, J. E, Erskine, L. L, Heika, A. A, Kuebler, S. M, Lee, I, Mccord-maughon, Y. S, Qin, D, Rockel, J, Rumi, H, Wu, M, Marder, X. -L, Perry, S. R, & Two-photon, J. W. polymerization initiators for three-dimensional optical data storage and microfabrication. *Nature*, , 398, 51-54.

[14] Denk, W, Strickler, J, & Webb, W. W. (1990). Two-photon laser scanning fluorescence microscopy. *Science*, , 248, 73-76.

[15] Devoe, R. J, Kalweit, H, Leatherdale, C. A, & Williams, T. R. (2002). Voxel shapes in two-photon microfabrication. *Proceeding of the SPIE*, , 4797, 310-316.

[16] Dittrich, P. S, Tachikawa, K, & Manz, A. (2006). Micro total analysis systems. Latest advancements and trends. *Analytical Chemistry*, , 78, 3887-3908.

[17] Emons, M, Obata, K, Binhammer, T, Ovsianikov, A, Chichkov, B, & Morgner, U. (2012). Two-photon polymerization technique with sub-50 nm resolution by sub-10 fs laser pulses. *Optical Materials Express*, , 2, 942-947.

[18] Hell, S. W, Jakobs, S, & Kastrup, L. (2003). Imaging and writing at the nanoscale with focused visible light through saturable optical transitions. *Applied Physics A: Materials Science & Processing*, , 77, 859-860.

[19] Fan, H. J, Werner, P, & Zacharias, M. (2006). Semiconductor Nanowires: From Self-Organization to Patterned Growth. *Small*, March 2006), , 2(700), 700-717.

[20] Fischer, J, Von Freymann, G, & Wegener, M. (2010). The Materials Challenge in Diffraction-Unlimited Direct-Laser-Writing Optical Lithography. *Advanced Materials*, , 22, 3578-3582.

[21] Fischer, J, Ergin, T, & Wegener, M. (2011). Three-dimensional polarization-independent visible-frequency carpet invisibility cloak. *Optics Letters*, , 36, 2059-2061.

[22] Fischer, J, & Wegener, M. optical laser lithography beyond the diffraction limit. *Laser & Photonics Reviews*, DOIlpor.201100046.

[23] Fischer, J, & Wegener, M. (2012b). Ultrafast Polymerization Inhibition by Stimulated Emission Depletion for Three-dimensional Nanolithography. *Advanced Materials*, , 24, OP65-OP69.

[24] Formanek, F, Takeyasu, N, Tanaka, T, Chiyoda, K, Ishikawa, A, & Kawata, S. (2006). Three-dimensional fabrication of metallic nanostructures over large areas by two-photon polymerization. *Optics Express*, , 14, 800-809.

[25] Goeppert-mayer, M. (1931). Über Elementarakte mit zwei Quantensprüngen, *Annalen der Physik*, , 401, 273-294.

[26] Gonsalves, K. E, Wang, M, Lee, C. T, Yueh, W, Tapia-tapia, M, Batina, N, & Henderson, C. L. (2009). Novel chemically amplified resists incorporating anionic photoacid generator functional groups for sub-50-nm half-pitch lithograph. *J. Mater. Chem.* March 2009), , 19, 2797-2802.

[27] Grotjohann, T, Testa, I, Leutenegger, M, Bock, H, Urban, N. T, Lavoie-cardinal, F, Willig, K. I, Eggeling, C, Jakobs, S, & Hell, S. W. (2011). Diffraction-unlimited all-optical imaging and writing with a photochromic GFP. *Nature*, , 478, 204-208.

[28] Harke, B, Bianchini, P, Brandi, F, & Diaspro, A. (2012). Photopolymerization Inhibition Dynamics for Sub-Diffraction Direct Laser Writing Lithography. *Chemical Physical Chemistry*, , 13, 1429-1434.

[29] Haske, W, Chen, V. W, Hales, J. M, Dong, W. T, Barlow, S, Marder, S. R, & Perry, J. W. (2007). nm feature sizes using visible wavelength 3-D multiphoton lithography. *Optics Express*, , 15, 3426-3436.

[30] He, G. S, Tan, L. S, Zheng, Q, & Prasad, P. N. (2008). Multiphoton absorbing materials: Molecular designs, characterizations, and applications. *Chemical Reviews*, , 108, 1245-1330.

[31] Hell, S. W, & Wichmann, J. (1994). Breaking the diffraction resolution limit by stimulated emission: stimulated emission depletion fluorescence microscopy. *Optics Letters*, , 19, 780-782.

[32] Henzie, J, Barton, J. E, Stender, C. L, & Odom, T. W. (2006). Large-area nanoscale patterning: chemistry meets fabrication. *Accounts of Chemical Research,* January 2006), , 39(4), 249-257.

[33] Im, S. G, Kim, B. S, Tenhaeff, W. E, Hammond, P. T, & Gleason, K. K. (2009). A directly patternable click-active polymer film via initiated chemical vapor deposition (iCVD). *Thin Solid Films*. January 2009), , 517(12), 3606-3611.

[34] Iosin, M, Scheul, T, Nizak, C, Stephan, O, Astilean, S, & Baldeck, P. (2011). Laser microstructuration of three-dimensional enzyme reactors in microfluidic channels. *Microfluidics and Nanofluidics*, , 10, 685-690.

[35] Jin, M, Malval, J. P, Versace, D. L, Morlet-savary, F, Chaumeil, H, Defoin, A, Allonas, X, & Fouassier, J. P. (2008). Two-photon absorption and polymerization ability of intramolecular energy transfer based photoinitiating systems. *Chemical Communications*, , 48, 6540-6542.

[36] Kaiser, W, & Garrett, C. G. B. (1961). Two-photon excitation in $CaF_2:Eu^{2+}$. *Physical Review Letters*, , 7, 229-232.

[37] Kato, J, Takeyasu, N, Adachi, Y, Sun, H. B, & Kawata, S. (2005). Multiple-spot parallel processing for laser micronanofabrication. *Applied Physics Letters*, , 86

[38] Kawata, S, Sun, H. B, Tanaka, T, & Takada, K. (2001). Finer features for functional microdevices- Micromachines can be created with higher resolution using two-photon absorption. *Nature*, , 412, 697-698.

[39] Klar, T. A, & Hell, S. W. (1999). Subdiffraction resolution in far-field fluorescence microscopy. *Optics Letters*, , 24, 954-956.

[40] Klein, F, Richter, B, Striebel, T, Franz, C. M, Von Freymann, G, Wegener, M, & Bastmeyer, M. (2011). Two-Component Polymer Scaffolds for Controlled Three-Dimensional Cell Culture. *Advanced Materials*, , 23, 1341-1345.

[41] Kodama, H. (1981). Automatic Method for fabricating a 3-dimensional plastic model with photo-hardening polymer. *Rev. Sci. Instrum.*, , 52, 1770-1773.

[42] Kovacs, G. T. A, Maluf, N. I, & Petersen, K. E. (1998). Bulk micromachining of silicon. *Proceeding of the IEEE* , 86, 1536-1551.

[43] Kumi, G, Yanez, C. O, Belfield, K. D, & Fourkas, J. T. (2010). High-speed multiphoton absorption polymerization: fabrication of microfluidic channels with arbitrary cross-sections and high aspect ratios. *Lab on a Chip*, , 10, 1057-1060.

[44] LaFrattaC. N.; Baldacchini, T.; Farrer, R. A.; Fourkas, J. T.; Teich, M. C.; Saleh, B. E. A.; Naughton, M. J. ((2004). Replication of two-photon-polymerized structures with extremely high aspect ratios and large overhangs. *Journal of Physical Chemistry B*, , 108, 11256-11258.

[45] LaFrattaC. N. ; Li, L. J.; Fourkas, J. T. ((2006). Soft-lithographic replication of 3D microstructures with closed loops. *Proceedings of the National Academy of Sciences of the United States of America*, , 103, 8589-8594.

[46] Lee, J-H, Singer, J. P, & Thomas, E. L. (2012). Micro-/Nanostructured Mechanical Metamaterials. *Advanced Materials*, , 24, 4782-4810.

[47] Lewis, J. A, & Gratson, G. M. (2004). Direct Writing in Three Dimensions. Materials Today, , 7, 32-39.

[48] Li, C, Luo, L, Wang, S, Huang, W, Gong, Q, Yang, Y, & Feng, S. (2001). Two-photon microstructure-polymerization initiated by a coumarin derivative/iodonium salt system. *Chemical Physics Letters*, , 340, 444.

[49] Li, X, Zhao, Y, Wu, J, Shi, M, & Wu, F. (2007). Two-photon photopolymerization using novel asymmetric ketocoumarin derivatives. *Journal of Photochemistry and Photobiology A: Chemistry*, , 190, 22-28.

[50] Li, L. J, Gattass, R. R, Gershgoren, E, Hwang, H, & Fourkas, J. T. (2009). Achieving lambda/20 Resolution by One-Color Initiation and Deactivation of Polymerization. *Science* , 324, 910-913.

[51] Lu, Y, Hasegawa, F, Goto, T, Ohkuma, S, Fukuhara, S, Kawazu, Y, Totani, K, Yamashita, T, & Watanabe, T. (2004). Highly sensitive measurement in two-photon absorption cross section and investigation of the mechanism of two-photon-induced polymerization. *Journal of Luminescence*, , 110, 1-10.

[52] Lu, W. E, Dong, X. Z, Chen, W. Q, Zhao, Z. S, & Duan, X. M. (2011). Novel photoinitiator with a radical quenching moiety for confining radical diffusion in two-photon induced photopolymerization. *Journal of Materials Chemistry*, , 21, 5650-5659.

[53] Malval, J. P, Jin, M, Morlet-savary, F, Chaumeil, H, Defoin, A, Soppera, O, Scheul, T, Bouriau, M, & Baldeck, P. L. (2011). Enhancement of the Two-Photon Initiating Efficiency of a Thioxanthone Derivative through a Chevron-Shaped Architecture. *Chemistry of Materials*, , 23, 3411-3420.

[54] Martineau, C, Anemian, R, Andraud, C. W, Bouriau, I, & Baldeck, M. P. L. ((2000). Efficient initiators for two-photon induced polymerization in the visible range. *Chemical Physics Letters,* , 362, 291-295.

[55] Maruo, S, Nakamura, O, & Kawata, S. (1997). Three-dimensional microfabrication with two-photon-absorbed photopolymerization. *Optics Letters,* , 22, 132-134.

[56] Maruo, S, & Ikuta, K. (2002). Submicron stereolithography for the production of freely movable mechanisms by using single-photon polymerization. *Sensors and Actuators, A: Physical,* A100, , 70-76.

[57] Maruo, S, & Inoue, H. (2006). Optically driven micropump produced by three-dimensional two-photon microfabrication. *Applied Physics Letters,* , 89, 144101.

[58] Mcleod, R. R, Kowalski, B. A, & Cole, M. C. (2010). Two-color photo-initiation/inhibition lithography. *Proceeding of the SPIE,* , 7591, 759102-759107.

[59] Moore, G. E. (1965). Cramming More Components Onto Integrated Circuits. *Electronics,* April 1965), , 38(8), 114-117.

[60] Ovsianikov, A, Viertl, J, Chichkov, B, & Oubaha, M. MacCraith, B.; Sakellari, L.; Giakoumaki, A.; Gray, D.; Vamvakaki, M.; Farsari, M.; Fotakis, C. ((2008). Ultra-Low Shrinkage Hybrid Photosensitive Material for Two-Photon Polymerization Microfabrication. *ACS Nano,* , 2, 2257-2262.

[61] Pao, Y-H, & Rentzepis, P. M. (1965). Laser-induced production of free radicals in organic compounds. *Applied Physics Letters,* , 6, 93-95.

[62] Park, S. H, Yang, D. Y, & Lee, K. S. (2009). Two-photon stereolithography for realizing ultraprecise three-dimensional nano/microdevices. *Laser & Photonics Reviews,* , 3, 1-11.

[63] Passinger, S, Saifullah, M. S. M, Reinhardt, C, Subramanian, K. R. V, Chichkov, B. N, & Welland, M. E. (2007). Direct 3D Patterning of TiO2 Using Femtosecond Laser Pulses. *Advanced Materials,* , 19, 1218-1221.

[64] Quake, S. R, & Scherer, A. (2000). From Micro- to Nanofabrication with Soft Materials. *Science,* , 290, 1536-1540.

[65] Reyes, D. R, Iossifidis, D, Auroux, P, & Manz, A. A. ((2002). Micro total analysis systems. 1. Introduction, theory, and technology. *Analytical Chemistry,* , 74, 2623-2636.

[66] Ridaoui, H, Wieder, F, Ponche, A, & Soppera, O. (2010). Direct ArF laser photopatterning of metal oxide nanostructures prepared by the sol-gel route. *Nanotechnology,* January 2010), , 21(6), 065303.

[67] Rill, M. S, Plet, C, Thiel, M, Staude, I, Von Freymann, G, Linden, S, & Wegener, M. (2008). Photonic metamaterials by direct laser writing and silver chemical vapour deposition. *Nature Materials,* , 7, 543-546.

[68] Ritschdorff, E. T, Nielson, R, & Shear, J. B. (2012). Multi-focal multiphoton lithography. *Lab on a Chip*, , 12, 867-871.

[69] Rittweger, E, Han, K. Y, Irvine, S. E, Eggeling, C, & Hell, S. W. (2009). STED microscopy reveals crystal colour centres with nanometric resolution. *Nature Photonics*, , 3, 144-147.

[70] Rumi, M, Ehrlich, J. E, Heikal, A, Perry, J. W, Barlow, S, Hu, Z. Y, Mccord-maughon, D, Parker, T. C, Rockel, H, Thayumanavan, S, Marder, S. R, Beljonne, D, & Bredas, J. L. (2000). Structure-property relationships for two-photon absorbing chromophores: Bis-donor diphenylpolyene and bis(styryl)benzene derivatives. *Journal of the American Chemical Society*, , 122, 9500-9510.

[71] Sakellari, I, Kabouraki, E, Gray, D, Purlys, V, Fotakis, C, Pikulin, A, Bityurin, N, Vamvakaki, M, & Farsari, M. (2012). Diffusion-Assisted High-Resolution Direct Femtosecond Laser Writing. *ACS Nano*, , 6, 2302-2311.

[72] Scott, T. F, Kowalski, B. A, Sullivan, A. C, Bowman, C. N, & Mcleod, R. R. (2009). Two-Color Single-Photon Photoinitiation and Photoinhibition for Subdiffraction Photolithography. *Science*, , 324, 913-919.

[73] Shevchenko, E. V, Talapin, D. V, Rogach, A. L, Kornowski, A, Haase, M, & Weller, H. (2002). Colloidal synthesis and self-Assembly of $CoPt_3$ nanocrystals. *Journal of the American Chemical Society*, , 104, 11480-11485.

[74] Shukla, S, Vidal, X, Furlani, E. P, Swihart, M. T, Kim, K. T, Yoon, Y. K, Urbas, A, & Prasad, P. N. (2011). Subwavelength Direct Laser Patterning of Conductive Gold Nanostructures by Simultaneous Photopolymerization and Photoreduction. *ACS Nano*, , 5, 1947-1957.

[75] Soppera, O, Croutxe-barghorn, C, & Lougnot, D. J. (2001). New insights into photoinduced processes in hybrid sol-gel glasses containing modified titanium alkoxides. *New Journal of Chemistry*, , 25, 1006-1014.

[76] Spangenberg, A, Malval, J. P, Akdas-kilig, H, Fillaut, J. L, Stehlin, F, Hobeika, N, Morlet-savary, F, & Soppera, O. (2012). Enhancement of Two-Photon Initiating Efficiency of a 4,4′-Diaminostyryl-2,2′-bipyridine Derivative Promoted by Complexation with Silver Ions. *Macromolecules*, , 45, 1262-1268.

[77] Steidl, L, Jhaveri, S. J, Ayothi, R, Sha, J, Mcmullen, J. D, Ng, S. Y. C, Zipfel, W. R, Zentel, R, & Ober, C. K. (2009). Non-ionic photo-acid generators for applications in two-photon lithography. *Journal of Materials Chemistry*, , 19, 505.

[78] Stocker, M. P, Li, L. J, Gattass, R. R, & Fourkas, J. T. (2010). Multiphoton photoresists giving nanoscale resolution that is inversely dependent on exposure time. *Nature Chemistry*, , 3, 223-227.

[79] Sun, H. B, & Kawata, S. (2003). Two-photon laser precision microfabrication and its applications to micro-nano devices and systems. *Journal of Lightwave* Technology, , 21, 624-633.

[80] Sun, H. B, Tanaka, T, & Kawata, S. (2002). Three-dimensional focal spots related to two-photon excitation. *Applied Physics Letters*, , 80, 3673-3675.

[81] Sun, Q, Juodkazis, S, Murazawa, N, Mizeikis, V, & Misawa, H. (2010). Freestanding and movable photonic microstructures fabricated by photopolymerization with femtosecond laser pulses. *Journal of Micromechanics and Microengineering*, , 20, 035004.

[82] Takada, K, Sun, H-B, & Kawata, S. (2005). Improved spatial resolution and surface roughness in photopolymerization-based laser nanowriting. *Applied Physics Letters*, , 86, 071122.

[83] Vienna (2012. http://amt.tuwien.ac.at/projekte/2pp/.

[84] Wang, I, Bouriau, M, Baldeck, P. L, Martineau, C, & Andraud, C. (2002). Three-dimensional microfabrication by two-photon-initiated polymerization with a low-cost microlaser. *Optics Letters*, , 27, 1348-1350.

[85] West, J, Becker, M, Tombrink, S, & Manz, A. (2008). Micro total analysis systems: Latest achievements. *Analytical Chemistry*, , 80, 4403-4419.

[86] Wolf, T. J. A, Fischer, J, Wegener, M, & Unterreiner, A-N. (2011). Pump-probe spectroscopy on photoinitiators for stimulated-emission-depletion optical lithography. *Optics Letters*, , 36, 3188-3190.

[87] Wu, E, Strickler, J. H, Harrell, W. R, & Webb, W. W. (1992). Two-photon lithography for microelectronic application. *Proc. SPIE* , 1674, 776-782.

[88] Xia, Y, & Whitesides, G. M. (1998). Soft Lithography. *Angewandte Chemie International Edition*, , 37, 550-575.

[89] Xia, R, Malval, J. P, Jin, M, Spangenberg, A, Wan, D, Pu, H, Vergote, T, Morlet-savary, F, Chaumeil, H, Baldeck, P, Poizat, O, & Soppera, O. (2012). Enhancement of Acid Photogeneration Through a Para-to-Meta Substitution Strategy in a Sulfonium-based Alkoxystilbene Designed for Two-Photon Polymerization. *Chemistry of Materials*, , 24, 237-244.

[90] Xing, J. F, Dong, X. Z, Chen, W. Q, Duan, X. M, Takeyasu, N, Tanaka, T, & Kawata, S. (2007). Improving spatial resolution of two-photon microfabrication by using photoinitiator with high initiating efficiency. *Applied Physics Letters*, , 90, 131106.

[91] Yanez, C. O, Andrade, C. D, & Belfield, K. D. (2009). Characterization of novel sulfonium photoacid generators and their microwave-assisted synthesis. *Chemical Communication*, , 7, 827.

[92] Zhang, X, Yu, X. Q, Sun, Y, Wu, Y, Feng, Y, Tao, X, & Jiang, M. (2005). Synthesis and nonlinear optical properties of a new D-π-A two-photon photopolymerization initiator. *Material Letters*, , 59, 3485-3488.

[93] Zhou, W, Kuebler, S. M, Braun, K. L, Yu, T, Cammack, J. K, Ober, C. K, Perry, J. W, & Marder, S. R. (2002). An efficient two-photon-generated photoacid applied to positive-tone 3D microfabrication. *Science*, , 296, 1106.

Chapter 3

Femtosecond Laser Lithography in Organic and Non-Organic Materials

Florin Jipa, Marian Zamfirescu, Alin Velea,
Mihai Popescu and Razvan Dabu

Additional information is available at the end of the chapter

http://dx.doi.org/10.5772/56579

1. Introduction

The lithography is a well established technology for fabrication of microelectronic components, integrated optics, microfluidic devices and Micro-Electro-Mechanical-Systems (MEMS) [1,2]. Using energy sources like ultra-violet (UV) photons or X-ray, various patterns are transferred from photo lithographic masks to photoresist materials. Developed for MEMS fabrications, the photoresists are light-sensitive materials. During exposure, chemical reactions are initiated in the irradiated volume, changing the chemical properties of the material. An imprinted pattern can be obtained when the exposed or unexposed material is removed by chemical solvents.

The size of the structures created through lithographic methods depends on the radiation wavelength. The optical diffraction limit represents the limiting factor for obtaining the minimum feature size. As a result, in the UV lithography the exposure wavelength was reduced initially from g-line (436 nm) to h-line (405 nm) and then to i-line (365 nm).

When smaller structures were required for the metal–oxide–semiconductor (MOS) circuits improvement, the UV lamps were replaced with excimer lasers. Deep UV lithography (DUV), based on 248 nm (KrF) and 194 nm (ArF) wavelengths [3,4], is used in semiconductor industry to produce transistors with 90 nm gate lengths. A further decrease of the radiation source to 157 nm wavelength (molecular fluorine) [5] was limited by the low transmission of fused silica material in this spectral range.

High quality crystalline calcium fluoride with low birefringence was grown for mask substrate and refractive lenses fabrication. Some technical difficulties, related to the mask protection, the requirement of special vacuum chambers and the growth of calcium fluoride material [6,7], have limited the implementation of this technology on large scale.

The 194 nm technology was reconsidered when a new technology, based on a liquid immersion between the last optical components and the photoresist material, was proposed [8]. In this case, the achieved resolution becomes similar to the 157 nm technology in air.

Despite some difficulties, like impurities from liquid immersion or bubbles formation [9,10], the immersion technology was used to manufacture 45 nm features in mass production. Using high-index fluids, it was tested for a 32 nm feature target too [11].

Other nano-technologies, like X-ray lithography [12] and ion/electron beam lithography [13,14], are used to obtain smaller feature size. Using exposure wavelengths under 100 nm, structures smaller than 30 nm were obtained [15]. The high cost of the mask fabrication for X-ray lithography limits the implementation of this technology on large scale production.

Another challenging technique, extreme ultraviolet lithography (EUV), uses a very short wavelength, 13.5 nm, to induce feature size down to 11 nm in photoresist materials [16]. The implementation of this technology is limited due to the complexity of installations for EUV radiation generation and photoresist characteristics like sensitivity and resolution.

Alternative methods like electron/ion beam lithography are used to produce nanometric structures in photoresist materials. A focused electron beam can generate chemical reactions in photoresist materials. This method does not require a photomask to create a pattern. The beam is focused directly in the photoresist material where structures with dimensions less than 30 nm can be obtained [17]. The implementation of this technology as mass volume production is limited by the high price of installations and the long exposure time for large area.

Alternative techniques like nano-imprint (soft) lithography can be used for low-cost pattern imprint. By this method, to create a pattern replica, a stamp or mold is pressed on a photoresist or thermoplastic material. The resulting mold replica can be used to multiply the original mold without complex and high cost production installations. Structures with dimensions under 100 nm were regularly imprinted, becoming the preferred method for high number microdevices fabrication.

All these lithographic methods which use masks or expose directly the photoresist material are limited to bi-dimensional structures. In some applications, such as photonic devices or microfluidics, three-dimensional (3D) structures with high-aspect ratio are needed. In order to create 3D structures, other exposure techniques, like laser holography or laser lithography with near infrared (NIR) femtosecond lasers, are required.

In case of holographic methods, based on the interference of several laser beams, the pattern is directly reconstructed in the photoresist volume. Using the holographic method [18,19], several research groups have obtained periodical structures with sub micrometer features on large surfaces. Nevertheless, the holographic method has some limitations when the design of the structure imposes localized defects in a complex 3D structure.

To overcome this drawback, infrared (IR) femtosecond laser installations with high repetition rate can be used to create 3D structures in photoresist materials. Due to the transparency in the NIR spectral band, the femtosecond laser pulses can be focused deeply in the photoresist volume. The high NIR photons density created in the focused spot induce two-photons or multi-photons absorption. Chemical reactions are initiated and structural modifications take place similar to the case of a single UV photon absorption.

In this paper we present an IR femtosecond laser processing method which can be used in organic and non-organic photoresist materials lithography. Section 2 is dedicated to micro-structures created by two-photon polymerization (TPP) in organic materials irradiated with femtosecond lasers. The TPP principle, the experimental set-up and processing procedure, as well as some applications of photoresist microstructures, are presented in this section. The non-organic photoresist materials and the main effects related to their interaction with the laser radiation are described in Section 3. Optical micro-lenses and photonic crystals, manufactured by photoresist laser lithography, are presented too. Conclusions are presented in Section 4.

2. Femtosecond laser lithography in organic photoresists

2.1. Principle of photo-polymerization mechanism in organic photoresist materials

Organic photoresist materials are composed by two main molecular components: monomers/oligomers (M) and photo-initiators (I). The monomers are unsaturated molecules and represent 90% from the photoresist material composition. These molecules can be bound in polymeric chains using the energy from a radiation source like photons. Two different photo-mechanisms, Chain Polymerizations and Step Polymerization (Photo-crosslinking), can be used to initiate this conversion.

When the light is used as an exposure energy source, the commonly encountered polymerization mechanism is based on chain propagation. Three stages: initiation, propagation and termination, are involved in this mechanism. The initiation stage starts when the excited photo-initiators (I*), produced in the presence of light, generate active species [20-22] like free radicals (R*), (Equation 1). Afterwards, these radicals interact with monomers, creating a new molecule which has an active termination (Equation 2). Once the process is started, a new monomer is added through this termination to the molecule (Equation 3). In this way, the polymeric chains grow rapidly without any supplementary photons excitations.

During propagation stage (Equation 4), the polymeric chains increase in length, until the active termination encounter a free radical (Equation 5) or an other active termination from a second polymeric chain (Equation 6). Through this coupling effect the formation of the polymeric chain comes to an end.

$$I \xrightarrow{h\upsilon} I^* \rightarrow \mathrm{R}^* \quad \text{(initiation)} \tag{1}$$

$$\mathrm{R}^*\mathrm{+M}_1 \rightarrow \mathrm{RM}_1{}^* \tag{2}$$

$$RM_1{}^{*}+M_2 \rightarrow RM_1M_2{}^{*} \tag{3}$$

$$RM_1...M_n{}^{*}+M_k \rightarrow RM_{(n+k)}{}^{*} \ \text{(propagation)} \tag{4}$$

$$RM_{(n+k)}{}^{*}+R^{*} \rightarrow RM_{(n+k)}R \ \text{(termination)} \tag{5}$$

$$RM_n{}^{*}+RM^{*} \rightarrow RM_nM_mR \ \text{(termination)} \tag{6}$$

Usually, depending on the concentration and the photo-initiator type, the commercially available photoresist materials have a high absorption for the UV photons and are highly transparent to NIR radiation. The photo-initiator molecule excitation can be realized using the energy from UV photons or by two IR photons absorption. The non-linear optical effect of the two photon absorption (TPA), represents the rule behind the Two Photon Photopolymerization (TPP) method, where 3D structures can be realized deeply in the photoresist volume.

2.2. Two-Photons Absorption (TPA) effect

Two-photons absorption process was early emphasized in 1965 [23] using a ruby laser focused in a styrene and p-isopropylstyrene photoresist. The importance of this method was proved when a 3D spiral coil microstructure was created in SCR 500 (urethane acrylate) by an infrared femtosecond laser [24].

Like other commercially available photoresist materials, the SCR 500 was developed for UV lithography. Under specific conditions, the photo-polymerization mechanism can be initiated by infra-red photons. The probability to obtain the photo-polymerization effect using IR photons strongly depends on the radiation intensity. A much higher photon density is necessary compared with the UV exposure. IR lasers which work in continuous wave (CW) regime can not achieve this high density without inducing thermal damages in material. For this reason, low energy short-pulses IR lasers are tightly focused in the photoresist volume. When a femtosecond laser pulse is focused in the photoresist volume, high power density can be reached. In the central area of the focused laser beam, the photo-polymerization effect is induced in a very tiny volume (voxel). The voxel dimensions, that can be smaller than the size of the focused spot, correspond to the TPA volume (Figure 1).

By translating the focused laser spot through the photoresist volume, complex 3D microstructures can be obtained (Figure 2a). The voxel size and geometry can be modified by controlling the laser intensity or by using focusing optics with different numerical apertures (NA). Depending on the geometry created in the photoresist material, the NA of the focusing optics can be selected in order to control the axial resolution. The voxel shape is defined by an ellipsoid with a long axis d_z, which represent the axial resolution, and a short axis d_{xy}, which is the voxel thickness (Figure 2b).

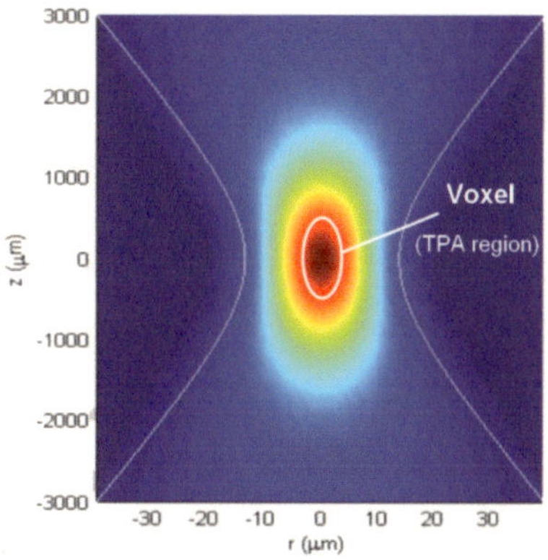

Figure 1. Longitudinal beam profile of the laser waist and the photopolymerized volume (voxel).

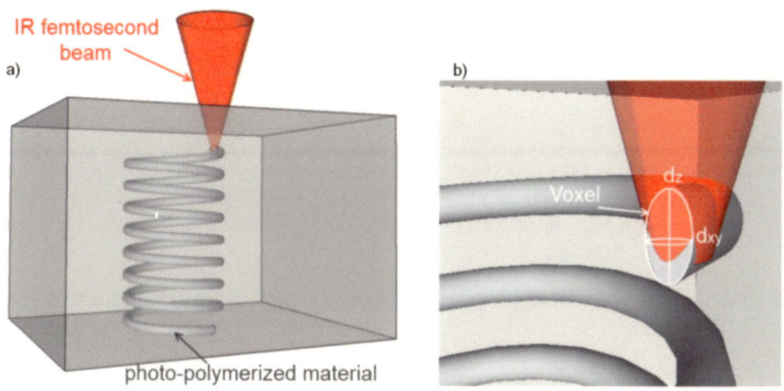

Figure 2. a) The sketch of the 3D structuring inside the transparent photopolymer. b) The geometry of the photo-polymerized voxel.

The voxel dimensions are also related to the photo-initiator used in the chemical composition of the photoresist material. Used in TPA processing method, the photo-initiator absorption efficiency is defined through a cross-section value (δ). Commercially available photoresists used in UV exposure are not created specially for the TPA processing. The low δ value of this processing method implies an increasing of laser energy and time exposure [25,26]. New photoresist materials with photo-initiators adapted for laser processing such as IP-L (Nano-scribe) were developed in order to obtain higher δ values.

This necessity generates new applications like up-converted lasing [27], where molecules with a large TPA cross-section are used. Several methods can be used to measure the δ value and to evaluate the TPA photo-initiators [28-30]. The elongation ratio value (d_z /d_{xy}) depends on the NA and can be reduced to 1.5 – 3, when focusing optics with high NA (~1.4 NA) are used. This NA value can be obtained only with immersion oil/water microscope objectives where the lateral spatial resolution can be less than 100 nm [31]. For smaller elongation value, shaded-

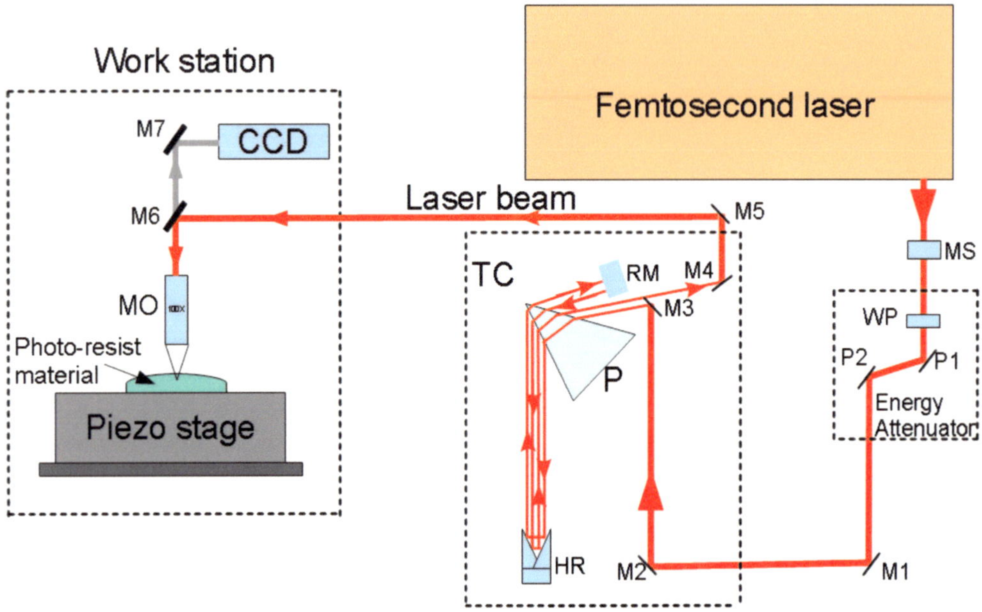

Figure 3. Schematic diagram of the DLW setup for 3D microprocessing using temporal compressor (TC): MS - Mechanical shutter; WP - Half wave-plate; P1 and P2 - Reflection Polarizers; P - BK7 prism; HR - Hollow Retro-reflector; RM - Roof Mirror; CCD - video camera; MO - microscope objective; M1 to M7 - Steering mirrors; M6 - Dichroic mirror.

rings [32] or annular binary filters [33] can be used to control the axial component (d_z) of the intensity distribution.

2.3. Experimental setup

A common Direct Laser Writing (DLW) setup for processing photoresist materials consists of the femtosecond laser source with high repetition rate, the attenuator for controlling the laser intensity, the focusing optics and translation stages with displacements at submicrometer resolution or better. For the 3D microstructuring experiments (Figure 3), a femtosecond laser oscillator with 5 nJ pulse energy, 15 fs pulse duration and 80 MHz repetition rate at 800 nm central wavelength was used. In order to preserve the laser pulse duration, the optical components of the experimental setup must be designed for a spectral bandwidth larger than 100 nm. Silver coated broadband optics were used for laser beam steering.

Positive group delay dispersion (GDD) generated by the glass in the optical path (focusing optics, polarizers, waveplates, beam colimators) produces a stretching of the pulse duration of the order of few hundreds of femtoseconds up to picosecond range. The temporal broadening of the pulses implies a considerable reduction of the laser peak power, leading to reduction of the photo-polymerization efficiency. In order to recover the pulse duration, for spectrally broad femtosecond laser sources, a dispersion compensator has to be used. The pulse chirp and pulse duration is controlled by tuning the GDD. In a standard dispersion compensator with prisms, four or two prisms are used, with total length of the beam path up to few meters. For minimizing the set-up foot-print, the compensator can be folded by a hollow corner cube reflector (HR) and a roof mirror (RM) [34]. The femtosecond laser beam passes four time through a BK7 single prism. By changing the distance between the dispersive prism and the hollow reflector, variable GDD is introduced. In this approach, the temporal compression can be adjusted in a flexible way, even if the focusing optics or other optical components are changed in the optical path. The laser intensity is adjusted by a variable attenuator, which consists in a half wave-plate and two reflection polarizers.

For processing the photoresist, the laser was coupled with a work station realized in a modular configuration. The work station consists in focusing optics, sample translations with piezo and motorized stages, and the visualization module. The focusing optics were selected depending on the resolution and the desired working distance. A 100x IR Mitutoyo microscope objective with 0.5 NA and 12 mm working distance was used for microfabrication of high-aspect-ratio structures [35]. For sub-micrometer feature size, a 100x Zeiss oil immersion microscope objective with NA 1.4 is more appropriate. However, the total high of the structures is limited by the short working distance of such objective of 0.17 mm.

The photoresist samples are translated by a piezoelectric translation stage with 300 μm displacement on all XYZ axes. A software controls the scanning path, scanning speed and the laser shutter according to a programmed design. Pre-defined geometries can be selected from a library, or can be imported in stereolithography (STL) format. A CCD camera is used for monitoring the laser focus on the sample surface and for direct observation of the irradiated area during the laser processing.

The CCD camera can be replaced by a photomultiplier, or by a spectrometer for spectroscopy measurements working with the focusing optics in confocal regime. Using the above presented DLW setup, both organic photoresist materials and non-organic materials (chalcogenide glasses) were processed.

2.4. Processing protocol for organic materials

When photoresist materials are used in micro-device fabrication, a processing protocol including some steps has to be followed. Substrate cleaning, photoresist deposition, material exposure, and development are the main stages (Figure 4).

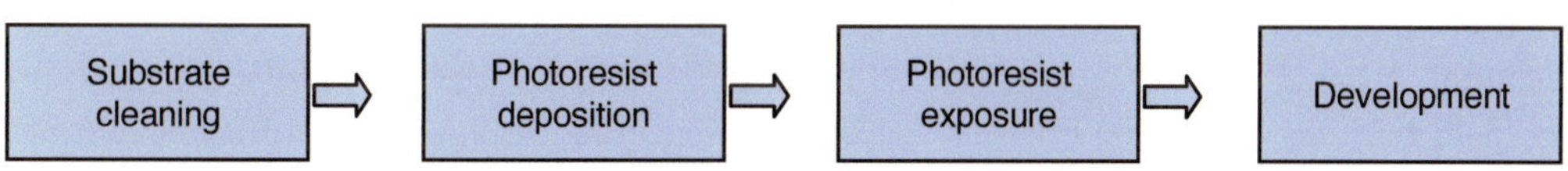

Figure 4. Usual process flow for the organic photoresist materials.

2.4.1. Substrate cleaning

Micro and nano processing technologies require a rigorous impurities control in order to prevent the sample contamination with dust particles or other chemical elements. The wafer substrate cleaning represents one critical stage. Here, several procedures have to be carried out for a good adhesion between photoresist material and substrate. First, a ultrasonic bath with chemicals solvents, like acetone or piranha solution, is used to remove the impurities from the substrate surface. Then, the substrate is rinsed in pure deionised water in order to eliminate the remaining solvent. After this stage, water molecules are left on the substrate surface, inducing a weak photoresist adhesion to the substrate. A water desorption step, which can be realized by heating the substrates to 150 °C for 30 minutes or using oxygen plasma cleaning, is necessary. The photoresist adhesion to different substrates can be poor, even if the water desorption was performed. Several methods can be used to eliminate this drawback. Reactive plasma etching, chemical corrosion or adding a promoter adhesion layer can be used. These treatments are usually recommended by the photoresist manufacturer.

2.4.2. Photoresist deposition

The photoresist material can be deposited on the substrate using various methods. Spray-coating, spin-coating, drop-cast or roller-coating are few methods which can be used. The most used common technique is spin-coating. Characterized through uniform and thin film thickness, this method uses the spinning speed to control the photoresist deposition. In order to use this deposition method, the photoresist must be in a liquid state. For this reason, the viscosity of the photoresist material is controlled using chemical solvents. For different viscosities, photoresist manufacturers provide informations about the evolution of the film thickness reported to spin rate (rpm) and time (s). Varying the ratio between the solid content and the solvent, photoresist materials can be deposited from 1 μm up to 1 mm.

One of the most popular negative photoresists, SU-8, is an epoxies based material where the polymerization is done by a cationic photo-polymerization mechanism. For this material a wide range of viscosities can be found. When the required thickness of the film is less than few micrometers, the manufacturer recommends to use a photoresist version which has 40% solid content and 60% solvent. For this viscosity, 1 ml of liquid SU-8 is necessary for every inch diameter of the substrate. The distribution of this quantity on the substrate surface begins with low spinning speed (500 rpm) for few seconds. Then, the speed is increased up to 8000 rpm in order to obtain 1 μm film thickness.

After deposition stage, for some photoresist materials a soft bake step is required. For SU-8 material, a thermal treatment was performed. A hot plate heated at 95^0 C is recommended to be used. During this step, the solvent is eliminated from the material, resulting a solidified material. A baking time ranging from 2-3 minutes for 4-10 microns photoresist thickness up to 15-40 minutes for 40-100 microns thickness is required. Because an insufficient thermal treatment could create a weak adhesion to the substrate and deformed structures, this step is very important.

2.4.3. Photoresist exposure by IR femtosecond lasers

The dimensions of the photoresist structures produced through TPP method depend on the focusing optics and processing parameters like fluence and scanning speed. The importance of the focusing optics, especially the NA, to voxel shape has been detailed in subsection 2.2. Here, the importance of the fluence and the scanning speed will be emphasized. These two parameters control the dimension and strength of the photoresist structures. Usually, in order to determine the most suitable conditions, a processing map is plotted for a given fluence F and different scanning speeds. The process is repeated for different fluence values (e.g. 2F, 3F, …, nF). In this way, through combinations between scanning speed and fluence, photoresist structures with different thickness are produced (Figure 5). After the imprinted pattern is investigated, the optimum exposure conditions are selected.

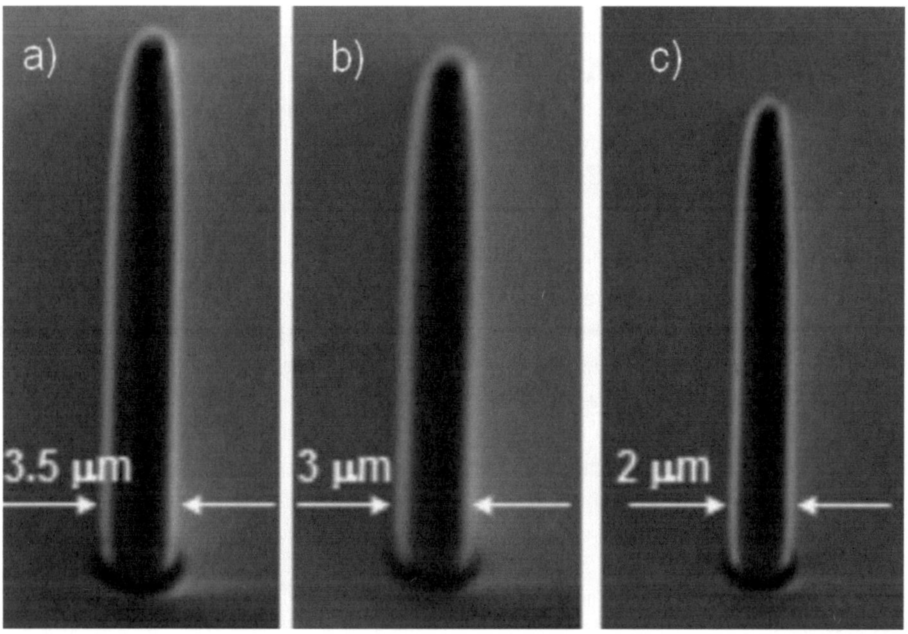

Figure 5. Pillars realised in OrmoComp photoresist by translating the IR focused beam (SynergyPro oscillator) in Z directions with different scanning speed. The fluence was fixed at 8.9 mJ/cm^2 and the scanning speed was : a) 5 µm/s b) 10 µm/s c) 50 µm/s.

In the SU-8 photoresist case, after exposure, the refractive index difference between the exposed and unexposed area is very low due to the fact that the polymerization is not realized instantaneously during exposure. In order to complete the formation of the polymeric chains, a supplementary post-exposure bake step is required.

Unlike the SU-8, the Ormocers and OrmoComp hybrid photoresists are laser photo-polymerized in a single irradiation step, without any additional thermal treatment. In these classes of inorganic-organic hybrid photoresists, the polymerization is instantly produced.

After exposure, for all photoresist materials a chemical etching stage is performed. Depending on the photoresist type, different chemical solvents are used to remove the exposed or unexposed photoresist material.

2.4.4. Development stage

When the photo-polymerization mechanism is accomplished, chemical solvents are used to develop the sample. Two classes of photoresists are identified: positive and negative. In the negative photoresists case, the irradiated material is transformed in polymer chains and becomes insoluble to solvents. Therefore, the non-irradiated material is removed. Unlike negative photoresists, the positive irradiated photoresists become soluble, being removed by the solvent.

2.4.5. Limiting factors in organic photoresist processing

Photoresist materials developed for nano-processing lithography have to satisfy several conditions in order to be used for MEMS mass production. The physical and chemical material properties, temperature conditions storage, chemical reactions with other materials, film thickness, minimum dot size, etching rate, are important parameters for semiconductor industry. Besides these parameters, other limiting factors are encountered in photoresist materials processing.

One limiting factor, which can induce some deviations from the initial design, is the shrinkage effect. Due to the photoresist chemical properties, the shrinkage effect appears after the etching stage where the structure becomes temporally a swollen gel. After complete removal of the solvent, the volume of the structure reduces and the initially designed opto-mechanical characteristics are changed. The shrinkage effect can be pre-compensated by adjusting the initial design of the structure. This way, a final photoresist structure with the desired symmetry and dimensions can result after the shrinkage. An attenuation of the shrinkage effect can be also obtained by developing new improved photoresist materials or by adjusting the photo-resist protocol [36].

Even if the shrinkage was compensated, another limiting factor can appear after chemical etching. Here, the microstructures created are stand alone, immersed in the solvent. During the sample extraction from solvent, the capillary forces can induce pattern distortions. To avoid this, methods like super-critical drying (SCD) [37] or the change of the contact angle between the structure and solvent during extractions were used [38].

2.5. Applications of the organic photoresist microstructures

2.5.1. High power laser targets

Several materials (e.g Zr, Mo, Au, Ag, Al) can be used as laser targets in experiments like fast ignition [39], X-ray emission [40], shock compression [41], ion acceleration [42] or hot electrons generation [43]. When a high power laser pulse interacts with a material, due to the strong electromagnetic field, hot dense, highly ionized plasma is generated. The resulted electrons, protons, ions or X-rays are analysed and used in many applications. Concepts of fusion or fast ignitions lead to new targets geometry, where the properties of the emitted particles like energy and directionality are analysed and improved. Using micro-cones target geometry, highly charged and collimated particles were produced through a better coupling between laser energy and material. Plasma structuring and optimization of radiation emission [44], represent two advantages of the new targets geometry. When the thickness [45] and lateral dimensions [46] of the target are reduced, the proton energy and the laser energy conversion efficiency increase.

With a sub-micrometer resolution, the photoresist materials offer an alternative to conventional high power laser targets fabrication methods. Using a microscope objective with 0.5 NA, capillary microstructures were created in Ormocore material through TPP method (Figure 6). Keeping the IR femtosecond beam focused on the substrate surface and by translating the sample, capillary structures with reduced height and thickness were obtained (Figure 6a). The capillary height can be increased by translating the focused beam along the Z axis through the photoresist volume (Figure 6b).

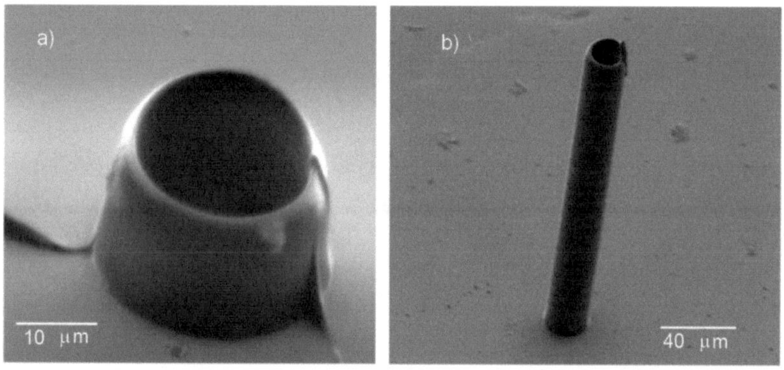

Figure 6. Capillary microstructures obtained in Ormocore photoresist by TPP method.

The photoresist structures can be used as template for more complex geometries. Other materials like metal or ceramics can be afterward deposited on the photopolimerised structures by different techniques, such as electroplating, thermal evaporation, pulsed laser deposition. After deposition, the photoresist material is eliminated by thermal treatment. The final target will reproduce the photoresist microstructure design.

2.5.2. *Light coupling masks*

Other applications of photoresist microstructures are in the optical contact lithography [47] and laser interference surface processing [48]. In both cases, a transparent photoresist mask with a periodic pattern is used to change the electromagnetic field propagation, generating a phase-shift during the propagation through the mask. Due to the periodic phase-shift introduced by the mask, an interference pattern is obtained. If the mask is placed in the proximity of a sample, the interference pattern and the localized high laser fluence induce modifications on the surface of the material or even inside the transparent photoresists. Initially used for photoresist processing, where sub-wavelength photoresist structure were produced, this method was implemented for semiconductor processing too.

We created a photoresist mask in a PMMA positive photoresist for light coupling mask (LCM) experiments. Using the spin-coating method, a photoresist layer with 300 nm thickness was deposited on a glass substrate. The photoresist mask, with a grid pattern periodicity of 2 μm, has been created by electron beam lithography (Figure 7a). We placed the photoresist mask in contact with a silicon wafer, without any additional pressure.

The mask was irradiated by an IR femtosecond laser pulse (775 nm central wavelength, 200 fs pulse duration) through a focusing lens with 75 mm focal length (Figure 7b). The laser fluency on silicon surface was 0.15 J/cm^2, below the silicon ablation threshold. Due to the electromagnetic field enhancement under photoresist mask, the laser fluency was enough to locally ablate the silicon wafer. Periodic grooves with 350 nm width and few nm depth, were obtained on silicon wafer surface (Figure 8). The imprinted pattern is identical with the configuration of the mask. In order to imprint a large area, the size of the spot was increased by changing the focus position with 1 cm below the photoresist mask position (Figure 7b).

A large silicon wafer surface was structured with a single IR laser pulse (Figure 9). Both the photoresist mask and the imprinted pattern were investigated by Atomic Force Microscopy (AFM).

Because the exposed area dimensions are limited only by the laser available energy and not by the photoresist mask, this method represents a fast and low cost surface processing technology.

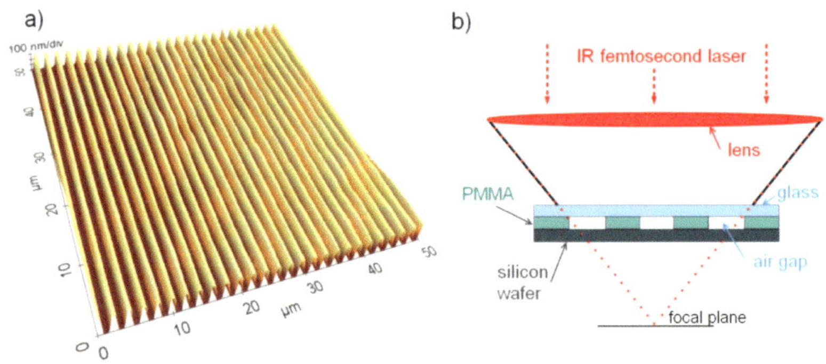

Figure 7. Large areas processing using LCM. a) The 3D AFM image of the PMMA photoresist mask. b) The experimental setup for large area processing.

2.5.3. Photonic nanostructures

The actual computing speed of the micro-processors is limited mainly by the resistive heating resulting from the electrons current in circuits and it is hard to be improved. To overcome this limitation a new technology which uses light instead of electrons is foreseen. A photonic device will provide a fast signal processing, transmission of higher information volume and lower power consumption.

http://dx.doi.org/10.5772/56579

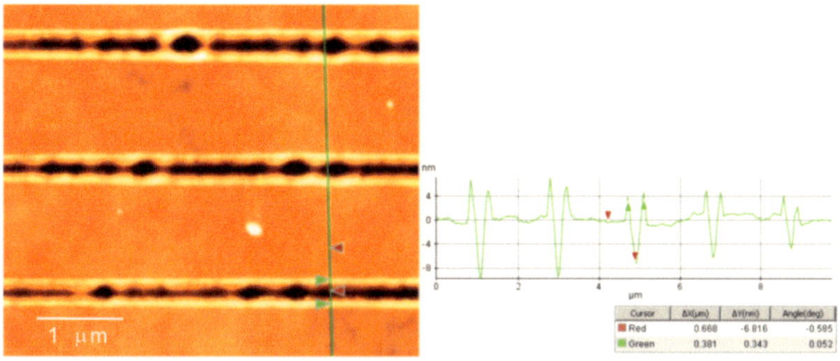

Figure 8. AFM image and profile of the periodic grooves imprinted in silicon wafer using LCM.

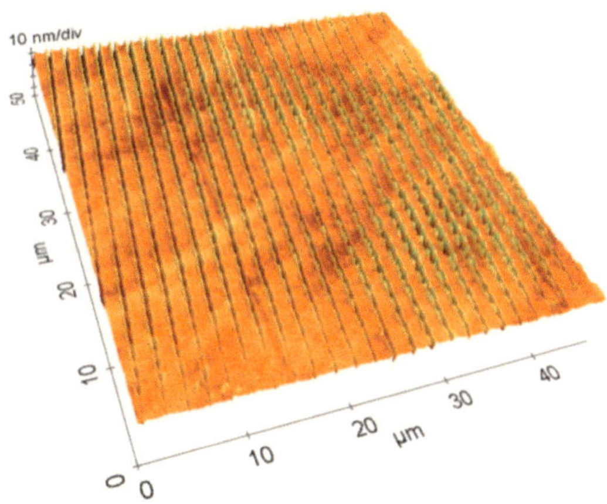

Figure 9. Extended area of imprinted pattern on silicon surface using LCM.

In such devices the propagation of the light is controlled by the so called photonic bandgap structures. These structures usually consist of periodic arrangements of photonic "atoms" with designed voids or defects such as cavities, or waveguides. The dimension of the photonic "atoms" and the period of the photonic structures are comparable with the light wavelength. Special designs and material configurations could even lead to non-conventional optical effects such as negative refraction index, negative refraction, cloaking at visible frequencies. The structures presenting such fascinating effect are widely known as negative index metamaterials (NIM). The dimension of an elementary cell in a metamaterial is several times smaller than the light wavelength and could decrease down to 100 nm for optical frequencies. For this reason, the fabrication of the photonic bandgap structures and metamaterials for visible spectral range requires complex processing techniques which are able to generate 3D structures with dimensions of the order of few hundreds nm and even below 100 nm. DLW is one of these techniques providing high accuracy in 3D. Such processing equipments are already commercially available for producing 2D and 3D structures with minimum linewidth of about

100 nm. Using a 3D lithography system (Photonic Professional - Nanoscribe) based on TPP, we have obtained structures with dimensions of 90 nm (Figure 10a). A scanning speed of 50 μm/s and a 0,1 nJ pulse energy at 80 MHz repetition rate were used to create a periodic pattern. An immersion oil microscope objective 100x and 1.4 numerical aperture was used for focusing the pulsed laser beam in the comercial photoresist IP-L (Nanoscribe), specially developed for TPP method.

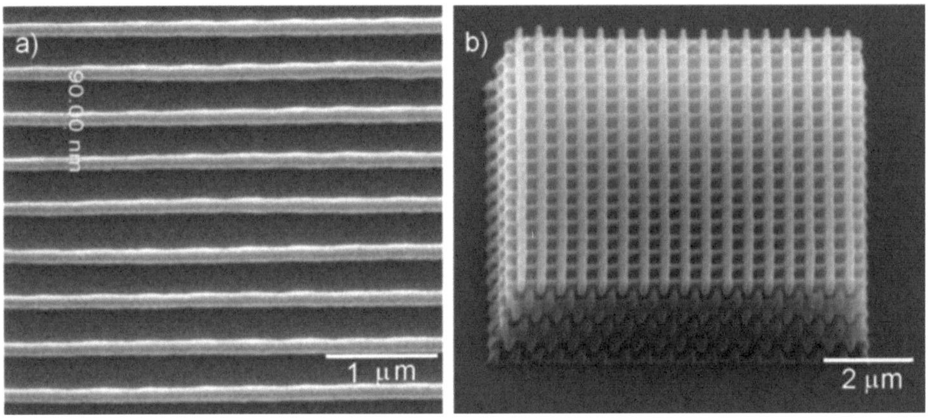

Figure 10. Nanostructures produced in IP-L photoresist. a) periodic lines; b) woodpile structure.

The organic materials have low refractive index, usually at the order of 1.4-1.6 at the visible wavelengths. In simple geometries, such as periodic arrangements of pillars in hexagonal or rectangular symmetry, the structures will not show photonic bandgaps. Complex geometries such as woodpiles has to be designed. 3D structures can be obtained by scanning the photoresist layer by layer. Recently, we have fabricated a woodpile structure with a period of 500 nm between lines and 5 μm tall (Figure 10b). For this value of the period, these structures could show photonic bandgaps for the visible spectral domain.

An alternative solution to these complex designs is to realize microstructures in non-organic photoresist materials, like chalcogenide glasses, with higher refractive index.

3. Laser processing of non-organic materials

3.1. Non-organic photoresist materials

The chalcogenides are materials either in crystalline or amorphous states, which are based on the chalcogen elements (Sulphur, Selenium and Tellurium) in combination with other elements (Arsenic, Phosphorus, Germanium, Tin, etc.). The structure of the chalcogenides is based on a network of covalent bonds, which gives the specific properties to these materials. While the crystalline state of the chalcogenide can be hardly obtained, the amorphous phase can be easily obtained by the melt cooling [49].

The properties of the amorphous (glassy) compositions are quite different from the crystalline counterparts. The basic structure at the atomic level consists in chains of atoms for elemental chalcogen and disordered layers for complex chalcogenides. A typical chalcogenide structure (Arsenic trisulphide, As_2S_3) in a bulk glass is shown in Figure 11a, while a disordered layer is presented in Figure 11b. Arsenic trisulphide glass is intensively studied due to its optical properties and versatility in structural modifications. It is characterized by a high nonlinear refractive index (Figure 12a), high transmission in infrared regions, and low phonon energies.

The study of the optical features of non-crystalline vitreous semiconductors near the absorption edge is of great interest. The absorption edge of non-crystalline materials is very sensitive to material composition and structure as well as external factors such as electric and magnetic fields, heat, light, and other radiation. Under the influence of these factors the optical properties of non-crystalline semiconductors are reversible or irreversible modified. They are suitable for optical investigations because their absorption edge is in the visible region of the spectrum. The absorption edge of As_2S_3 is at 2.4 eV (Figure 12b). Moreover, vitreous materials can be easily obtained in bulk samples, thin film, and optical fibers. These properties make them suitable to be used as materials for components in optoelectronic devices, as solid electrolites, photonic crystals, IR-transmitters, optical and electrical phase change memories or rewritable memory materials for CD's and DVD's.

Due to their sensitivity to different radiation wavelengths, chalcogenide glasses are suitable for laser lithography. They have small molecular units thus having the possibility to obtain higher resolution. They are much harder than polymers and can maintain the shape.

Due to the metal photodissolution in chalcogenide glasses, dry grayscale lithography can be done [50,51]. Chalcogenide glasses can also be used for wet lithography. Much simpler etching can be done without any treatments and they are resistant to acids. Thereby, there is easy to transfer patterns in substrates using reactive ion etching.

Depending on the type of glasses, both positive and negative resists can be obtained. Besides currently available organic photoresists, chalcogenide glasses offer a new powerful class of photoresists for a versatile lithography.

3.2. Light interaction with chalcogenide glasses

Many applications of the glassy chalcogenides are based on their interaction with light. Recently, different effects produced in chalcogenide glasses and thin films were investigated [52]. The main effects related to the interaction of the light with the glass are:

- Photodarkening and photobleaching
- Photoexpansion of the material
- Change of composition with the elimination of one chalcogen
- Vaporization of the material

The photodarkening [53 - 57] and photobleaching [58] effects were the first optical effects discovered in chalcogenide glasses. The optical absorption edge can be shifted toward higher

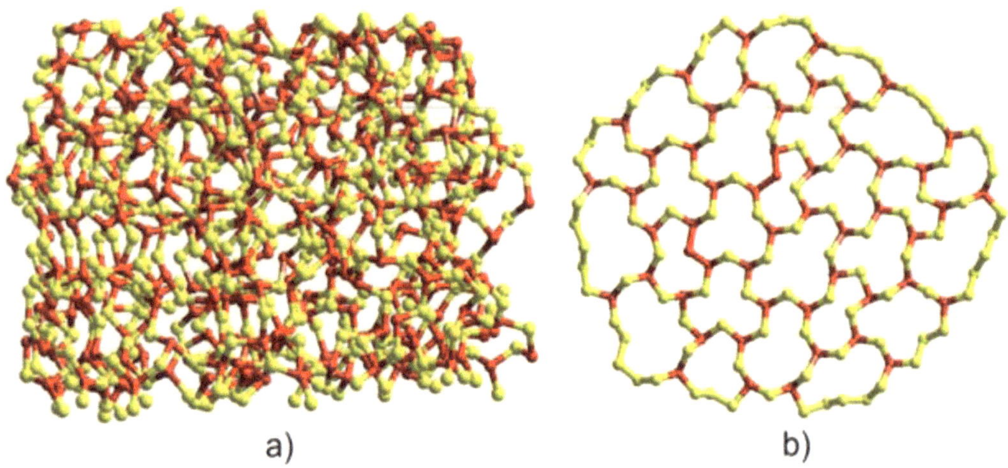

Figure 11. a) The structure of As_2S_3 glass in a bulk glass. b) A thin disordered film. (As – red, S – yellow)

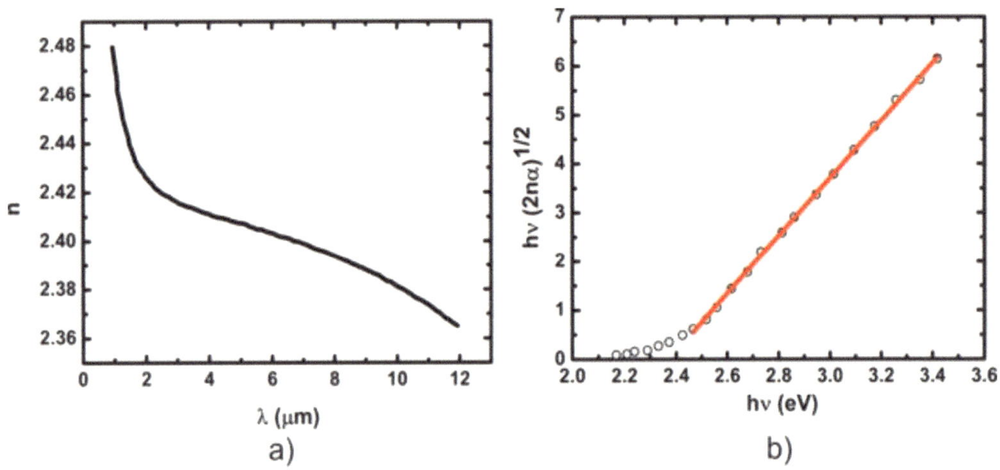

Figure 12. Optical properties of As_2S_3 glass. a) The variation of the refractive index with the wavelength. b) The absorption edge.

wavelengths (photodarkening) or toward lower wavelengths region (photobleaching) under laser light illumination with photon energy near absorption edge (2.4 eV). These processes are reversible as a function of heat treatment (under the glass transition temperature (T_g)).

The photoexpansion is one of the main phenomena which produce photoinduced volume change. The irradiation of amorphous chalcogenides films with bandgap light produces an increase in thickness and is termed photoexpansion [59, 60]. It has been demonstrated that amorphous As_2S_3 thin films irradiated with bandgap and sub-bandgap light expand with about 0.5 % and 4 % (giant photoexpansion) [61], respectively. Both photoexpansion and giant photoexpansion vanish after annealing close to the glass transition temperature. On the other hand, when the material is irradiated with super-bandgap light, photocontraction effects ascribed to ablation or photovaporization [62] are induced. Photoexpansion and photocontraction are usually produced in chalcogenides, the sign of the effect depending on the glass composition [63 – 68].

One important effect occurring during interaction of high power light with the chalcogenide material is the vaporization in the chalcogenide mass. The vaporization occurs by elimination of clusters of different composition and size [69]. Vaporization is preceded by photofluidity effect discovered by Hisakuni and Tanaka [70]. The standard processing protocol of a chalcogenide photoresist is described in the next section.

3.3. Processing protocol for chalcogenide glasses

As shown in the previous section, chalcogenide materials have specific properties that make them suitable for lithography. The steps of the lithographic process in chalcogenide glasses are presented in Figure 13. First, a chalcogenide thin layer is deposited on a glass substrate using the pulsed laser deposition method (PLD). As_2S_3 films were deposited on glass substrate by PLD using a KrF* excimer laser, with 80 mJ pulse energy and 25 ns pulse duration at 248 nm wavelength. Homogeneous films were obtained with thickness of around 2 micrometers. Secondly, the layer is irradiated using femtosecond laser pulses (λ = 800 nm).

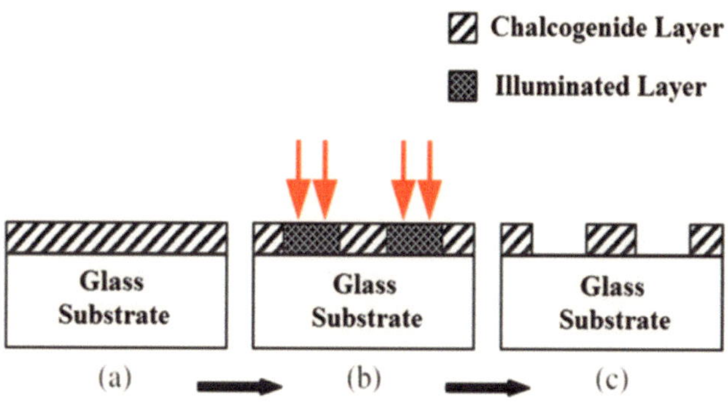

Figure 13. The lithographic process in As_2S_3 chalcogenide glass.

In this case the chalcogenide glass acts like a negative photoresist. In the final step, the sample is etched using an amine based aqueous etchant. The removal of the irradiated regions by etching leaves behind the selected regions of the photoresist. In the following subsections we present the main methods used for chalcogenide glasses processing using laser irradiation: direct laser writing by pulsed femtosecond laser irradiation and interference lithography.

3.3.1. Pulsed femtosecond laser irradiation. Two photon absorption in chalcogenide glasses

It is known that the chalcogenide glasses exhibits a characteristic one-photon absorption spectrum. The absorption edge consists of three functional curves, i.e., a square-root dependence, the so-called Urbach edge, and an exponential weak-absorption tail. The weak-absorption tail limits optical transparency of chalcogenide glasses [71]. The two-photon absorption spectrum of As_2S_3 glass has an exponential form $\beta=\exp(h\omega/E_\beta)$, where $E_\beta \sim 150$ meV [72].

This exponential form implies that the two photon process is enhanced by the gap states, which cause the weak-absorption tail. When the incident light intensity is less than 10 MW/cm^2, one-photon excitation of the gap states occurs more frequently than two-photon excitation of free carriers, and accordingly, the former could be responsible for the phenomena photoinduced by sub bandgap photons.

From the studies of the 3D optical data storage into As_2S_3 blocks via photodarkening with 800-nm femtosecond laser pulses [73] it was shown that two photon absorption can be achieved using relatively low energy laser pulses. The two-photon absorption cross-section was found to be 6.2 ± 0.5 GM (Goeppert-Mayer units, where 1 GM is 10^{-50} cm^4 s photon^{-1}) at around 800 nm wavelength.

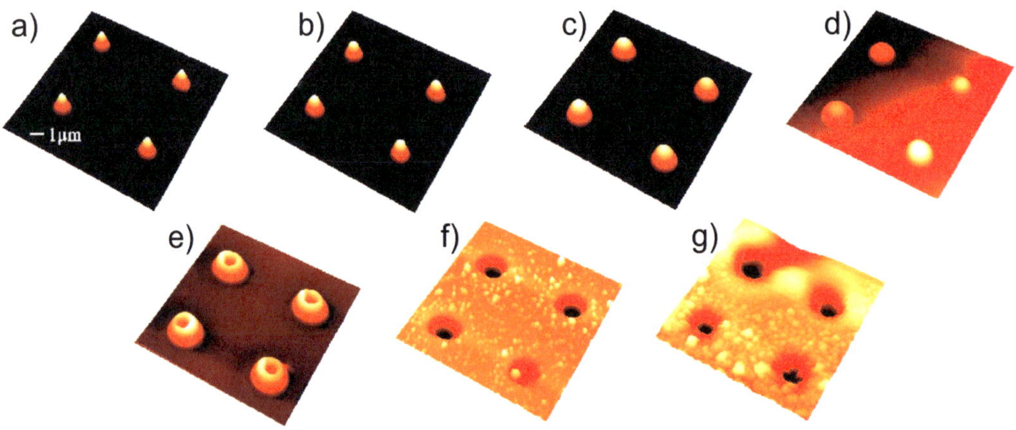

Figure 14. AFM images of laser irradiated As_2S_3 thin film surface using an average laser power of a) 8 mW, b) 12 mW, c) 18 mW, d) 20 mW, e) 25 mW, f) 30 mW, g) 50 mW

Some results of femtosecond laser imprints on As_2S_3 film surface have been reported in reference [74].

Using a femtosecond laser with 80 MHz repetition rate and 15 femtosecond pulse durations, a network of nano-lenslets was created by local exposure of individual sites separated by 5 μm × 5 μm. The irradiation was performed for 300 ms in each point. The average laser power was varied from 8 to 50 mW, corresponding to femtosecond pulse laser energy from 0.1 nJ to 0.63 nJ.

In Figure 14, one can see the formation of hillocks and/or holes on the surface of a thin amorphous As_2S_3 film by direct laser writing method using femtosecond laser pulses (central wavelength, $\lambda_c = 800$ nm). The shape of the modified surface is found to be a function of the laser power. Thus, a network of nano-lenslets could be imprinted at appropriate laser power and might be used in planar optoelectronic circuits.

It can be observed a boundary between the low energy laser pulses and high energy laser pulses. For low energy laser pulses the main evidenced effect is the photoexpansion. For higher energy pulses, over 0.25 nJ, a process of material ablation takes place and determines the holes formation into the film.

3.3.2. Interference lithography

Interference lithography (IL) is widely used for the fabrication of one dimensional nanostructures [75], production of the master mold for nano-imprinting lithography [76], formation of grating structures on semiconductor surfaces [77,78], pre-patterning of the substrate before the formation of photonic crystals by electrochemical etching [79] or vacuum deposition [80] etc.

We used As_2S_3-As_2Se_3 as an inorganic photoresist for the fabrication of submicrometer periodic relief on silicon wafers using interference lithography [81]. A 300 nm thick photoresist of As_2S_3-As_2Se_3 was vacuum evaporated on a (100) silicon substrate on which a 50 nm thick chromium layer was previously deposited. The obtained samples were exposed to an interference pattern that was generated by an argon laser ($\lambda = 488$ nm) using a holographic setup. To generate interference fringes, light beam has to be divided into two waves which afterwards are recombined. In an amplitude-division system, a beam splitter is used to divide the light into two beams travelling in different directions, which are then superimposed to produce the interference pattern. The laser fluence was around 0.5 J/cm^2. For the formation of bi-gratings each exposure can be 1.5-2 times reduced. The two-dimensional periodic structures on Si (100) surface were formed by double exposure on two perpendicular orientations of the Si wafers.

During the first exposure, Si (100) wafers were aligned by a base cut in parallel to interference grating lines and during the second irradiation the wafers were rotated with 90°. The size of the exposed part on the substrate reached up to 0,075 mm ▪ x 0,075 mm. ▪

After exposure, the samples were chemically treated in non-water alkaline organic solutions (negative etching) to form a relief pattern. The removal of Cr layer using water solution of HCl through a chalcogenide mask was the next step. Thus, the obtained two-layer resistive mask As_2S_3-As_2Se_3-Cr was used to form a corresponding relief on Si surface. Anisotropic etching of silicon was carried out using ethylene diamine solutions.

As ethylene diamine actively dissolves chalcogenides, etching of silicon occurred, mainly, through a Cr resistive mask that is neutral to alkaline solutions.

Figure 15a shows the AFM image of a diffraction grating formed on the silicon (100) surface by the anisotropic etching through As_2S_3-As_2Se_3-Cr resistive mask (grating period is near 1.0 μm). Figure 15b shows the bi-grating structure that was formed using double exposure of 0.3 J/cm^2. Symmetrical photoresist islands were obtained, with the ratio of the island diameter to the interval width between islands closed to unity. Time of the silicon etching was 15 s. Depth of the obtained relief is 0.15 micrometers. The size of photoresist islands depends on the value of exposure, and the form of islands depends on the ratio of exposures in two mutually perpendicular directions.

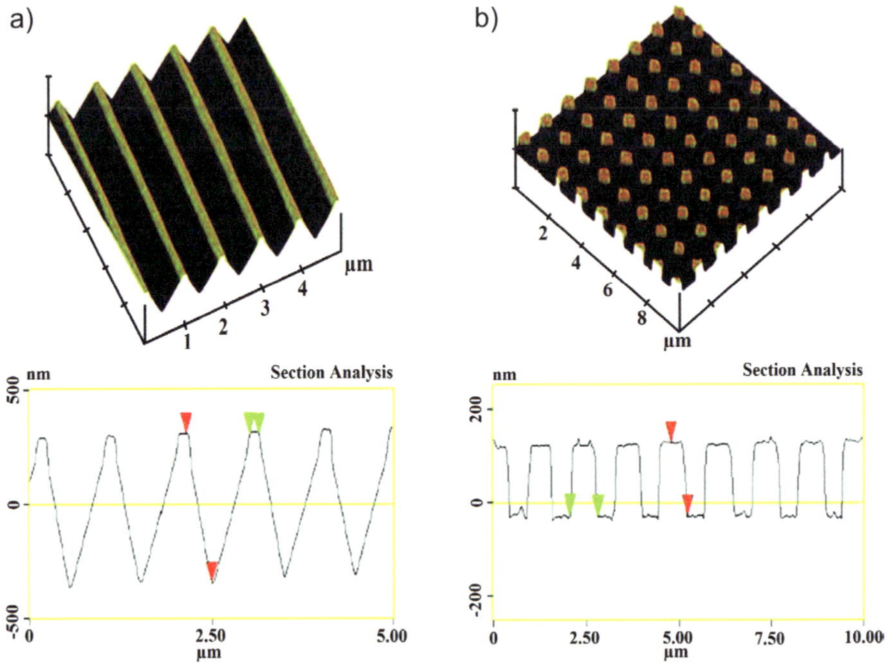

Figure 15. AFM images of the created relief on the surface of Si (100). a) Relief and groove profile of a grating obtained on by 50 s silicon etching. b) Profile of bi-grating with symmetrical elements obtained by a 15 s etching time.

Two applications of the lithographic process in chalcogenide glasses are presented in the next section.

3.4. Chalcogenide glasses applications

3.4.1. Optical microlenses

A first application is related to the formation of microlenslets on the surface of the chalcogenide film. By irradiating a thin chalcogenide film with the above mentioned femtosecond laser (Section 3.3.1), As_2S_3 microlenses were obtained. The profile of the lenslet measured from the AFM data was fitted with an asymmetric double sigmoidal curve (Figure 16). The fitting curve is given by equation (7), where y_0 is the offset, A is the amplitude, x_c is centroid and w_1, w_2, w_3 are width parameters.

$$y = y_0 + \frac{A \cdot \left(1 - \frac{1}{1 + e^{-\left(\frac{x - 0.5 \cdot w_1}{w_3}\right)}}\right)}{1 + \frac{1}{1 + e^{-\left(\frac{x - 0.5 \cdot w_1}{w_2}\right)}}} \tag{7}$$

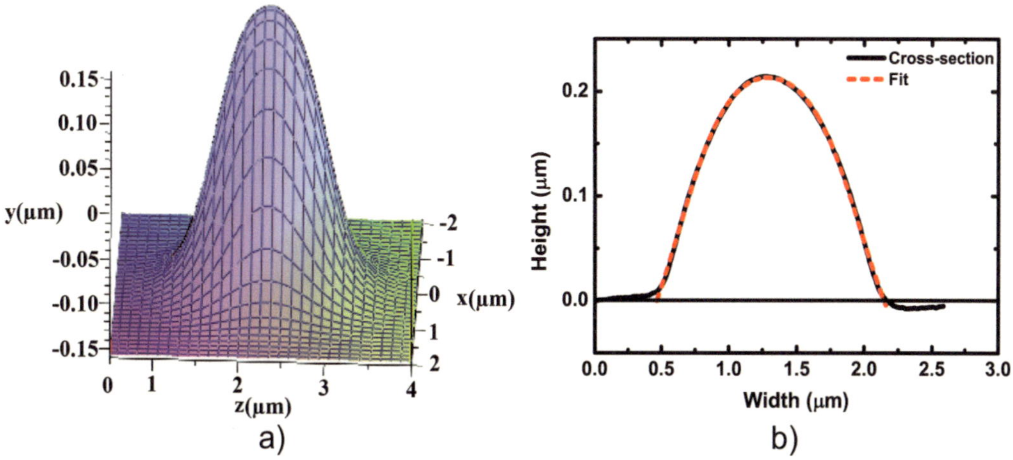

Figure 16. Chalcogenide microlenses. a) A lens represented by a 3D-plot. b) The shape of the lens fitted by an asymmetric double sigmoidal curve.

Fitting parameter	Value	Standard error
y_0	-0.09925	0.00382
x_c	8.30668	0.00035
A	0.33291	0
w_1	1.34572	0.00763
w_2	0.18998	0.00267
w_3	0.20191	0

Table 1. Fitting parameters

The lenslet profile is very well fitted to the parameters of the fitting curve presented in Table 1. The geometrical characteristics of the lenslet are: a diameter of 2.03 μm and a height of 0.21 μm. The focal length was between 1.21 μm at λ = 650 nm and 1.37 μm at λ = 5 μm (Figure 17). For computing the focal length we used the values of the refractive indices at different wavelengths from [82]. The transmission of light through the lenslets is limited by the As_2S_3 optical absorption edge of 2.41 eV.

3.4.2. Photonic crystals

Photonic structures are important components of the optoelectronic circuits used in telecommunications and in non-linear optics, as lossless guiding [83] and tightly bent 90° waveguides [84]. They can combine optical waveguides, resonators, dispersive devices and modulators for on-chip integration. Recently, it was shown that various 2D or 3D structures can be inscribed on the surface and bulk of an arsenic sulphide glass by the action of femtosecond laser pulses followed by etching in alkali or amine based etchants [85,86]. The laser installation is presented in section 2.3.

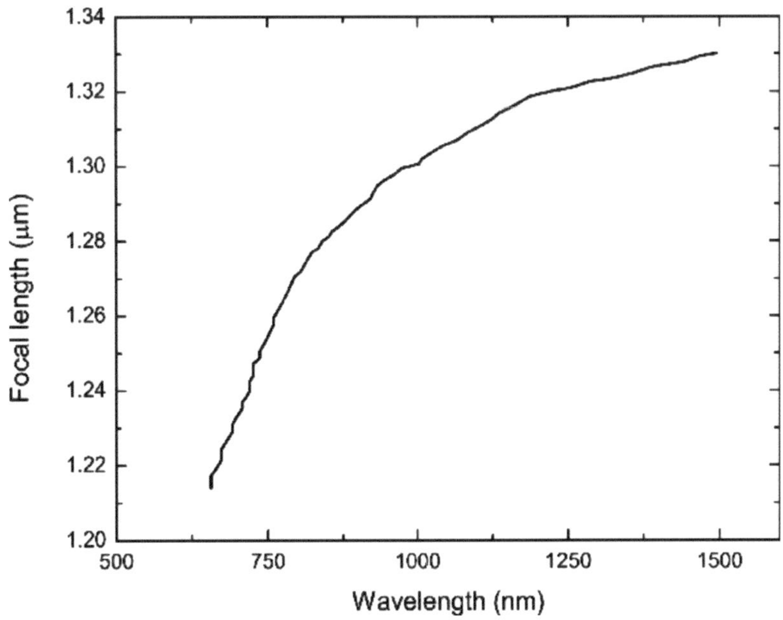

Figure 17. The variation of the focal length of the microlenslets with the radiation wavelength.

A 2D photonic crystal structure was imprinted on the surface of bulk As_2S_3 chalcogenide glass. Regular bumps obtained by photoexpansion of the glassy material have the height of 150 – 200 nm (Figure 18a) [87].

After etching, using an amine based etchant, a hexagonal lattice of holes having the diameter of about 2 micrometers was obtained (Figure 18b).

In order to obtain photonic devices for the visible light domain, further investigations are in progress to improve the processing parameters.

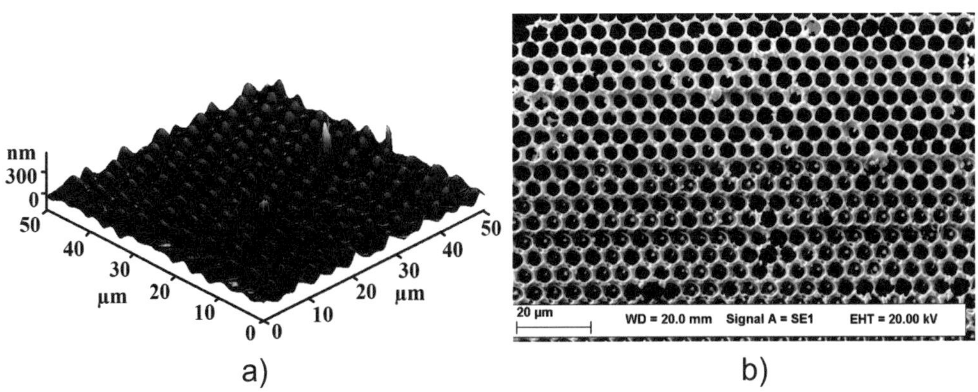

Figure 18. photonic crystal. a) Before etching. b) After etching.

4. Conclusions

The laser lithography can be considered an alternative to the classical lithography methods. In this chapter we emphasize the possibility to obtain nano/micro-structures in organic and non-organic materials by femtosecond laser lithography.

The organic and non-organic photoresist materials properties are presented. Photo-chemical reactions induced by femtosecond laser irradiation of photoresists are described. Due to the transparency at IR wavelengths, the laser pulses can be focused deeply in the photoresist volume. A Direct Laser Writing station coupled with a high repetition rate femtosecond laser was used to process the photoresist materials.

Various geometry structures were obtained in organic photoresists and chalcogenide glasses. When photoresist materials specially developed for the TPP method were used, nanostructures with dimensions under 100 nm were obtained.

Using the TPP method, high power laser targets were fabricated in organic photoresist materials. These structures can be used as template for other materials. An other domain where the organic photoresist structures can be used is optical contact lithography. Based on the electromagnetic field enhancement produced by the photoresist masks, structures with 350 nm width were created in a silicon wafer.

Besides the organic photoresists, the femtosecond laser lithography can be used to process other photoresist materials, like chalcogenide glasses. Taking advantage of their higher refractive index in comparison with organic photoresists, the chalcogenide glasses are suitable for visible and near-IR micro-optical devices fabrication. They act either as negative or positive photoresists. Optical microlenses and photonic crystals structures where produced in As_2S_3 chalcogenide glasses using DLW method. The microlenses imprinted on the surface of PLD deposited As_2S_3 thin films could be used to focus the red-infrared laser light transmitted through optical fibers. Transition from the bump to the hole configuration has been revealed when the laser pulse power was increased. Bi-dimensional photonic structures characterized by a hexagonal assembly of bumps or gratings with traces of micrometer width have been obtained.

The possibility of structure direct writing by translating the focused spot through the photoresists volume, recommends the femtosecond laser lithography technique as a fast, cheap and flexible processing method.

Author details

Florin Jipa[1], Marian Zamfirescu[1], Alin Velea[2], Mihai Popescu[2] and Razvan Dabu[1*]

*Address all correspondence to: razvan.dabu@inflpr.ro

1 National Institute for Laser, Plasma and Radiation Physics, Magurele, Romania

2 National Institute of Materials Physics, Magurele, Romania

References

[1] Pilnam K. Soft Lithography for Microfluidics. A Review, BIOCHIP JOURNAL 2008; 2(1) 1-11.

[2] Unger M. A, Chou H, Thorsen T, Scherer A, Quake S. Monolithic Microfabricated Valves and Pumps by Multilayer Soft Lithography. Science 2000; 288 (5463) 113-116.

[3] Matsuyama T, Ishiyama T, Ohmura Y. Nikon projection lens update. Proc. SPIE Optical Microlithography 2004; 5377 730.

[4] Hsu S. H, Fang S. P, Huang I.H, Lin B.S. M, Hung K.C Extension of ArF lithography for poly gate patterning of 65-nm generation and beyond. Proc. SPIE 2004; 5377 1214.

[5] Bloomstein T.M, Horn M.W, Rothschild M, Kunz R.R, Palmacci S.T, Goodman R.B. Lithography with 157 nm lasers. J. Vac. Sci. Technol. 1997; (B15) 2112.

[6] Grenville A, Liberman V, Rothschild M, Sedlacek J.H.C, French R.H, Wheland R.C, Zhang X, Gordon J. Behavior of candidate organic pellicle materials under 157-nm laser irradiation, Proc. SPIE 2002; (4691) 1644.

[7] French R, Wheland R.C, Qiu W, Lemon M.F, Blackman G.S, Zhang E, Gordon J, Liberman V, Grenville A, Kunz R.R, Rothschild M. 157-nm pellicles: polymer design for transparency and lifetime. Proc. SPIE 2002; 4691 576.

[8] Switkes M, Rothschild M. Resolution enhancement of 157-nm lithography by liquid immersion. Proc. SPIE 2002; 4691 459.

[9] Owa S, Nagasaka H, Nakano K, Ohmura Y. BCurrent status and future prospect of immersion lithography. Proc. SPIE 2006; 6154 (61) 5408-1.

[10] Sewell H, Mulkens J, McCafferty D, Markoya L, Streefkerk B, Graeupner P. B. The next phase for immersion lithography. Proc. SPIE 2006, 6154 615.

[11] Peng S, French R.H, Qiu W, Wheland R.C, Yang M, Lemon M.F, Crawford M.K. Second generation fluids for 193 nm immersion lithography. Proc. SPIE 2005; 5754 427.

[12] Ehrfeld W, Lehr H. Deep X-ray lithography for the production of three-dimensional microstructures from metals, polymers and ceramics. Radiation Physics and Chemistry 1995; 45, (3) 349–365.

[13] Broers A. N, Hoole A. C. F, Ryan J. M. Electron beam lithography--Resolution limits. Microelectronic Engineering 1996; 32 (131 EOF 142)1-4.

[14] Liu K, Avouris P, Bucchignano J, Martel R, Sun S, Michl J. Simple fabrication scheme for sub-10 nm electrode gaps using electron-beam lithography. Applied Physics Letters 2002; 80 (5) 865-867.

[15] Yu-Tung C, Tsung-Nan L, Yong S C, Jaemock Y, Chi-Jen L, Jun-Yue W, Cheng-Liang W,Chen-Wei C, Tzu-En H, Yeukuang H, Qun S, Gung-Chian Y, Keng S L, Hong-

Ming L, Jung Ho J, Giorgio M. Full-field hard x-ray microscopy below 30 nm: a challenging nanofabrication achievement.Nanotechnology 2008; 19 395302.

[16] Tichenor D. A, Kubiak G, Malinowski M, Stulen R. H, Haney S, Berger K. W, Brownet L. A, Freema R. R, Mansfield W M, Wood R, Tennant D. M, Bjorkholm J.E, MacDowell A. A, Bokor J, Jewell T. E, White D. L, Windt D. L, Waskiewicz W K. Diffraction-limited soft-x-ray projection imaging using a laser plasma source. Optics Letters 1991; 16 (20) 1557.

[17] Bilenberga B, Jacobsena S, Schmidta M.S, Skjoldinga L.H.D, Shib P, Bøggilda P, Tegenfeldtc J.O, Kristensena A. High resolution 100 kV electron beam lithography in SU-8. Microelectronic Engineering 2006; 83 (4–9) 1609–1612.

[18] Miklyaev Yu. V, Meisel D. C, Blanco A, Freymann von G, Busch K, Koch W, Enkrich C, Deubel M, Wegener M. Three-dimensional face-centered-cubic photonic crystal templates by laser holography: fabrication, optical characterization, and band-structure calculations APPLIED PHYSICS LETTERS VOLUME 2003; 82 (8).

[19] Liyun Z, Wenbi Z, Bingheng X, Qiming Y, Aibing G "Making a photoresist relief hologram with expection groove depth"Proc. SPIE 2866, International Conference on Holography and Optical Information Processing (ICHOIP '96), 316 1996 doi: 10.1117/12.263101

[20] George Odian. Principles of Polymerization. John Wiley & Sons, Inc 2004 ISBN 0-471-27400-3

[21] Pollino J, Nair K, Stubbs L, Adams J, Weck M "Cross-linked and functionalized 'universal polymer backbones' via simple, rapid, and orthogonal multi-site self-assembly" Tetrahedron 2004 (60) 7205–7215.

[22] Varadan V.K, Jiang X, Varadan VV. Microstereolithography and other fabrication techniques for 3D MEMS. Chichester: Wiley 2001.

[23] Pao Y.H, Rentzepis P.M. LASER-INDUCED PRODUCTION OF FREE RADICALS IN ORGANIC COMPOUNDS. Appl Phys Letters 1965; 6 (5) 93.

[24] Maruo S, Nakamura O, Kawata S.Three-dimensional microfabrication with two-photon-absorbed photopolymerization.Optics Letters 1997; 22 (2) 132-134.

[25] Martineau C, Anemian R, Andraud C, Wang I, Bouriau M, Baldeck P.L. Efficient initiators for two-photon induced polymerization in the visible range. Chem. Phys. Letters 2002; 362 (3-4) 291–295.

[26] Belfield K.D, Schafer K.J, Liu Y.U, Liu J, Ren X.B, Van Stryland E.W. Multiphoton-absorbing organic materials for microfabrication, emerging optical applications and non-destructive three-dimensional imaging. J. Phys. Org. Chem 2000; 13 (12) 837–849.

[27] Yu-fang Z, Sheng-yu F, Xiao-mei W. Investigation of TPP up-conversion lasing of stilbene derivatives. Journal of Molecular Structure 2002; 613 (1–3) 91–94.

[28] Ajami A, Husinsky W, Liska R, Pucher N. Two-photon absorption cross section measurements of various two-photon initiators for ultrashort laser radiation applying the Z-scan technique. Journal of the Optical Society of America B 2010; 27(11) 2290.

[29] Kumpfmu°ller J, Pucher N, Stampfl J, Liska R. Optical Application in Polymer Photochemistry. New York:Springer Publishers; 2011.

[30] Clay G. O, Schaffer C. B, Kleinfeld D. Large two-photon absorptivity of hemoglobin in the infrared range of 780–880 nm. J. Chem. Physics 2007; 126, (2) 025102.

[31] Tanaka T, Sun HB, Kawata S. Rapid sub-diffraction-limit laser micro/nanoprocessing in a threshold material system. Appl Phys Letters 2002; 80 (2) 312.

[32] Martínez-Corral M, Ibáñez-López C, Saavedra G. Axial gain resolution in optical sectioning fluorescence microscopy by shaded-ring filters. Optics Express 2003; 11 (15) 1740-1745.

[33] Martinez C. M, Pons A. Axial apodization in 4Pi-confocal microscopy by annular binary filters. J. Opt. Soc. Am. A 2002; 19 (8) 1782-1789.

[34] Akturk S, Gu X, Kimmel M, Trebino R. Extremely simple single-prism ultrashort-pulse compressor. OPTICS EXPRESS 2006; 14 (21) 10101-10108.

[35] Teha W. H, Dürig U, Drechsler U, Smith C. G, Güntherodt H. J. Effect of low numerical-aperture femtosecond two-photon absorption on SU-8 resist for ultrahigh-aspect-ratio microstereolithography. Journal of Applied Physics 2005; 97 (5) 054907.

[36] Chung C K, Hong Y Z. Surface modification of SU8 photoresist for shrinkage improvement in a monolithic MEMS microstructure. J. Micromech. Microeng. 2007; 17 207–212.

[37] McHugh M.A, Krukonis V.J. Supercritical Fluid Extraction. Butterworth-Heinemann, Boston, 2nd ed., 1994.

[38] Jipa F, Zamfirescu M, Luculescu C, Dabu R. High-aspect-ratio structures produced by two-photon photopolimerization. J Optoelectron. Adv. Materials 2010;12 (1) 124-128.

[39] Tabak M, Hinkel D, Atzeni S, Campbell E. M, Tanaka K. Fast Ignition: Overview and Background. Fusion Science and Technology 2006; 49 (3) 254-277.

[40] Quéré F, Thaury C, Monot P, Dobosz S, Martin Ph, Geindre J.-P, Audebert P. Coherent Wake Emission of High-Order Harmonics from Overdense Plasmas. Phys. Rev. Lett. 2006; 96 (12) 125004.

[41] Tahir N. A, D. Hoffmann H. H, Kozyreva A, Shutov A, Maruhn J. A, Neuner U, Tauschwitz A, Spiller P, Bock R. Shock compression of condensed matter using intense beams of energetic heavy ions. Phys. Rev. E 2000; 61 1975–1980.

[42] Fuchs J, Audebert P, Borghesi M, Pepin H, Willi O. Laser acceleration of low emittance, high energy ions and applications. C. R. Physique 2009; 10 176.

[43] Dewald E.L, Suter L.J, Thomas C, Hunter S, Meeker D, Meezan N, Glenzer S.H Bond E, Kline J, Dixit S, Kauffman R.L, Kilkenny J, Landen O.L. First hot electron measurements in near-ignition scale hohlraums on the National Ignition Facility. Journal of Physics: Conference Series 2010; 244 022074.

[44] Galloudec N. R, D'Humieres E. New micro-cones targets can efficiently produce higher energy and lower divergence particle beams. Laser and Particle Beams 2010; 28 513–519.

[45] Buffechoux S, Psikal J, Nakatsutsumi M, Romagnani L, Andreev A, Zeil K, Amin M, Antici P, Burris-Mog T, Compant-La-Fontaine A, d'Humie`res E, Fourmaux S, Gaillard S, Gobet F, Hannachi F, Kraft S, Mancic A, Plaisir C, Sarri G, Tarisien M, Toncian T, Schramm U, Tampo M, Audebert P, Willi O, Cowan T. E, Pe´pin H, Tikhonchuk V, Borghesi M, Fuchs J. Hot Electrons Transverse Refluxing in Ultraintense Laser-Solid Interactions 2010; PHYSICAL REVIEW LETTERS 105 015005.

[46] Ceccotti T, Le´vy A, Popescu H, Re´au F, D'Oliveira P, Monot P, J. Geindre P, Lefebvre E, Martin Ph. Proton Acceleration with High-Intensity Ultrahigh-Contrast Laser Pulses. PHYSICAL REVIEW LETTERSPRL 2007; 99 185002.

[47] Schmid H, Biebuyck H, Michel B, Martin O. J. F. Light-coupling masks for lensless, subwavelength optical lithography. Appl. Phys. Lett.:1998 72 (19) 2379-2381.

[48] Borchers B, Bekesi J, Simon P, Ihlemanna J. Submicron surface patterning by laser ablation with short UV pulses using a proximity phase mask setup. JOURNAL OF APPLIED PHYSICS 2010; 107 063106

[49] Popescu M. Non-Crystalline Chalcogenides. Dordrecht: Kluwer Academic Publishers; 2000.

[50] Jain H., Vlcek M. Glasses for lithography. Journal of Non-Crystalline Solids 2008; 354 1401-1406.

[51] Kovalskiy A., Cech J., Vlcek M., Waits CM, Dubey M., Heffner WR, Jain H. Chalcogenide glass e-beam and photoresists for ultrathin grayscale patterning. J. Micro/Nanolith. MEMS MOEMS 2009; 8(4) 043012.

[52] Popescu M., Disordered chalcogenide optoelectronic materials: Phenomena and applications J. Optoelectron. Adv. Mater. 2005; 7(4) 2189-2210.

[53] De Neufville JP, Moss SC, Ovshinsky SR. Photostructural transformations in amorphous As2Se3 and As_2S_3 films. J. Non-Cryst. Solids 1974; 13(2) 191-223.

[54] Tanaka K. Reversible photostructural change: Mechanisms, properties and applications. J. Non-Cryst. Solids 1980; 35-36 1023-1034.

[55] K. Tanaka, A. Odajima, Photodarkening in amorphous selenium. Solid State Commun. 1982; 43(12) 961–964.

[56] Katayama Y., Yao M., Ajiro Y., Inui M., Endo H. Photo-Induced Phenomena in Isolated Selenium Chains. J. Phys. Soc. Jpn. 1989; 58 1811-1822.

[57] Tanaka K., Ohtsuka Y. Transient characteristics of photochromic processes. J. Opt. 1977; 8 121.

[58] Averyanov VL, Kolobov AV, Kolomiets BT, V. Lyubin M. Thermal and optical bleaching in darkened films of chalcogenide vitreous semiconductors. Phys. Status Solidi (a) 1980; (a) 57 81-88.

[59] Hamanaka H., Tanaka K., Matsuda A., Iizima S. Reversible photo-induced volume changes in evaporated As_2S_3 and As4Se5Ge1 films. Solid State Commun. 1976; 19 499-501.

[60] Vateva E., Arsova D., Skordeva E., Pamukchieva V. Irreversible and reversible changes in band gap and volume of chalcogenide films. J. Non-Cryst. Solids 2003; 326-327 243-247.

[61] Hisakuni H., Tanaka K. Giant photoexpansion in As_2S_3 glass. Appl. Phys. Lett. 1994; 65 2925-2927.

[62] Tanaka K. Spectral dependence of photoexpansion in As_2S_3 glass. Phil. Mag. Lett. 1999; 79 25-30.

[63] Ramachandran S., Pepper JC, Brady DJ, Bishop SG, Micro-optical lenslets by photoexpansion in chalcogenide glasses. J. of Lightwave Technol. 1997; 15 1371-1377.

[64] Saliminia A., Galstian TV, Villeneuve A. Optical Field-Induced Mass Transport in As_2S_3 Chalcogenide Glasses. Phys. Rev. Lett. 2000; 85, 4112-4115.

[65] Tanaka K., Saitoh A., Terakado N. Giant photo-expansion in chalcogenide glass. J. Optoelectron. Adv. Mater. 2006 2058-2065.

[66] Saitoh A., Gotoh T., Tanaka K. Chalcogenide-glass microlenses for optical fibers. J. Non-Cryst. Solids 2002; 299-302 983-987.

[67] Saitoh A., Tanaka K. Self-developing aspherical chalcogenide-glass microlenses for semiconductor lasers. Appl. Phys. Lett. 2003; 83 1725-1727.

[68] Messaddeq SH, Siu Liu M., Lezal D., Ribeiro SJL., Messaddeq Y. Above bandgap induced photoexpansion and photobleaching in Ga–Ge–S based glasses. J. Non-Cryst. Solids 2001; 284 282-287.

[69] Andriesh AM, Buzdugan AI, Dolghier V, Iovu MS. Modeling of chalcogenide glass structures before and after laser illumination, based on mass spectroscopy data. ROMOPTO '97: Fifth Conference on Optics, SPIE 1998; 3405 258.

[70] Hisakuni H., Tanaka K. Optical Microfabrication of Chalcogenide Glasses. Science 1995; 270, 974-975

[71] Tanaka K. The charged defect exists? J. Optoelectron. Adv. Mater. 2001; 3 189-198.

[72] Tanaka K. Two-photon absorption spectroscopy of As_2S_3 glass. Appl. Phys. Lett 2002; 80(2) 177-179.

[73] Juodkazis S., Kondo T., Misawa H. Photo-structuring of As_2S_3 glass by femtosecond irradiation. Optics Express 2006; 14(17) 7751-7756.

[74] Velea A., Popescu M., Sava F., Lőrinczi A., Simandan ID, Socol G, Mihailescu IN, Stefan N., Jipa F., Zamfirescu M., Kiss A., Braic V.Photoexpansion and nano-lenslet formation in amorphous As2S3 thin films by 800 nm femtosecond laser irradiation. J. Appl. Phys. 2012; 112 033105.

[75] Zaidi SH, Chu AS, Brueck SRJ. Optical properties of nanoscale, one-dimensional silicon grating structures. J.Appl. Phys. 1996; 80 6997-7000.

[76] Yu Z., Chen L., Wu W., Ge H., Chou SY. J. Vac. Sci. Technol. B 2003; 21 2089-2092.

[77] Popov E., Hoose J., Frankel B., Keast C., Fritze M., Fan TY, Yost D., Rabe S. Low polarization dependent diffraction grating for wavelength demultimlexing. Optics Express 2004; 12 269-275.

[78] Hoose J., Frankel R., Popov E., Neviere M., US Patent 2005: 6,958,859 B2.

[79] Geppert T., Schweizer SL, Gosele U, Wehrspohn RB. Deep trench etching in macroporous silicon. Appl. Phys. A-Mater 2006; 84 237-242.

[80] Gish DA, Summers MA, Brett MJ. Morphology of periodic nanostructures for photonic crystals grown by glancing angle deposition. Photonics Nanostruct. 2006; 4 23-27.

[81] Indutnyy I. Z., Popescu M., Lőrinczi A., Sava F., Min'ko VI, Shepeliavyi PE. Fabrication of submicrometer periodic structures using interference lithography and two-layer chalcogenide photoresist. J. Optoelectron. Adv. Mater. 2008; 10(12) 3188 – 3192.

[82] [82 Popescu A., Savastru D., Miclos S. Refractive index anisotropy in non-crystalline As_2S_3 films. J. Optoelectron. Adv. Mater. 2010; 12 (5) 1012-1018

[83] Bayindir M., B. Temelkuran, Ozbay E. Tight-Binding Description of the Coupled Defect Modes in Three-Dimensional Photonic Crystals. Phys. Rev. Lett. 2000; 84 2140-2143

[84] Noda S., Tomoda K., Yamamatu N., Chutinan A. Full Three-Dimensional Photonic Bandgap Crystals at Near-Infrared Wavelengths. Science 2000; 289 604-606

[85] Popescu A., Micloş S., Savastru D., Savastru R., Ciobanu M., Popescu M., Lőrinczi A., Sava F., Velea A., Jipa F., Zamfirescu M. Direct laser writing of two-dimensional photonic structures in amorphous As_2S_3 thin films. J. Optoelectron. Adv. Mater. 2009; 11(11) 1874-1880.

[86] Popescu M., Sava F., Lőrinczi A., Velea A., Vlček M., Jain H. Modelling of dissolution kinetics of thin amorphous chalcogenide films. Phil. Mag. Lett. 2009; 89(6) 370-376.

[87] Popescu M., Velea A., Lőrinczi A., Zamfirescu M., Jipa F., Micloş S., Popescu A., Ciobanu M., Savastru D. Two-dimensional photonic structures based on As-S chalcogenide glass. Dig. J. Nanomater. Bios. 2010; 5(4) 1101-1105.

Chapter 4

Three-Dimensional Lithography Using Combination of Nanoscale Processing and Wet Chemical Etching

Noritaka Kawasegi and Noboru Morita

Additional information is available at the end of the chapter

http://dx.doi.org/10.5772/56354

1. Introduction

Nanofabrication technology is an important field of research, and numerous attempts have been made to improve this technology in recent years. This technology is demanded in various industrial fields such as electronic, photonic, and biomedical engineering to miniaturize machine components. Through miniaturization, the integration and parallelization of components in a machine device are possible, leading to reductions in dead space and fabrication cost. A high-performance device can be fabricated using the changes in physicochemical behavior due to scaling effects; the resonance frequency, thermal response, and chemical reaction all increase with miniaturization. These characteristics are particularly effective in devices such as sensors and reactors. Additionally, Newton's laws of motion do not hold for distances less than 10 nm owing to increased quantum mechanical effects, so that certain types of functions only occur in miniaturized devices. Devices that maximize these effects include microelectromechanical systems and nanoelectromechanical systems, which consist of micro/nanometer-scale mechanical components integrated on a silicon surface.

The conventional fabrication methods of these components are based on photolithography [1], which is a fabrication method used for semiconductor devices. This technique is suitable for mass producing micro/nanostructures because it is a high-throughput fabrication process. It was developed based on Moore's law [2], and it can fabricate sub-100-nm patterns by using deep ultraviolet light with wavelengths of 248 nm and 193 nm. Other major approaches include reactive ion etching [3] and LIGA (lithographie, galvanoformung, abformung) [4], which are mainly used to fabricate high-aspect-ratio structures. However, these fabrication processes use a photomask, which is a high-cost and time-consuming step. Layering and etching techniques are used to fabricate complex structures, but it is difficult to fabricate complex three-dimensional structures. However, several successful attempts have been made to fabricate nano/

micrometer-scale structures using scanning probe microscopy (SPM), focused ion beam (FIB), and electron beam (EB) lithography. Though these methods are effective for fabricating nanometer-scale structures, time-consuming steps are necessary for fabricating sub-micrometer to micrometer-scale structures. Higher and wider structures can be fabricated by taking advantage of the combination of the high resolution of these lithographic techniques and the high-speed material removal of wet chemical etching. Three-dimensional structures can also be fabricated by using this combination of lithography and etching.

Three processing techniques are introduced for three-dimensional fabrication using a combination of nanoscale processing and wet chemical etching. The first is tribo-nanolithography (TNL), which forms an amorphous silicon phase on a silicon substrate by using direct machining, similar to mechanical scratching. The second technique is FIB irradiation, which forms an amorphous phase on a silicon substrate by irradiating it with accelerated ions. The last is EB irradiation, which causes the formation of a thin hydrocarbon layer that has an etch resistance. The etch rate of the processing layer is different from the original material surface, enabling structures to be fabricated after the etching. Additionally, three-dimensional structures can be fabricated using these techniques. The fundamental characteristics and possible applications of these methods are described in the following sections.

2. Three-dimensional fabrication using tribo-nanolithography

2.1. SPM-based lithography

SPM, which includes scanning tunneling microscopy (STM) [5], atomic force microscopy (AFM) [6], and scanning near-field optical microscopy (SNOM) [7], was developed to observe surface characteristics via detecting interactions between a probe and solid materials. The most used instruments in the SPM family are STM and AFM. STM is used to measure the surface topography with atomic resolution by detecting the tunneling current, and AFM measures the topography by detecting minute forces between the probe and the sample. These instruments have a probe and a scanning stage, which means they can be used as a micro tool with a precise stage, similar to a machine tool. Therefore, several micro/nanostructuring technologies arose after the development of these instruments; this technique enabled researchers to manipulate even single atoms [8]. Thus, an SPM can be used as not only a measuring tool but also as a nanostructuring tool, making it appropriate for nanoscale lithography.

SPM lithography can be used to fabricate nanoscale structures by various methods such as electrochemical oxidation [9–11], material transfer [12–14], mechanical removal [15–19] and thermal reaction [21, 22]. This technique is effective for fabricating nanometer-scale structures. However, for fabricating larger structures, time-consuming steps are necessary. The chemical properties of the patterned area change, although the mechanism is different for each method. The patterned area can then be used as an etching mask or it can be selectively etched, and micro/nanometer structures can then be fabricated [9–11, 14, 19]. SPM lithography is used only for the patterning; the structures are fabricated with subsequent etching, which can remove a large amount of material at a time. Thus, SPM lithography with etching is suitable for fabricating structures with a height over several tens of nanometers, whereas SPM lithography alone can only be used for structures that are a few nanometers high. Additionally, three-dimensional structures can be fabricated by controlling the etching mask. Herein we describe a method of three-dimensional fabrication using TNL [19, 22–28] with wet chemical etching.

2.2. Fabrication method

TNL uses AFM as a machine tool for nanopatterning, which enables one to measure the machining forces as well as the machine materials. The technique can be used to machine a material surface with large normal loads by employing a specially designed diamond tip cantilever. The machining setup, based on AFM, is schematically shown in Figure 1(a). The machining forces can be measured from the deflection and torsion of the cantilever, which are both detected by a four-segment photodetector. The normal load is kept constant by a feedback control from a piezo scanner. In the machining process, the specially designed cantilever [18, 29, 30] shown in Figure 1(b) is installed in the system instead of conventional Si_3N_4 or silicon cantilevers. This cantilever has a diamond tip attached to a silicon lever, which has a very high stiffness of more than 1000 times that of conventional cantilevers.

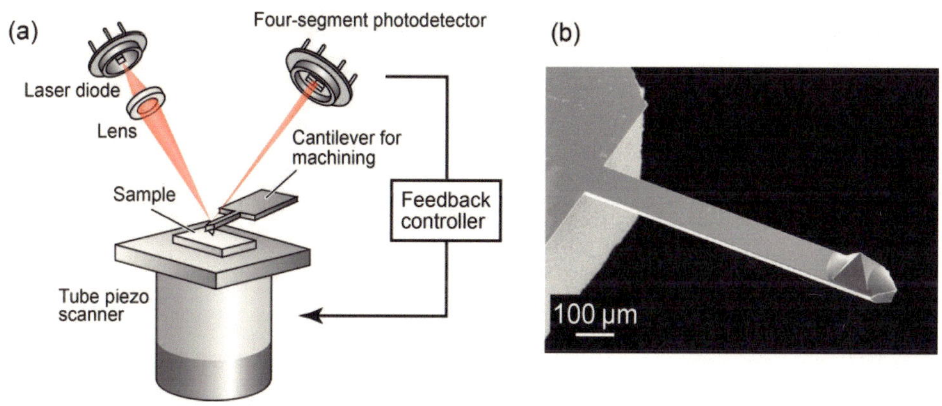

Figure 1. a) Schematic diagram of the experimental setup for TNL based on AFM. (b) SEM image of the cantilever for machining. Reprinted with permission from [29]. Copyright 2006, ASME.

The TNL method forms a modified area on the surface of the silicon substrate via a direct nanomachining method, similar to mechanical scratching. The machining characteristics of the cantilever change significantly owing to the shape of the diamond tip. Figure 2 shows a scanning electron microscopy (SEM) image of cutting tips with different radii. Figure 3 shows AFM topography images of silicon surfaces machined with these cantilevers [30]. When a sharper tip is used, the machined surface is removed and a concave pattern is fabricated, as shown in Figure 3(a) [29, 30]. In this case, continuous cutting chips are formed because the cutting was conducted in a ductile mode owing to the small cutting depth. A dull tip tends not to remove material from the machined area, as shown in Figure 3(b), but instead introduces a high-pressure region to the substrate. Additionally, a 1–2 nm high protuberance was formed, induced by the volume expansion of the machined area [19, 30]. A dull tip is used in the TNL method because it is more suitable for fabricating precise low-linewidth structures [23].

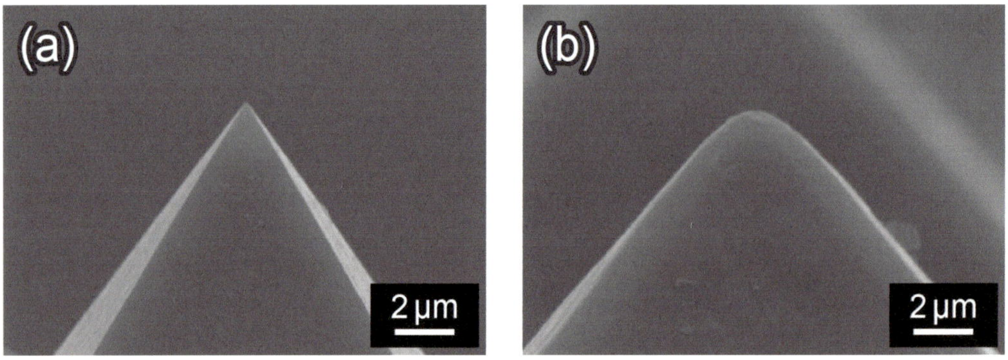

Figure 2. SEM image of diamond tips with (a) sharp and (b) dull cutting edges.

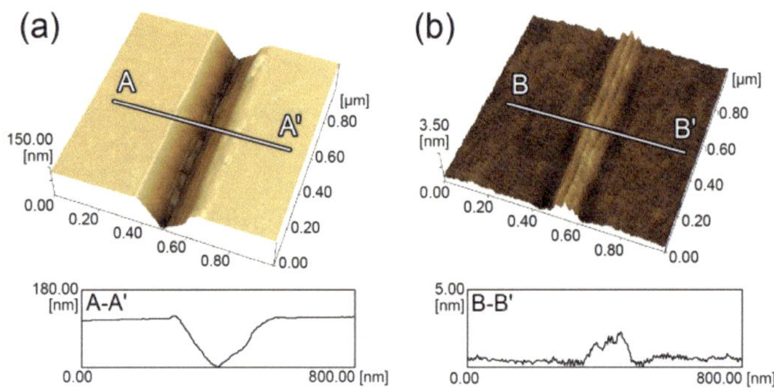

Figure 3. AFM topography images of machined areas. Silicon surfaces were machined with (a) sharp and (b) dull cutting edges at normal loads of 331 μN and 340 μN, respectively.

The fabrication of a structure using a combination of TNL and wet chemical etching is shown in Figure 4 [19]. Figure 4(a) shows an AFM image of the topography of a machined area (15 × 7.5 μm^2 area) prepared using TNL at a normal load of 310 μN and a scanning pitch of 59 nm. The minute protuberance, which is only 1–2 nm high, was formed on the machined area. Figure 4(b) shows an AFM topography image of the same area after etching in a 10 mass% potassium hydroxide (KOH) for 5 min. The machined area was able to resist etching in KOH, whereas the nonmachined area was etched. Thus, a protruding structure with a height of 110 nm was fabricated on the machined area. However, this machined area could also be selectively dissolved in hydrogen fluoride (HF), with the nonmachined silicon surface being barely etched [1]. In this way, a concave structure with a depth of several nanometers to several tens of nanometers could be fabricated from the machined area.

http://dx.doi.org/10.5772/56354

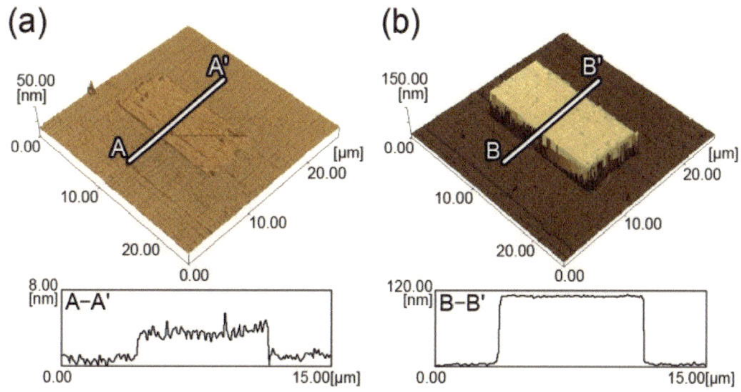

Figure 4. Structure fabrication using a combination of TNL and wet chemical etching. (a) AFM topography image of the machined area (15 × 7.5 μm^2 area) prepared using TNL at a normal load of 310 μN. (b) AFM topography image of the same area after etching in 10 mass% KOH for 5 min. Reprinted with permission from [19]. Copyright 2005, IOP Publishing.

The TNL-induced etch resistance of the machined area is due to the formation of an amorphous phase. Cross-sectional transmission electron microscopy (TEM) images and nano-electron diffraction (nano-ED) patterns of the machined areas prepared by TNL are shown in Figure 5 [19]. Figure 5(a) shows a cross-sectional TEM image of the silicon substrate after machining a single line at a normal load of 350 μN. This image shows that an affected layer measuring 100 nm wide and 15 nm deep was formed. This indicates that the volume expansion of the machined area shown in Figure 4(a) resulted from this layer. The nano-ED pattern of the affected layer in Figure 5(c) shows a diffuse ring pattern, whereas that of the nonmachined area in Figure 5(d) shows a silicon crystal pattern, meaning the machined area was converted to an amorphous phase. Figure 5(b) shows cross-sectional TEM images of the machined area prepared at a scanning pitch of 50 nm. The amorphous phase was thicker than that of a machined single line, and the thickness was observed to be 20 nm. Furthermore, a concave-convex pattern of the silicon crystal structure, which has the same pitch as the scanning pitch, is formed under the amorphous phase. Secondary ion mass spectrometry and Auger electron spectroscopy analyses have shown that the amorphous phase consists entirely of silicon. Therefore, the etch stop effect of the machined area results from the formation of the amorphous phase, rather than from the formation of a chemical compound such as silicon oxide or silicon hydroxide. The TNL-induced amorphous phase was also formed via the removal method, shown in Figure 3(a), by applying the sharp tip with a higher normal load [20]. In this case, a thicker amorphous phase was formed under the machined area as well as large dislocations. The dislocations induced a KOH etching enhancement rather than an etch stop effect. Therefore, in the material removal method, a convex or concave structure can be fabricated by the etching in KOH from the etch stop and etching enhancement effects, which is decided by the concentration of KOH.

The phase transition of the silicon from crystalline to amorphous is due to the high pressure induced by the TNL. During machining, a significantly high pressure is introduced at the contact region of the probe and the silicon substrate, which creates the amorphous silicon phase. Machining-induced phase transitions have been observed during various types of machining at nano to submicron scales, such as turning [31, 32], grinding [33], indentation [34, 35], and scratching [36, 37]. Silicon, which has a diamond crystal structure, converts to a β-Sn structure at a pressure of approximately 12 GPa, and then converts to an amorphous phase during the pressure release process [37]. This is the mechanism by which the area machined by TNL is converted to an amorphous phase.

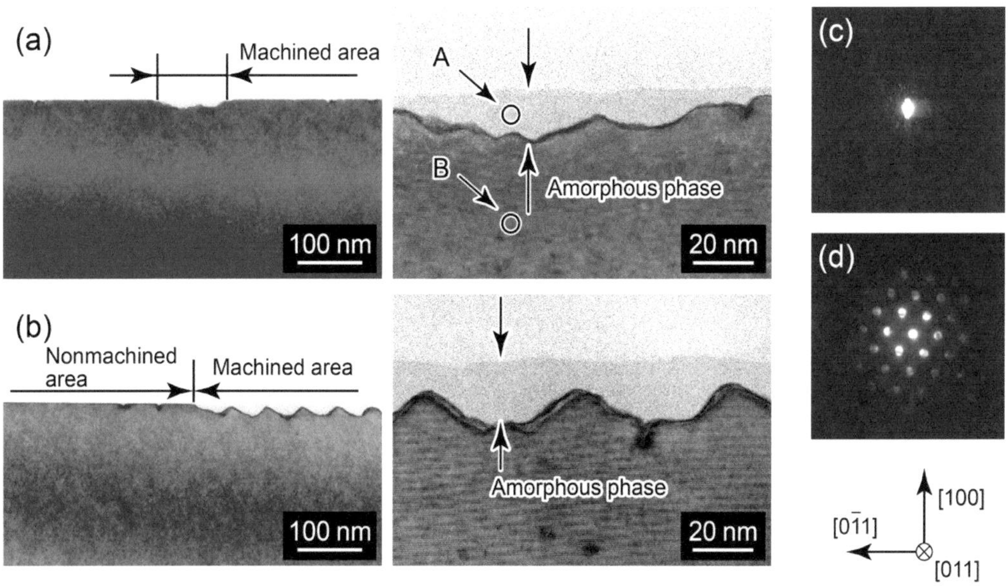

Figure 5. Bright field cross-sectional TEM image of machined area prepared by TNL. (a) Machined area of a single line at a normal load of 350 μN. (b) Machined area prepared at a normal load of 350 μN with a scanning pitch of 50 nm. Nano-ED patterns of a machined area A in (c), and a nonmachined area B in (d). Reprinted with permission from [19]. Copyright 2005, IOP Publishing.

2.3. Three-dimensional fabrication using TNL and wet chemical etching

The morphology of the amorphous phase depends on the machining-induced pressure, and therefore, it can be controlled by the machining conditions. Thus, the change in etch resistance of the amorphous phase can be used to fabricate a three-dimensional structure. The height dependence of the etch rate, the height of the protuberance, and the height of the structure on the normal load is shown in Figure 6 [27]. The silicon surface was machined at various normal loads and then etched in 10 mass% KOH for 10 min. The machined area protruded for normal loads less than 372 μN, whereas the machined area was removed and a concave structure was fabricated for normal loads greater than 372 μN. Therefore, the machining mode is divided into protuberance and removal regions, as shown in the figure. The etch rate of the machined area decreased and the structure height increased with increasing normal loading for the protuberance region. However, the etch rate was nearly constant in the removal region owing to the dislocations formed by the removal machining at higher normal loads. The dislocations enhanced the silicon etching in KOH [22]. Therefore, a constant etch rate at the removal region resulted from the interaction between the amorphous phase (etch stop effect) and dislocation (etching enhancement effect) formed by the machining. These results demonstrate that the etch resistance can be controlled by the normal load in the protuberance region, whereas it is constant in the removal region. The etch resistance can also be increased by other machining conditions, such as the overlap ratio and number of times the area is machined [23].

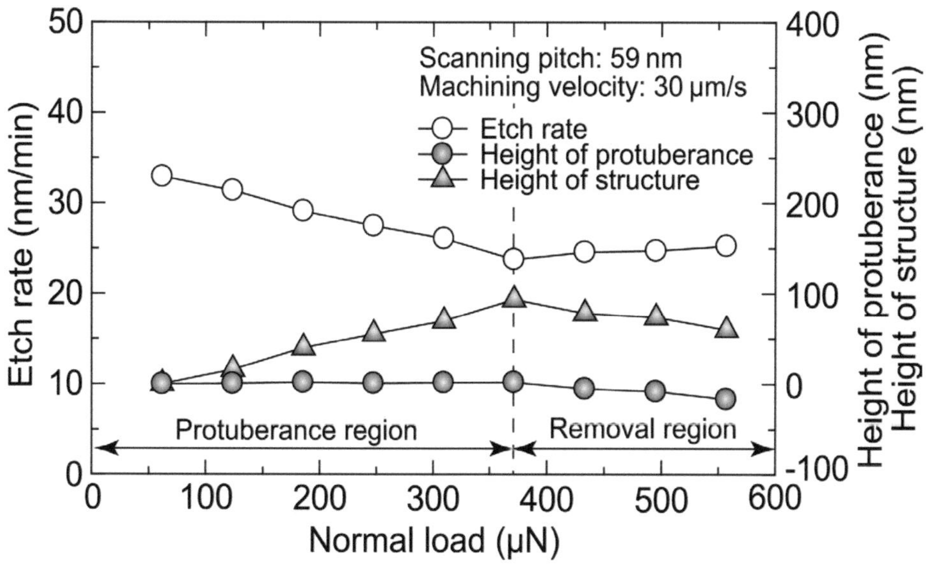

Figure 6. Etch rate of the machined area when the silicon was machined at various normal loads and then etched in 10 mass% KOH for 10 min.

The change in the etch rate of the machined area owing to the machining conditions is caused by the morphology of the amorphous phase. To study this, the morphology of the amorphous phase is measured by etching in HF. Figure 7 shows a cross-sectional TEM image of the machined area prepared at a normal load of 350 μN and a scanning pitch of 50 nm, the same conditions as Figure 5(b), after etching in 25 mass% HF for 10 min [27]. The amorphous phase formed in the machined area was removed, and the concave-convex pattern of the silicon crystal phase remained. The single-crystal silicon surface was scarcely etched in HF [1]. Therefore, the concave structure shown in Figure 7 resulted from the selective removal of the TNL-induced amorphous phase by etching in HF.

Figure 8 shows the thickness of the amorphous phase for various normal loads [27]. The maximum height of protuberance after machining was observed at a normal load of 278 μN. After etching, the depth of the machined area increased with increasing normal load. The amorphous phase was formed by the high-pressure phase transition induced by the TNL [19], and increased in thickness with increasing normal load. Hence, for higher normal loads, a deeper region tended to be transformed to the amorphous phase owing to the high pressure, forming a thicker amorphous phase and resulting in a higher etch resistance against KOH. The thickness of the amorphous phase is directly correlated to the changes in the etch resistance of the machined area. Alternatively, changes in the etch resistance against KOH due to the scanning pitch and the number of times the area is machined result from the density of the amorphous phase rather than its thickness [27].

Figure 7. Bright-field cross-sectional TEM image of the machined area after etching in 25 mass% HF for 10 min. The silicon was machined at a normal load of 350 μN and a scanning pitch of 50 nm.

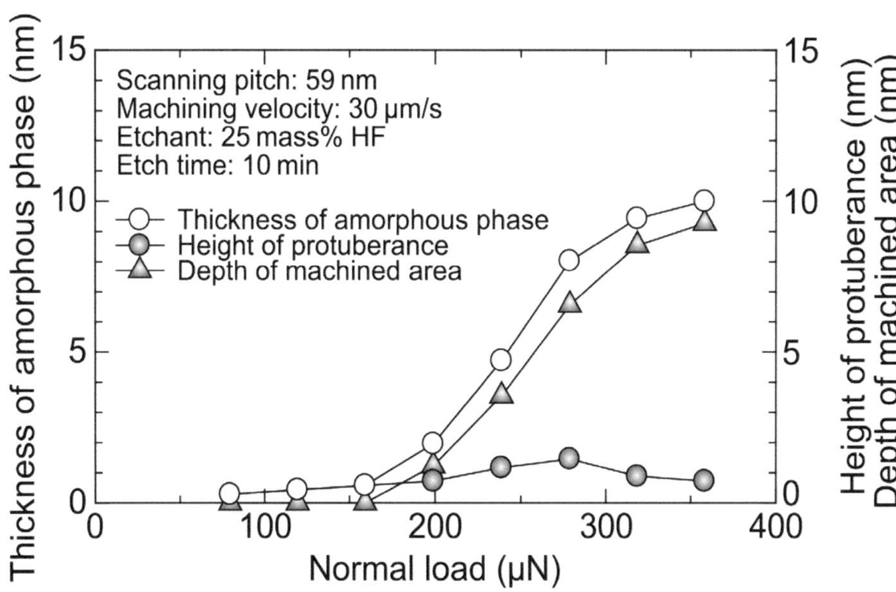

Figure 8. Changes in the thickness of the amorphous phase after the silicon was machined at various normal loads and etched in 25 mass% HF for 10 min.

Figure 9 shows three-dimensional structures with uniform height fabricated by TNL and wet chemical etching. Figures 9(a) shows a "Toyama Prefecture" (located in Japan) pattern.

http://dx.doi.org/10.5772/56354

Structures that are several tens to hundreds of nanometers high can be fabricated using this simple method by machining under constant machining conditions [19]. Figure 9(b) shows a structure with a high aspect ratio fabricated on a (110)-oriented silicon surface. The silicon surface was machined along the <112> direction to take advantage of the anisotropic etching of silicon, and then etched in KOH [28]. This produced structures 150 nm wide and 800 nm high, with an aspect ratio of 5.3. Figure 10 shows a three-dimensional structure fabricated by exploiting the change in the etch resistance with the normal load [23]. The silicon surface was machined using five different normal loads in the range of 124 to 372 μN and protruded to a height of several nanometers after machining, as shown in Figure 10(a). After etching, a stepped structure with five different heights was produced, as shown in Figure 10(b).

This method can be used to fabricate three-dimensional structures with varying heights that cannot be fabricated via conventional photolithographic processes. A three-dimensional sloped structure can also be fabricated by machining while etching in KOH, owing to the simultaneous formation of an amorphous phase with etching [24, 25]. Therefore, this method is effective for various industrial fields in which three-dimensional structures are required.

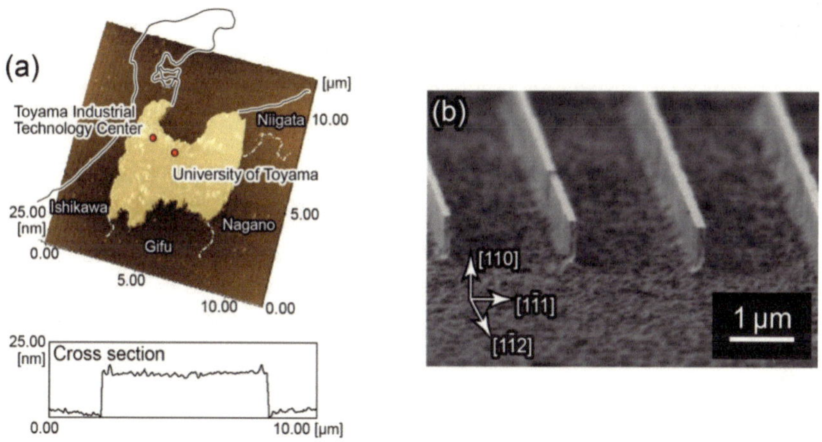

Figure 9. Structures fabricated by TNL and wet chemical etching. (a) AFM topography image of a "Toyama Prefecture" pattern fabricated at a normal load of 350 μN followed by wet chemical etching in KOH for 1 min. (b) SEM image of a high-aspect-ratio structure. The (110)-oriented silicon surface was machined along the <112> direction and then etched in KOH for 12 min.

3. Three-dimensional fabrication using FIB irradiation

3.1. FIB-based lithography

FIB is an instrument that irradiates ions focused over a range of a few nanometers to a few micrometers, accelerating them to an ion energy of 5 to 150 keV [38]. Using FIB, nanostructures can be fabricated by using the interactions of the irradiated ions with substrate atoms and/or introduced gases. The interaction causes sputtering [39–41], deposition [42–44], and implan-

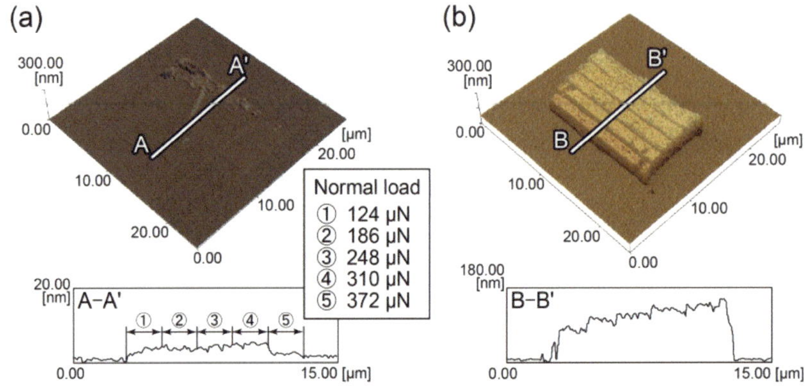

Figure 10. Stepped structure fabricated by TNL and wet chemical etching. (a) AFM topography image of a silicon surface machined using five different normal loads spanning the range of 124 to 372 μN. (b) The same area after etching in 10 mass% KOH for 10 min [27].

tation [45, 46] effects. Sputtering and deposition are used to fabricate nanometer-scale structures, whereas implantation is used to control the electrical properties of the material. FIB methods can be used to machine a sample surface to atomic scales and are thus suitable for fabricating structures that are a few tens of nanometers in size. The most used application of this method is the preparation of TEM samples, for which a thin sample is necessary.

A silicon surface irradiated with ion beams resists some etchants such as KOH [47–55], tetramethylammonium hydroxide [58], sodium hydroxide [50], and hydrazine [56, 57], so that protruding structures can be fabricated via etching. This phenomenon is not dependent on the species of the irradiated ions and has been reported after irradiating Ga [47–53], Si [49, 51, 54–56], Au [49], BF_2 [56], Ni [57], and P [56, 58] ions. The "etch stop" effect of the ion-irradiated area is caused by the formation of an amorphous phase due to ion irradiation [55]. This method is effective for fabricating large structures because the ion irradiation time is significantly shorter than what is required for sputtering and deposition processes. A wet chemical etching process can be used to remove a large amount of material in a short amount of time. The height of the fabricated structure is uniform because the ion-irradiation-induced etching masks have sufficient etch resistance, similar to conventional photolithographic techniques. However, by controlling the etch resistance of the etching mask, the height can be controlled, and therefore, three-dimensional structures can be fabricated. Herein, a three-dimensional fabrication method using FIB irradiation and wet chemical etching [59–61] is described.

3.2. Fabrication method

A fabrication method using a combination of FIB irradiation and wet chemical etching is shown in Figure 11 [59]. Figure 11(a) shows an AFM topography image of a silicon surface area after 30 keV Ga^+ ion irradiation ($5 \times 5\ \mu m^2$ area) at a dose of 13.0 $\mu C/cm^2$. This dose value is significantly lower than that used for structure fabrication via sputtering. This image shows a minute protuberance of the irradiated area, which is only 1–2 nm in height. This phenomenon results from the formation of an amorphous phase induced by FIB irradiation, similar to the effect of TNL shown in Figure 4(a). Figure 11(b) shows an AFM topography image of the irradiated area after etching in 20 mass% KOH for 5 min. The irradiated area resists etching in KOH, whereas the nonirradiated area is etched, resulting in a protruding structure with a height of 169 nm. The etching characteristics of the irradiated area are different depending on the etchant species. Figure 11(c) shows an AFM topography image of the irradiated area after etching in 46 mass% HF for 20 min. The irradiated area is selectively etched in HF, whereas the nonirradiated area is scarcely etched [1]. Therefore, a concave structure with a depth of 33 nm is fabricated on the irradiated area. This result indicates that the shape of the structure (convex or concave) can be selected by using different etchants.

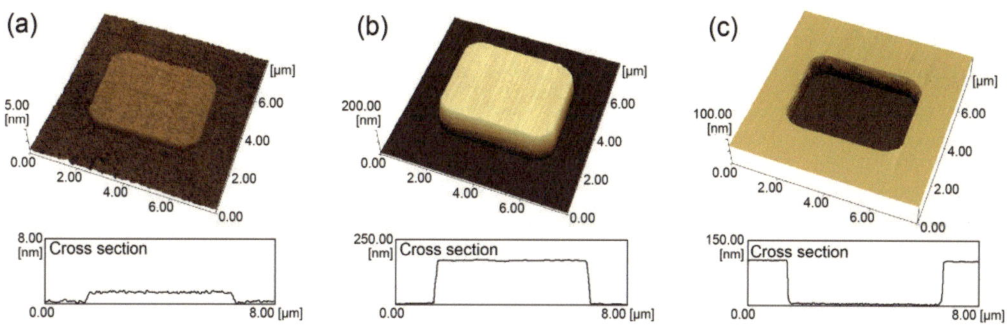

Figure 11. Etching characteristics of a FIB-irradiated area [59, 61]. (a) AFM topography image of a silicon surface after irradiation with 30 keV Ga^+ ions at a dose of 13.0 μC/cm^2. AFM topography image of an irradiated area (b) after etching in 20 mass% KOH for 5 min, and (c) after etching in 46 mass% HF for 60 min.

The difference in the etch resistance against KOH is caused by the thickness and density of the irradiation-induced amorphous phase. Figure 12 shows the cross-sectional TEM images and nano-ED patterns of 30 keV Ga^+ ion-irradiated areas at different doses before and after etching in KOH. An amorphous phase formed in the irradiated area at lower doses, as indicated by the TEM and nano-ED images in Figures 12(a) and (e), respectively. The center of the amorphous phase was 20–30 nm deep, expanding to a depth of 70 nm. The amorphous phase was completely etched, forming a concave–convex pattern on the surface, as shown in Figure 12(b). A thicker and wider amorphous phase formed at a higher dose, as shown in Figure 12(c). The amorphous phase remained after etching, indicating that the amount of silicon etching decreased significantly in the amorphous region. Hence, the etch stop of the irradiated area was caused by the amorphization of silicon. A higher etch resistance resulted from the higher doses owing to the resulting expansion and higher density of the amorphous phase. However, by etching in HF, the amorphous phase was selectively dissolved and a concave structure was fabricated. The depth of the concave structure was determined by the longitudinal expansion of the amorphous phase when HF was used as an etchant.

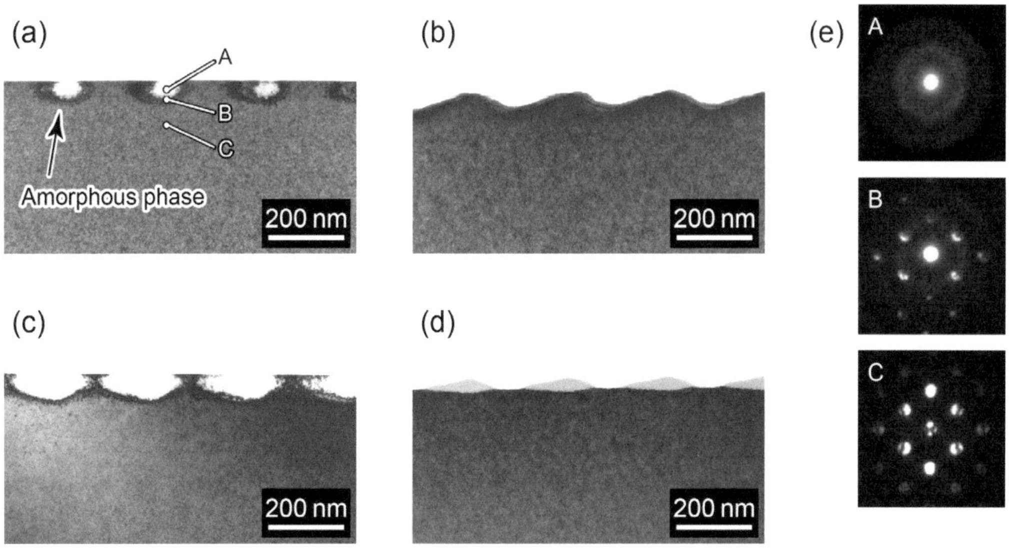

Figure 12. Bright-field cross-sectional TEM images of a 30 keV irradiated area (a) before and (b) after etching in KOH at a low dose, and the irradiated area (c) before and (d) after etching in 20 mass% KOH at a higher dose. Doses per dot in (a) and (c) were 0.9 pC and 1.4 pC, respectively. (e) Nano-ED pattern of the irradiated area denoted by A, B, and C in (a) [59].

As shown in Figure 12, the morphology and the etch rate of the amorphous phase were different according to the irradiation conditions. The height of the structure can be controlled with this technique. Figure 13 demonstrates the relationship between the height of the structure and the ion dose [59]. The silicon surface was irradiated by Ga^+ ions at various doses and then etched in KOH. The etch rate of the irradiated area decreased with increasing ion doses up to 10.0 μC/cm^2. Therefore, higher structures were fabricated using higher doses. For doses over 10.0 μC/cm^2, the etch rate and height were nearly constant at approximately 24 nm/min and 85 nm, respectively. For doses less than 10.0 μC/cm^2, the amorphous phase induced by the ion irradiation was etched, due to the thinner and lower density amorphous phase at low dose condition. The etch resistance at low dose condition varied according to the ion doses, due to the differences in thickness and density of the amourphous phase. Therefore structures of various heights were fabricated owing to the time lag of the dissolution in KOH. A thick and highly dense amorphous phase formed when higher doses of irradiation were used, resulting in a high etch resistance against KOH. The resulting structure was somewhat also etched in KOH because the height of structure was approximately 35 nm lower than the etched depth of the non-irradiated area. This indicates that the amorphous phase formed in the interior of silicon. Therefore, the maximum etch resistance occurred in the interior of silicon, and the resulting height was somewhat less than that of the non-etched area.

The height of the structure can be controlled also by the ion energy. Figure 14 shows the relationship between the ion energy and the height of a structure fabricated after etching in KOH for 5 min [61]. The height of the structure decreased with increasing ion energy at an ion dose of 25 μC/cm^2, whereas the height remained constant at an ion dose of 150 μC/cm^2. For

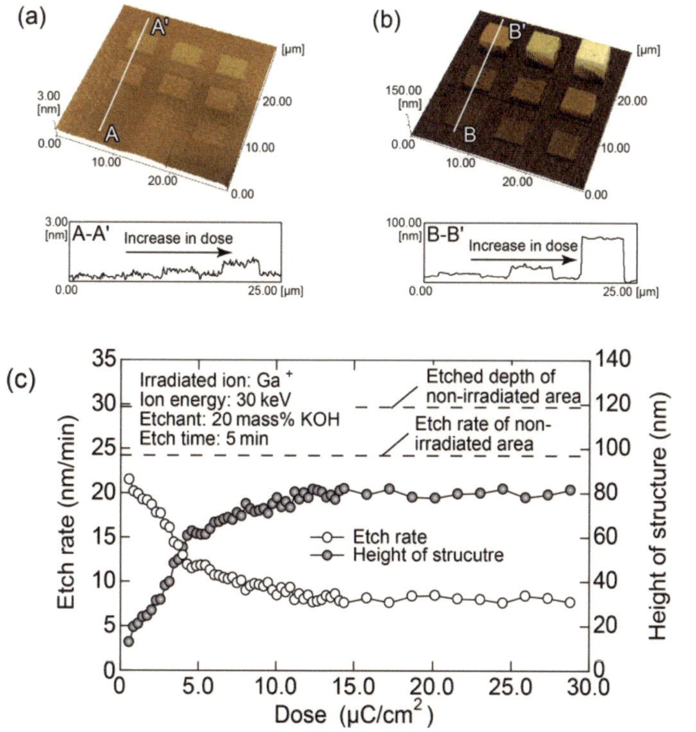

Figure 13. AFM topography image of a FIB-irradiated area at doses of 0.6 to 9.1 μC/cm^2 and an ion energy of 30 keV. (b) The same area after etching in 20 mass% KOH for 5 min. (c) Change in the height of the structure for various ion doses [59].

the lower dose, an amorphous phase formed in the interior of the silicon and the depth of amorphous region increased with the ion energy owing to the dissolution of the non-damaged silicon above the amorphous phase. A lower structure formed at the higher ion energy owing to the increase in the projected range of the irradiated ion. However, an amorphous phase formed from the surface at the higher dose so that the resulting height of the structure was constant for all ion energies. These results indicate that the height of the structure can be controlled by adjusting the ion energy and taking advantage of the difference in depth of the amorphous phase produced by ion irradiation.

A protruding three-dimensional structure can be fabricated by using the methods described above [59]. Figure 15(a) shows an AFM topography image of the stepped structures fabricated using three different doses of FIB irradiation followed by wet chemical etching in KOH. The etch resistance increased with the ion dose, resulting in a higher structure under the high dose condition. Therefore, a single structure with multiple heights can be fabricated by changing the ion dose. Smooth and sloped three-dimensional dome-shaped structures can also be fabricated using this method by continuously changing the ion dose, as shown in Figure 15(b).

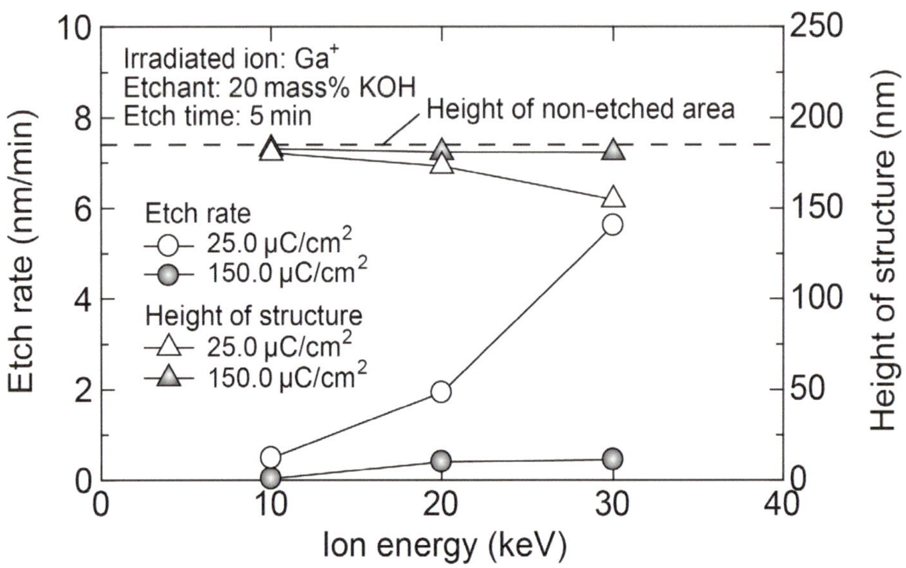

Figure 14. Height change of the structure as a function of the ion energy.

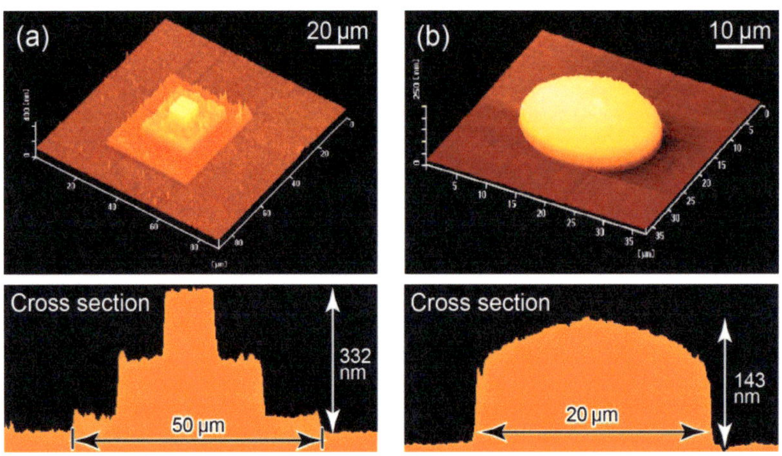

Figure 15. Application of FIB and wet chemical etching to three-dimensional fabrication. AFM topography images of a (a) stepped structure and (b) dome structure [59].

3.3. Three-dimensional fabrication using the etching enhancement

In this section, we describe a three-dimensional fabrication method based on a FIB-induced etching enhancement. In the etching enhancement, the shape of the structure was decided by the morphology of the amorphous phase because the structure was fabricated by the selective dissolution of the amorphous phase. Therefore, a three-dimensional structure was fabricated using the change in morphology of the amorphous phase according to the ion irradiation conditions.

Figures 16(a) and (c) show AFM topography images of the silicon surface (5 × 5 μm² area) irradiated with 30 keV Ga^{+} ions at doses of 0.2 to 27.1 μC/cm² and 224.0 to 2016.0 μC/cm², respectively [61]. The irradiated area protruded at a height of 1 to 2 nm under the lower dose conditions. Under the higher dose conditions, the irradiated area was sputtered and concave structures were fabricated. A burr-like structure also formed around the edge of the irradiated area owing to the reattachment of the sputtered ions. Figures 16(b) and (d) show the AFM topography images of the same areas after etching in 46 mass% HF for 20 min. The irradiated areas were selectively etched in HF, and concave structures were fabricated. In addition, the burr-like structure shown in Figure 16(c) was entirely removed, leaving a smooth surface around the irradiated area. Thus, precise structures could be fabricated in spite of the sputtering that occurs owing to the simultaneous etching of the reattached atoms. However, for lower doses, the irradiated area was scarcely etched.

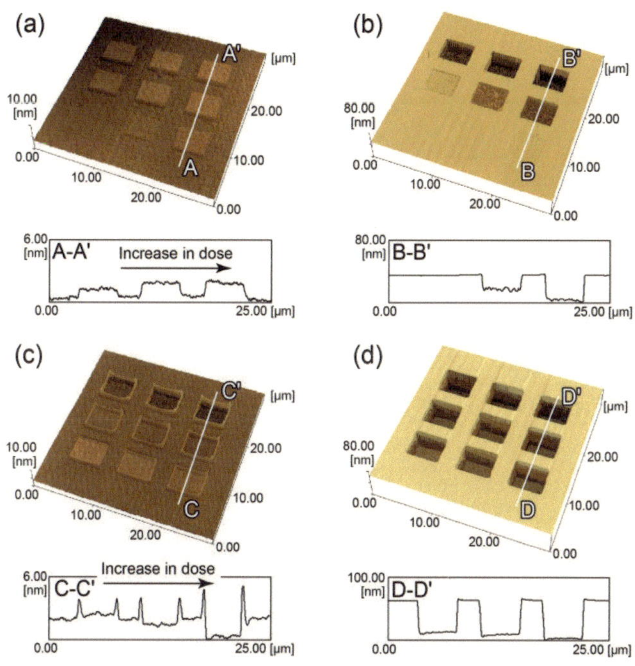

Figure 16. AFM topography image of a silicon surface showing the change in the irradiated area. The areas were irradiated at doses of (a) 0.2 to 27.1 μC/cm² and (b) 224.0 to 2016.0 μC/cm². (c) and (d) The same areas after etching in 46 mass% HF for 20 min [61].

Figure 17 shows the dependence of the depth of the structure on the dose after etching in HF for 20 min [61]. The irradiated area protruded at a dose of less than 1120 μC/cm². At this value, the irradiated area was sputtered and a concave structure was fabricated. The irradiated area was scarcely etched by HF when the dose was less than 6.9 μC/cm². The depth of the irradiated area rapidly increased above this value, whereas a more gradual increase was observed at doses greater than 20.4 μC/cm². Amorphization initially occurs near the most heavily damaged region, where most of the irradiated ions have slowed and have low energies, which is shown in Figure 12. Therefore, at lower doses, an amorphous phase is formed in the interior of the silicon near the range of the irradiated ions, while the surface area recrystallizes or is not transformed to an amorphous phase. This crystalline surface layer causes the surface of the irradiated area to be scarcely etched by HF. An amorphous phase forms on the surface at the higher doses owing to the expansion, resulting in dissolution of the irradiated area in HF. A deep structure is fabricated when the dose increases because of the longitudinal expansion of the amorphous phase. Hence, the depth of the structure can be controlled by the ion dose.

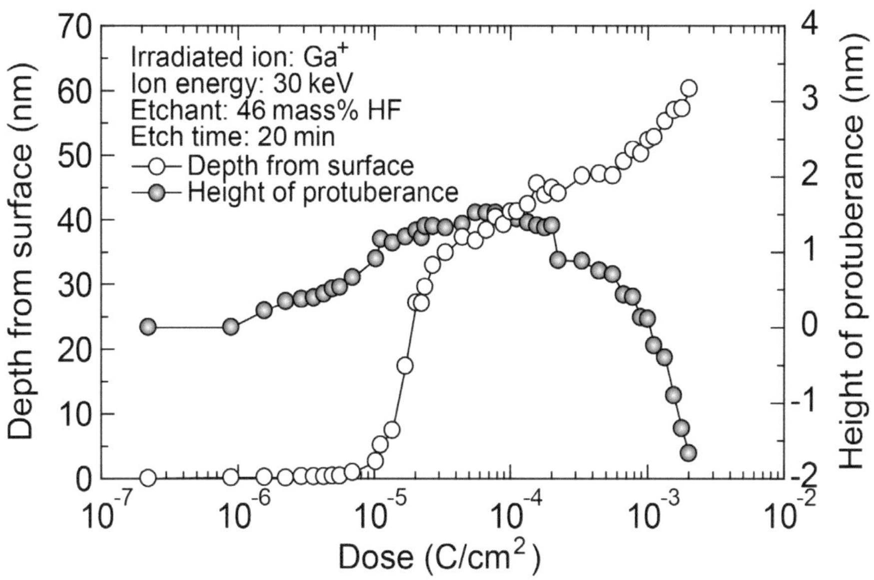

Figure 17. Relationship between the dose and depth from the surface.

The structure depth can also be controlled by the ion energy. Figure 18 shows the relationship between the ion energy and the depth of the structure after etching in HF for 20 min [61]. The depth of the irradiated area was proportional to the ion energy for both doses. The projected range of the irradiated ions increases with the ion energy, resulting in the formation of a thick amorphous phase and therefore a deep structure. Though the maximum depth of the structure was approximately 100 nm owing to the ion energy limitations of conventional FIB instruments, deeper structures over several hundreds of nanometers deep can be fabricated by using a high-energy ion irradiation facility, which can irradiate ions at several hundreds of keV [60].

Because the depth of the structure can be controlled by the ion irradiation conditions such as the dose and the ion energy, a complex three-dimensional structure can be fabricated. Figure 19(a) shows an AFM topography image of a silicon surface irradiated with four different doses

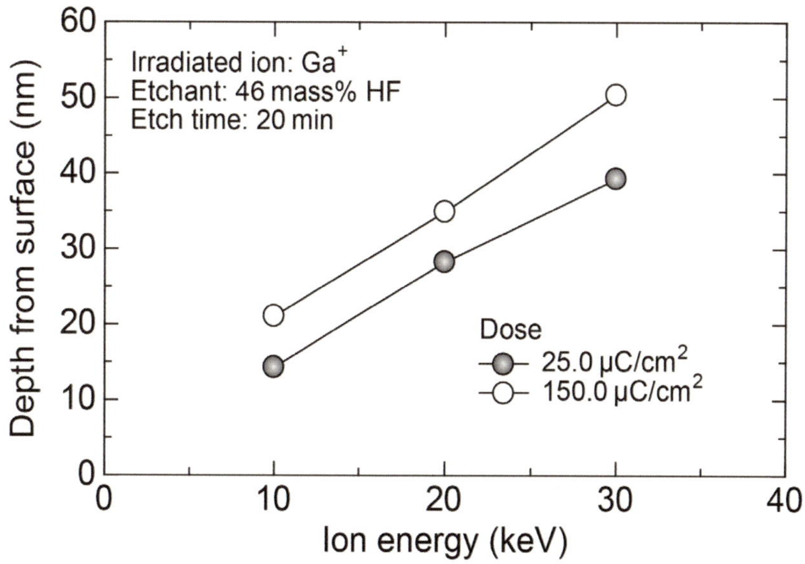

Figure 18. Relationship between the ion energy and depth from the surface.

[61]. The irradiated area protruded owing to the formation of an amorphous phase, but the height difference was only a few nanometers. Figure 19(b) shows an AFM topography image of the same area after etching in 46 mass% HF for 20 min. The depth increased with increasing dose values owing to the change in the thickness of the amorphous phase, and consequently, a stepped structure with four different depths was fabricated from the irradiated area.

Figure 20 shows an AFM topography image of a Fresnel lens pattern structure fabricated by gradually changing the ion dose. This is a three-dimensional structure with a smooth curved surface [61].

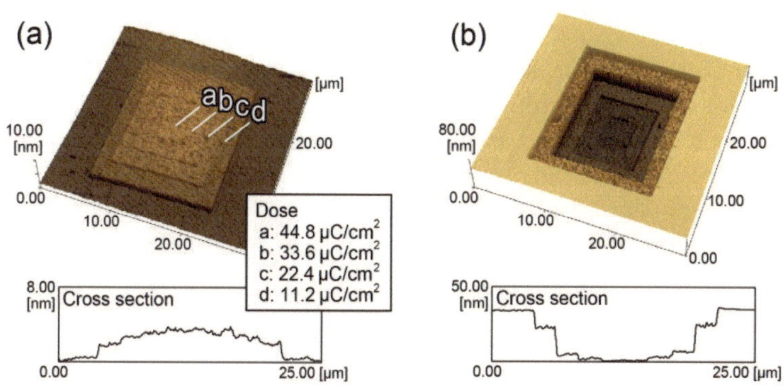

Figure 19. AFM topography image of a three-dimensional structure fabricated by using the change in the etching depth with the ion dose [61].

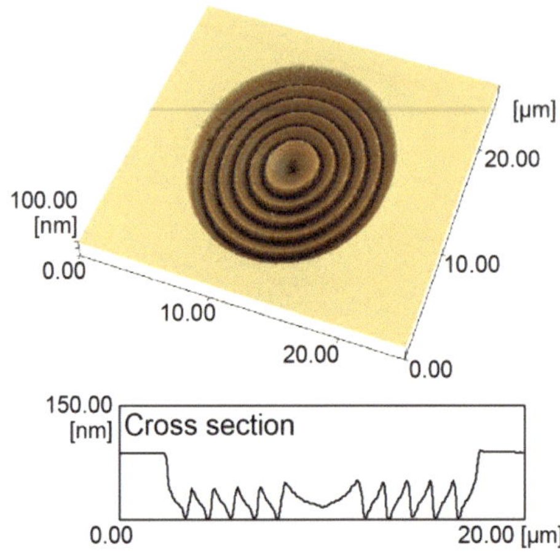

Figure 20. AFM topography image of a Fresnel lens pattern structure fabricated using FIB irradiation with various ion doses followed by wet chemical etching in 46 mass% HF for 20 min [61].

4. Three-dimensional fabrication using electron beam irradiation

4.1. EB-based lithography

The EB is an instrument used for various purposes such as high-resolution surface observation (SEM observation), and resist exposure. In particular, the exposing technique can fabricate fine line patterns in what is known as EB lithography [62], used in a similar way to mask fabrication in photolithography. These methods permit the fabrication of nanometer- or micrometer-scale patterns. However, a more productive method is necessary to fabricate deep and wide structures or complex structures because the direct machining approach requires a time-consuming step. Generally considered contamination, it is known that the EB irradiation causes the formation of a thin hydrocarbon layer on the surface [64–67]. The carbon is introduced by the residual gas and pump into the vacuum chamber. It reacts with the EB and is then deposited on the surface [63]. Because the etching characteristics of the irradiated material change because of EB irradiation, structures can be fabricated efficiently by combining these methods with wet chemical etching, which effectively overcomes the problems of solo-irradiation methods. This method is effective because EB irradiation facilities are used worldwide, and precise patterning, with a minimum line width of several nanometers, is possible using a very simple process. In this session, the three-dimensional fabrication technique using EB-induced carbon deposition [68] is described.

4.2. Fabrication method

Figure 21 shows SEM and AFM images of a GaAs area irradiated using the EB at a dose of 60 mC/cm^2 [68]. The irradiated area appears as a dark area in the SEM image, which indicates the formation of a hydrocarbon layer. This area was raised by 1–2 nm, as shown in Figure 21(b). The hydrocarbon layer has an etch resistance against AH solution, which consists of ammonia

http://dx.doi.org/10.5772/56354

(NH_3), hydrogen peroxide (H_2O_2), and water (H_2O) in a ratio of NH_3:H_2O_2:H_2O = 4:1:3312 by weight. By using this phenomenon, a protruding structure can be fabricated from the irradiated area.

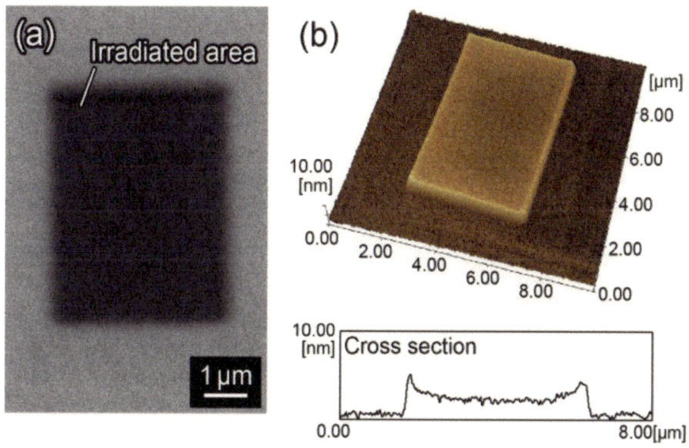

Figure 21. a) SEM and (b) AFM images of the area irradiated at an EB dose of 60 mC/cm^2. Reprinted with permission from [68]. Copyright 2008, IOP Publishing.

4.3. Three-dimensional fabrication using EB irradiation and wet chemical etching

The etch resistance of the irradiated area changes owing to the irradiation conditions, and therefore, the height of the fabricated structure can be controlled via hydrocarbon layers with different etch resistances. By controlling the etch resistance, structures with different heights can be fabricated by taking advantage of the difference in the dissolution time of the hydrocarbon layers. The change in the height of the irradiated area before and after etching in AH solution for 15 s is plotted as a function of the dose in Figure 22 [68]. The height of the hydrocarbon layer ranged from 1–3 nm and increased with the dose. After etching, the irradiated area resisted etching in the AH solution, whereas the nonirradiated area was selectively etched. Consequently, a structure higher than that present before etching was fabricated on the irradiated area. The height of the structure increased with the irradiation dose. With a high dose, a thicker hydrocarbon layer formed on the irradiated area, resulting in a high etch resistance against the AH solution.

Figure 23 shows the change in the height of the structure while etching in AH solution [68]. The increase ratio of the height in the figure denotes the average value between two plots. The height of the structure increased rapidly after etching for 15 s. It remained constant for etch times over 30 s, indicating that the etch-resistant hydrocarbon layer had completely dissolved at 15 s. Additionally, after etching for 15 s, the relative increase in height at 89 mC/cm^2 was greater than the relative increase at 30 mC/cm^2. The values were similar for both dose conditions for etch times over 30 s. Therefore, the difference in the height of a structure according to the irradiation dose results from the difference in the etch rate in the early stages of the etching process. Because the etch resistance of the hydrocarbon layer depends on the concentration of the AH solution, the maximum height of the structure can be controlled by the AH solution concentration [68].

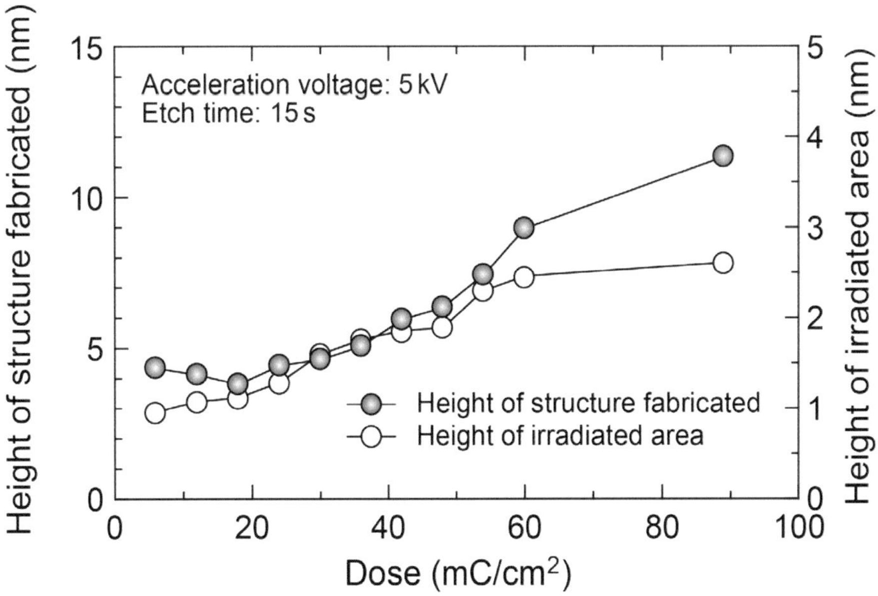

Figure 22. Change in the height of the irradiated area before and after etching in AH solution at various irradiation doses. Reprinted with permission from [68]. Copyright 2008, IOP Publishing.

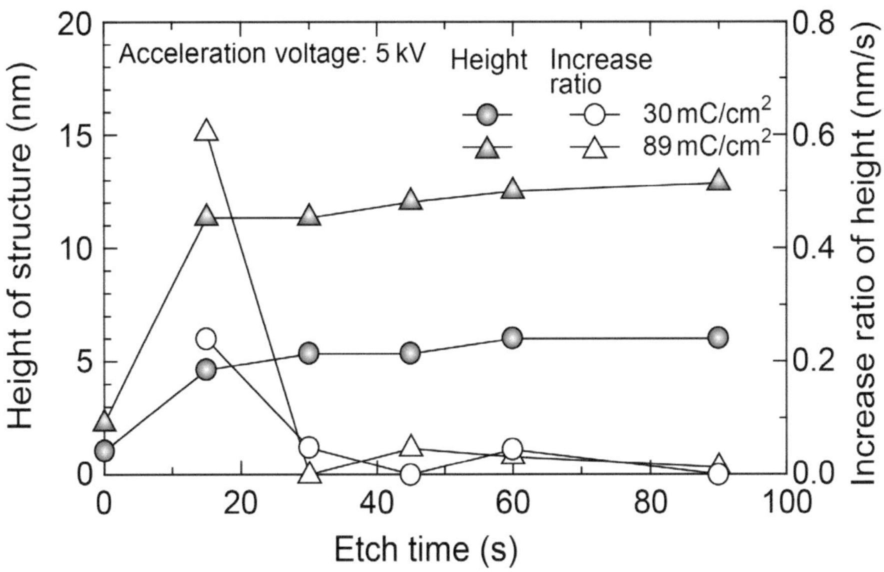

Figure 23. Relationship between the etch time and the height of the structure. The structures were fabricated at different doses. Reprinted with permission from [68]. Copyright 2008, IOP Publishing.

From the results described above, the height of a structure can be controlled by the irradiation condition. Figure 24 shows an AFM topography image of a grating pattern structure with a uniform height, fabricated using constant irradiation conditions [68]. The GaAs surface was irradiated by an EB with a spot size of 1 μm and then etched for 30 s. As a result, structures with a constant height of 12 nm and pitch of 5 μm were fabricated. Figure 24(b) shows an AFM topography image of a three-dimensional structure fabricated by changing the EB dose. The GaAs surface was irradiated with four different doses and then etched for 30 s, resulting in a step structure. The area irradiated with the highest dose is the highest, and the area irradiated with the lowest dose is the lowest.

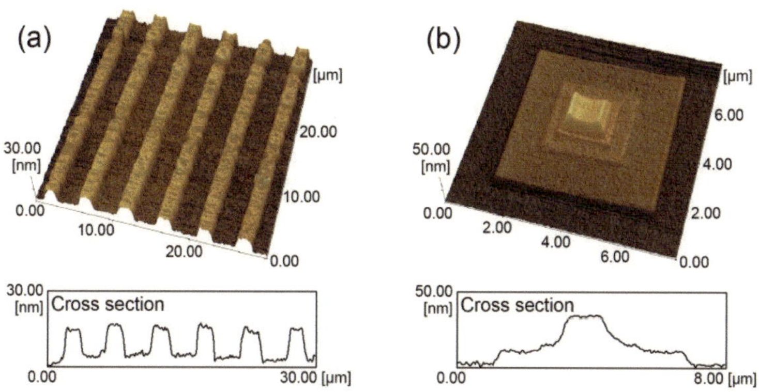

Figure 24. AFM topography images of structures fabricated using EB irradiation and wet chemical etching. (a) A grating pattern structure fabricated using a constant EB dose. (b) A step structure fabricated using the change in etch resistance of the irradiated area with the EB dose. Reprinted with permission from [68]. Copyright 2008, IOP Publishing.

5. Conclusions

We described three-dimensional fabrication techniques using nanoscale processing and wet chemical etching. Three types of processing methods were introduced. We indicated the superiority of these methods for fabricating structures with several tens to hundreds of nanometer high (deep) structures in comparison to conventional photolithographic techniques because they are simpler and more precise. Each of the three processing methods—TNL, FIB, and EB—have distinctive characteristics. Therefore, the processing method should be decided by the desired structure shape, resolution, patterning time, cost, and other factors. TNL simply forms an etching mask because the process is operated in air. The shape is decided by the machining parameters and the tip shape, which is the key technology for this method. FIB forms an etching mask via a rapid process because the dose value needed for the mask fabrication is significantly lower than that needed for a sputtering process. Deeper structures can be fabricated using the FIB-induced etching enhancement. The drawback of this method is the lateral expansion of the irradiated ions and the high-cost instrument. EB forms a high-resolution etching mask using its low-linewidth patterning ability. The maximum height of the structure was limited to several tens of nanometers and therefore can be improved by patterning and/or etching conditions. The combination of nanoscale processing and wet chemical etching is expected to become an essential tool for emerging nanotechnology and nanoscience applications related to electronic, photonic, biomedical, and nanosystem engineering.

Author details

Noritaka Kawasegi[1*] and Noboru Morita[2]

*Address all correspondence to: kawasegi@itc.pref.toyama.jp

1 Central Research Institute, Toyama Industrial Technology Center, Takaoka, Japan

2 Graduate School of Engineering, Chiba University, Chiba, Japan

References

[1] Elwenspoek, M, & Jansen, H. V. (1998). Silicon Micromachining, Cambridge University Press, Cambridge.

[2] Intel. Moore's Law Inspires Intel Innovation: http://www.intel.com/content/www/us/en/silicon-innovations/moores-law-technology.html (accessed 3 September (2012).)

[3] Kovacs, G. T. A, Maluf, N. I, & Petersen, K. E. Bulk micromachining of silicon. Proc. IEEE (1998). , 86-1536.

[4] Becker, E. W, Ehrfeld, W, Hagmann, P, Maner, A, & Münchmeyer, D. Fabrication of Microstructures with High Aspect Ratio and Great Structural Heights by Synchrotron Radiation Lithography, Galvanoformung and Plastic Moulding (LIGA process). Microelectoron. Eng. (1986). , 4, 35-56.

[5] Binnig, G, & Rohrer, H. Gerber Ch. and Weibel E. Surface Studies by Scanning Tunneling Microscopy. Phys. Rev. Lett. (1982). , 49-57.

[6] Binnig, G, & Quate, C. F. and Gerber Ch. Atomic Force Microscope. Phys. Rev. Lett. (1986). , 56, 930-933.

[7] Pohl, D. W, Denk, W, & Lanz, M. Optical stethoscopy: Image recording with resolution $\lambda/20$ Appl. Phys. Lett. (1984). , 44, 651-53.

[8] Eigler, D. M, & Schweizer, E. K. Positioning single atoms with a scanning tunnelling microscope. Nature (1990). , 344, 524-526.

http://dx.doi.org/10.5772/56354

[9] Dagata, J. A, Schneir, J, Harary, H. H, Evans, C. J, Postek, C. J, & Bennett, J. Modification of hydrogen-passivated silicon by a scanning tunneling microscope operating in air. Appl. Phys. Lett. (1990). , 56, 2001-3.

[10] Chien, F. S. S, Wu, C. L, Chou, Y. C, Chen, T. T, Gwo, S, & Hsieh, W. F. Nanomachining of (110)-oriented silicon by scanning probe lithography and anisotropic wet etching. Appl. Phys. Lett. (1999). , 75-2429.

[11] Snow, E. S, Campbell, P. M, & Shanabrook, B. V. Fabrication of GaAs nanostructures with a scanning tunneling microscope. Appl. Phys. Lett. (1993). , 63, 3488-90.

[12] Kolb, D. M, Ullmann, R, & Will, T. Nanofabrication of Small Copper Clusters on Gold(111) Electrodes by a Scanning Tunneling Microscope. Science (1997). , 275, 1097-99.

[13] Piner, R. D, Zhu, J, Xu, F, Hong, S, & Mirkin, C. A. Dip-Pen" Nanolithography. Science (1999). , 283, 661-3.

[14] Weinberger, D. A, Hong, S, Mirkin, C. A, Wessels, B. W, & Higgins, T. B. Combinatorial generation and analysis of nanometer- and micrometer-scale silicon features via "dip-pen" nanolithography and wet chemical etching. Adv. Mater. (2000). , 12, 1600-3.

[15] Mccord, M. A, & Pease, R. F. Scanning tunneling microscope as a micromechanical tool. Appl. Phys. Lett. (1987). , 50, 569-71.

[16] Magno, R, & Bennett, B. R. Nanostructure patterns written in III-V semiconductors by an atomic force microscope. Appl. Phys. Lett. (1997). , 70, 1855-7.

[17] Lee, H. T, Oh, J. S, Park, S. J, Park, K. H, Ha, J. S, Yoo, H. J, & Koo, J. Y. Nanometer-scale lithography on H-passivated Si(100) by atomic force microscope in air. J. Vac. Sci. Technol. A (1997). , 15, 1451-4.

[18] Ashida, K, Morita, N, & Yoshida, Y. Study on nano-machining process using mechanism of a friction force microscope. JSME Int. J. Ser. C (2001). , 44, 244-53.

[19] Kawasegi, N, Morita, N, Yamada, S, Takano, N, Oyama, T, & Ashida, K. Etch stop of silicon surface induced by tribo-nanolithography. Nanotechnology (2005). , 16, 1411-4.

[20] Ashida, K, Chen, L, & Morita, N. New maskless miro-fabrication technique of single crystal silicon using the combination of nanometer-scale machining and wet etching. Balsamo A., Evans C., Frank A., Knapp W., Mana G., Mortarino M., Sartori S. Thwaite E.G. (Eds.) Proceedings of 2nd european society for precision engineering and nanotechnology, May 27th-31st, 2001, Turin, Italy; (2001).

[21] Vettiger, P, Despont, M, Drechsler, U, Durig, U, Haberle, W, Lutwyche, M. I, Rothuizen, H. E, Stutz, R, Widmer, R, & Binnig, G. K. The 'millipede'-more than one thousand tips for future AFM data storage. IBM J. Res. Develop. (2000). , 44, 323-40.

[22] Bae, J. H, Ono, T, & Esashi, E. Scanning probe with an integrated diamond heater element for nanolithography. Appl. Phys. Lett. (2003). , 82, 814-6.

[23] Kawasegi, N, Morita, N, Yamada, S, Takano, N, Oyama, T, Ashida, K, & Micro-fabrication, D. using Combination Technique of Nano-scale Processing and Chemical Etching (1st Report, Possibility of 3D micro-fabrication using the mechanism of friction force microscope). Trans. Jpn. Soc. Mech. Eng. C (in Japanese) (2004).

[24] Kawasegi, N, Park, J. W, Morita, N, Yamada, S, Takano, N, Oyama, T, & Ashida, K. Nanoscale fabrication in aqueous KOH solution using tribo-nanolithography. J. Vac. Sci. Technol. B (2005). , 23, 2471-5.

[25] Park, J. W, Kawasegi, N, Morita, N, & Lee, D. W. Tribonanolithography of silicon in aqueous solution based on atomic force microscopy. Appl. Phys. Lett. (2004). , 85, 1766-8.

[26] Park, J. W, Kawasegi, N, Morita, N, & Lee, D. W. Mechanical approach to nanomachining of silicon using oxide characteristics based on tribo nanolithography (TNL) in KOH solution. ASME J. Manuf. Sci. Eng. (2004). , 126, 801-6.

[27] Kawasegi, N, Morita, N, Yamada, S, Takano, N, Oyama, T, & Ashida, K. Morphological control of a tribo-nanolithography-induced amorphous silicon phase for three-dimensional lithography. J. Adv. Mecha. Des. Sys. Manuf. (2007). , 1, 283-93.

[28] Kawasegi, N, & Morita, N. High-aspect ratio structure fabrication on (110)-oriented silicon surfaces using tribo-nanolithography. J. Nanosci. Nanotechnol. (2010). , 10, 2394-400.

[29] Kawasegi, N, Takano, N, Oka, D, Morita, N, Yamada, S, Kanda, K, Takano, S, Obata, T, & Ashida, K. Nanomachining of silicon surface using atomic force microscope with diamond tip. ASME J. of Manuf. Sci. Eng. (2006). , 128, 723-9.

[30] Kawasegi, N, Fukase, T, Takano, N, Morita, N, Yamada, S, Kanda, K, Takano, S, Obata, T, & Ashida, K. Development and its applications of diamond array tool using silicon mold (2nd report)-Fabrication of machining cantilever with arbitrary tip shape-J. Jpn. Soc. Prec. Eng. (2006). in Japanese).

[31] Shibata, T, Ono, A, Kurihara, K, Makino, E, & Ikeda, M. Cross-section transmission electron microscope observations of diamond-turned single-crystal Si surfaces. Appl. Phys. Lett. (1994). , 65, 2553-5.

[32] Yan, J. Laser micro-Raman spectroscopy of single-point diamond machined silicon substrates. J. Appl. Phys. (2004). , 95, 2094-101.

[33] Puttick, K. E, Whitmore, L. C, Chao, C. L, & Gee, A. E. Transmission electron microscopy of nanomachined silicon crystals. Phil. Mag. (1994). , 69, 91-103.

[34] Clarke, D. R, Kroll, M. C, Kirchner, P. D, Cook, R. F, & Hockey, B. Amorphization and conductivity of silicon and germanium induced by indentation. J. Phys. Rev. Lett. (1998)., 60, 2156-9.

[35] Bradby, J. E, Williams, J. S, Wong-leung, J, Swain, M. V, & Munroe, P. Transmission electron microscopy observation of deformation microstructure under spherical indentation in silicon. Appl. Phys. Lett. (2000)., 77, 3749-51.

[36] Gogotsi, Y, Baek, C, & Kirscht, F. Raman microspectroscopy study of processing-induced phase transformations and residual stress in silicon. Semicond. Sci. Tech. (1999)., 14, 936-44.

[37] Gogotsi, Y, Zhou, G, Ku, S, & Cetinkunt, S. S. Raman microspectroscopy analysis of pressure-induced metallization in scratching of silicon. Semicond. Sci. Technol. (2001)., 16, 345-52.

[38] Orloff, J, Utlaut, M, & Swanson, L. High resolution focused ion beam: FIB and Its Application. New York: Kluwer Academic / Plemium Pulication; (2003).

[39] Efremow, N. N, Geis, M. W, Flanders, D. C, Lincoln, G. A, & Economou, N. P. Ion-beam-assisted etching of diamond. J. Vac. Sci. Technol. B (1985)., 3, 416-8.

[40] Hopkins, L. C, Griffith, J. E, Harriott, L. R, & Vasile, M. J. Polycrystalline tungsten and iridium probe tip preparation with a Ga+ focused ion beam. J. Vac. Sci. Technol. B (1995)., 13, 335-7.

[41] Russell, P. E, Stark, T. J, Griffis, D. P, Phillips, J. R, & Jarausch, K. F. Chemically and geometrically enhanced focused ion beam micromachining. J. Vac. Sci. Technol. B (1988)., 16, 2494-8.

[42] Gamo, K, Takakura, N, Samoto, N, Shimizu, R, & Namba, S. Ion beam assisted deposition of metal organic films using focused ion Beams. Jpn. J. Appl. Phys. (1984). L, 293-5.

[43] Matsui, S, Kaito, K, Fujita, J, Komuro, M, Kanda, K, & Haruyama, Y. Three-dimensional nanostructure fabrication by focused-ion-beam chemical vapor deposition. J. Vac. Sci. Technol. B (2000)., 18, 3181-4.

[44] Fujita, J, Ishida, M, Ichihashi, T, Ochiai, Y, Kaito, T, & Matsui, S. Growth of three-dimensional nano-structures using FIB-CVD and its mechanical properties. Nucl. Instr. and Meth. B (2003)., 206, 472-7.

[45] Reuss, R. H, Morgan, D, Greeneich, E. W, Clark, W. M, & Rensch, D. B. Vertical npn transistors by maskless boron implantation. J. Vac. Sci. Technol. B (1985)., 3, 62-6.

[46] Evason, A. F, Cleaver, J. R. A, & Ahmed, H. Focused ion implantation of gallium arsenide metal-semiconductor field effect transistors with laterally graded doping profiles. J. Vac. Sci. Technol. B (1988)., 6, 1832-5.

[47] Berry, I. L, & Caviglia, A. L. High resolution patterning of silicon by selective gallium doping. J. Vac. Sci. Technol. B (1983)., 1, 1059-61.

[48] La Mache PH., Levi-Setti R. and Wang Y.L. Focused ion beam microlithography using an etch-stop process in gallium-doped silicon. J. Vac. Sci. Technol. B (1983)., 1, 1056-8.

[49] Steckl, A. J, Mogul, H. C, & Mogren, S. Localized fabrication of Si nanostructures by focused ion beam implantation. Appl. Phys. Lett. (1992)., 60, 1833-5.

[50] Schmidt, B, Bischoff, L, & Teichert, J. Writing FIB implantation and subsequent anisotropic wet chemical etching for fabrication of 3D structures in silicon. Sensors Actuators A (1997)., 61, 369-73.

[51] Fuhrmann, H, Döbeli, M, & Mühle, R. Focused ion-beam structuring of Si and Si/ $CoSi_2$ heterostructures using adsorbed hydrogen as a resist. J. Vac. Sci. Technol. B (1999)., 17, 945-8.

[52] Edenfeld, K. M, Jarausch, K. F, Stark, T. J, Griffis, D. P, & Russell, P. E. Force probe characterization using silicon three-dimensional structures formed by focused ion beam lithography. J. Vac. Sci. Technol. B (1994)., 12, 3571-5.

[53] Kawasegi, N, Morita, N, Yamada, S, Takano, N, Oyama, T, Ashida, K, Taniguchi, J, Miyamoto, I, & Momota, S. Etching characteristics of a silicon surface induced by focused ion beam irradiation. Int. J. Manuf. Technol. Manag. (2006).

[54] Fuhrmann, H, Döbeli, M, Kötz, R, Mühle, R, & Schnyder, B. Thin oxides on passivated silicon irradiated by focused ion beams. J. Vac. Sci. Technol. B (1999)., 17, 3068-71.

[55] Fuhrmann, H, Candel, A, Döbeli, M, & Mühle, R. Minimizing damage during focused-ion-beam induced desorption of hydrogen. J. Vac. Sci. Technol. B (1999)., 17, 2443-6.

[56] Koh, M, Sawara, S, Goto, T, Ando, Y, Shinada, T, & Ohdomari, I. New process for si nanopyramid array (NPA) fabrication by ion-beam irradiation and wet etching. Jpn. J. Appl. Phys. (2000)., 39, 2186-8.

[57] Koh, M, Goto, T, Sugita, A, Tanii, T, Iida, T, Shinada, T, Matsukawa, T, & Ohdomari, I. Novel Process for high-density buried nanopyramid array fabrication by means of dopant ion implantation and wet etching. Jpn. J. Appl. Phys. (2001)., 40, 2837-9.

[58] Masahara, M, Matsukawa, T, Ishii, K, Liu, Y, Nagao, M, Tanoue, H, Tanii, T, Ohdomari, I, Kanemaru, S, & Suzuki, E. Fabrication of ultrathin Si channel wall for vertical double-gate metal-oxide-semiconductor field-effect transistor (DG MOSFET) by using ion-bombardment-retarded etching (IBRE). Jpn. J. Appl. Phys. (2003)., 42, 1916-8.

[59] Kawasegi, N, Morita, N, Yamada, S, Takano, N, Oyama, T, Ashida, K, & Micro-fabrication, D. using combination technique of nano-scale processing and chemical etch-

ing (2nd report, possibility of 3D micro-fabrication using focused ion beam process). Trans. Jpn. Soc. Mech. Eng. C (2004). in Japanese).

[60] Kawasegi, N, Morita, N, Yamada, S, Takano, N, Oyama, T, Momota, S, Taniguchi, J, & Miyamoto, I. Depth control of a silicon structure fabricated by 100q keV ar ion beam lithography. Appl. Surf. Sci. (2007)., 253-3284.

[61] Kawasegi, N, Morita, N, Yamada, S, Takano, N, Oyama, T, Ashida, K, Taniguchi, J, & Miyamoto, I. Three-dimensional nanofabrication utilizing selective etching of silicon induced by focused ion beam irradiation. JSME Int. J. Ser. C (2006)., 49, 583-9.

[62] Yamazaki, K, Yamaguchi, T, & Namatsu, H. Three-Dimensional Nanofabrication with nm Resolution. Jpn. J. Appl. Phys. (2004). L1111-3., 10.

[63] Djenizian, T, Santinacci, L, & Schmuki, P. Factors in electrochemical nanostructure fabrication using electron-beam induced carbon masking. J. Electrochem. Soc. (2004). G, 175-80.

[64] Miura, N, Ishii, H, Shirakashi, J, & Yamada, A. and Konagai M Electron-beam-induced deposition of carbonaceous microstructures using scanning electron microscopy. Appl. Surf. Sci. (1997)., 114-269.

[65] Djenizian, T, Salhi, B, Boukherroub, R, & Schmuki, P. Bulk micromachining of silicon using electron-beam-induced carbonaceous nanomasking. Nanotechnology (2006)., 17-5363.

[66] Djenizian, T, Santinacci, L, Hildebrand, H, & Schmuki, P. Electron beam induced carbon deposition used as a negative resist for selective porous silicon formation. Surf. Sci. (2003)., 524-40.

[67] Chen, I. C, Chen, L. H, Orme, C, Quist, A, Lal, R, & Jin, S. Fabrication of high-aspect-ratio carbon nanocone probes by electron beam induced deposition patterning. Nanotechnology (2006)., 17-4322.

[68] Morita, N, Kawasegi, N, & Ooi, K. Three-dimensional fabrications on gaas surfaces using electron beam-induced carbon deposition followed by wet chemical etching Nanotechnology (2008).

Chapter 5

Combination of Lithography and Coating Methods for Surface Wetting Control

Athanasios Milionis, Ilker S. Bayer, Despina Fragouli,
Fernando Brandi and Athanassia Athanassiou

Additional information is available at the end of the chapter

http://dx.doi.org/10.5772/56173

1. Introduction

In recent years, many different lithographic approaches have been applied in order to fabricate micro and nano-structures which are successfully used for the formation of surfaces with special wetting properties [1,2]. In fact, the modification of the surface roughness in the micron, sub-micron and nano-scale with or without chemical treatment, results in surfaces with controlled wetting properties exhibiting the extreme limits (e.g. superhydrophobic, superhydrophilic surfaces) [3-5]. In this work, we examine different approaches in order to achieve such patterned surfaces with tunable wetting characteristics. In particular, we fabricate micro-rough substrates by using different lithographic techniques. As an additional step, these substrates are coated with different types of sub-micrometer particles or nanoparticles (NPs) (organic or inorganic) in order to achieve the desired chemical modification and to induce submicron or nano-roughness to the surfaces. Specifically, the coatings consist of polytefluoroethylene (PTFE) sub-micrometer particles and iron oxide colloidal NPs and they are applied by using triboelectric (fricition induced) or spray deposition. Apart from their interesting wetting properties from the theoretical point of view, such type of surfaces can be used in various applications such as in biological scaffolds, microfluidics, lab-on-a-chip devices and aerospace vehicles [6-9].

The first part of this study refers to the use of the appropriate lithographic technique so as to obtain a controlled rough surface on soft polymeric or hard substrates. Common lithographic techniques that have been used successfully for this purpose are UV lithography, laser-micromachining, electron-beam lithography, soft-lithography, X-ray lithography, plasma etching etc [10-17]. The main advantage of the lithographic techniques is the creation of well-defined patterned structures with controllable geometrical characteristics.

The second step is to induce additional sub-micron or nano-roughness and to chemically modify the patterned surfaces by adding layers of organic or inorganic particles. This provides an important aspect for tailoring the wetting characteristics of the surface since the chemistry is being modified together with the topography. In order to do this a thin coating layer can be deposited on top of the surface by different coating methods such as drop casting, chemical vapor deposition, spray coating, triboelectric deposition, etc.

Herein, we present two different methods for constructing surfaces with tunable wetting properties, both of them based on the concept of combining a lithographic technique to create an initial micro-rough surface and a deposition technique in order to add sub-micron or nano-rough features using particles with different chemical properties. In particular, the approaches are:

1. UV lithography with spray coating: At first, a pattern of equally sized square micro-pillars is fabricated by photopolymerizing an SU-8 photoresist and as a second step different types of particles are sprayed on top by using a spray coating setup. The particles are PTFE sub-micrometer particles and iron oxide NPs that are sprayed in successive coating layers. The different size and chemistry of the particles are responsible for the tuning of the wetting characteristics of the substrates.

2. Laser micromachining with triboelectric deposition: Silicon surfaces are micro-textured by laser ablation with a nanosecond pulsed UV laser and subsequently triboelectrically charged PTFE particles are deposited on them. Again the wetting properties are examined and they are compared with the micro-textured silicon carbide (SiC) surfaces which have been processed in a similar manner.

The morphology of the surfaces is characterized by means of atomic force microscopy (AFM) and scanning electron microscopy (SEM). In order to characterize the wetting properties, the apparent water contact angles (APCA) and the water contact angle hysteresis (CAH) are measured. The fabrication processes of the patterns are relatively quick and simple, making them good candidates for commercial production. Possible uses of such kind of materials can expand from self-cleaning surfaces, microfluidic devices and smart surfaces to biotechnological applications.

2. Combination of UV lithography and spray coating

Lithographic patterning using UV light is a very attractive technology due to the relatively low energy consumption, room temperature operation, rapid curing, spatial control and the ability to expose in a single step large surface areas [18]. For these reasons UV lithography is nowadays the most widespread used method for microfabrication.

The UV lithography technique involves various steps and it is suitable for processing materials that are called photoresists and absorb in the UV region of light. The common fabrication steps can be summarized as resist coating, pre-bake, mask alignment, exposure and development. Initially the substrate has to be cleaned with different chemicals to avoid the presence of dust, lint, bacteria, water and oil. Then the photoresist is applied by spin-coating, forming a uniform layer on the substrate. Its thickness depends on the speed of the spinning. Subsequently, the photoresist is baked on a hot plate or in a convection oven in order to evaporate its solvent. After this, the sample is aligned with the UV light source by using a mask aligner and it is exposed through a photomask that contains the design of the pattern to be transferred. The role of the photomask is to allow the UV light to pass only from some specific regions and to block it from others. When the sample is exposed, the development stage takes place. In this stage, chemicals are applied to the surface causing a positive or a negative photoresist reaction. In the case of the negative reaction the molecules that were subjected to UV light are bonded strongly, causing the polymerization of the material. The use of appropriate chemicals will remove the non-polymerized sections. In the positive reaction, the sections of the photoresist that were exposed are chemically altered and decomposed and after chemically washing them, the non-exposed parts remain on the substrate. In some occasions, in the development stage, a post exposure bake is required before the use of the chemicals in order to complete the chemical reaction that started during the exposure of the photoresist.

Spray coating is a standard method to apply thin coatings on surfaces. A spray coating setup consists of a high-pressure airstream flow that is able to break a liquid into miniature aerosol droplets through a confined nozzle head. The liquid is the coating to be deposited on the substrate and it is stored on a glass vial that is connected with the nozzle head. The main advantage of the spray coating compared to other coating techniques, is that the sprayed solution dries immediately after its deposition on the target, preventing thus possible aggregations and coating inhomogeneities that can occur during the evaporation of the solvent as in other methods (e.g. drop casting). In this way, the sprayed coatings are uniform while the preparation time is reduced.

The combination of the UV photolithography with the spray coating technique can be a very efficient way to create surfaces with hierarchical roughness in the micro, sub-micron and nano-scale that are necessary in order to obtain surfaces with tunable wetting characteristics [4,19]. Both techniques are fast and simple and are successfully used in the industry in the last decades.

2.1. Fabrication procedure and materials used

The material used as a polymeric substrate for the UV lithography, is the SU-8 3050 from Microchem, USA. SU-8 is a commercial biocompatible epoxy-based photoresist that is suitable for the microfabrication of high aspect-ratio hierarchical structures with smooth and vertical sidewalls, and it is used widely in the fabrication of MEMS devices, waveguides, microfluidic devices and stamps. It absorbs in the UV range of the spectrum with a maximum efficiency at 365 nm. When exposed to UV light, its molecular chains crosslink causing the polymerization of the material that becomes strong, stiff and chemically resistant. SU-8 has been also used for the fabrication of dual-scale rough structures. Such hierarchical structures produced with SU-8 have been coated with titanium dioxide, PTFE submicrometer particles and fluorocarbons in order to obtain surfaces with special wetting properties [19-21].

To fabricate the SU-8 patterns initially we dispensed a small quantity of the SU-8 3050 onto a silicon wafer. Then the spin-coating of the SU-8 was accomplished in two subsequent steps: (a) at 500 rpm for 10 s with spinning acceleration of 100 rpm/s and (b) at 4.000 rpm for 30 s with spinning acceleration of 300 rpm/s. The samples were pre-baked at 100 °C for 20 min on a hotplate. The thickness of the film obtained was 33 μm. A sodalime mask of square-shaped patterns (42 μm side) from Deltamask, Netherlands, with inter-square distance of 28 μm, was used for the exposure of the spin-coated samples. Patterning was performed by exposing the spin-coated material to UV radiation with a Karl-Suss MA6 mask aligner in hard contact mode with an i-line mercury lamp. An exposure dose of 323 mJ was used to fully polymerize the 33-μm-thick SU-8 layer. The exposure was followed by a post-exposure bake on a hotplate at 65 °C for 1 min and at 95 °C for 5 min, in order to achieve complete cross-linking of the resist. The samples were then allowed to cool down in order to improve the adhesion of SU-8 to the silicon wafer. Subsequently, the samples were washed for a couple of minutes with SU-8 developer followed by rinsing with 2-propanol. Following this process square SU-8 pillar structures of 33 μm height and 28 μm inter-pillar distance were obtained. In Figure 1 there is a schematic representation with the steps followed for the fabrication of the patterned surfaces.

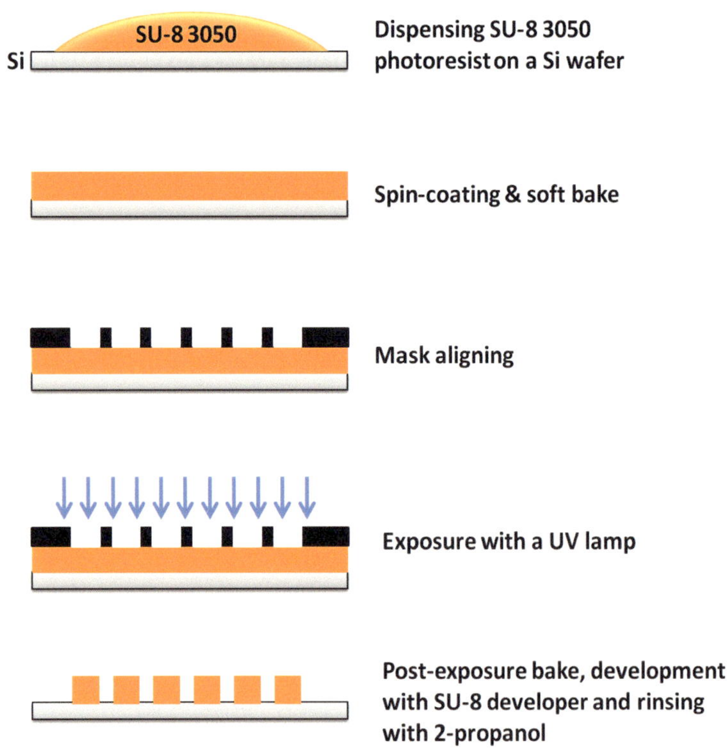

Figure 1. Schematic illustrating the various steps followed for the fabrication of the SU-8 micropillars.

In order to modify the surface chemistry and to induce different scales of roughness to the hierarchical SU-8 micro-patterns, the spray-coating technique was used. In particular, colloidal solutions of PTFE sub-micrometer particles and iron oxide NPs are used to form a first and a second coating layer respectively. Like this, sub-micron and nano- roughness is added on top of the SU-8 micro-patterns.

Regarding the preparation of the colloidal solutions, the PTFE powder, purchased from Sigma-Aldrich with an average particle diameter of ~ 150 nm, was dispersed in acetone (3% wt.) and sonicated for about 20 minutes in order to form a uniform and well-dispersed suspension. The nearly spherical colloidal iron oxide NPs (average diameter: 24±3 nm) are dispersed in chloroform (0.06% wt.) and they are synthesized in our lab [4]. The colloidal solutions were sprayed, as shown in Figure 2, at a distance of 10 cm from the substrate while the pressure of the airstream was 150 kPa. The resulting surface, after all this treatment, exhibits a triple-scale roughness in micro-, submicron- and nano-scale. Moreover, the specific treatment induces significant changes in the surface chemistry, since the PTFE particles are well-known for their hydrophobic properties. The iron oxide NPs are also hydrophobic due to the oleic acid (OLAC) molecules that are used as surfactants [22]. Thus, the sprayed particles have a dual role in the system, which is to increase the hydrophobicity and also to change the roughness scale of the patterns.

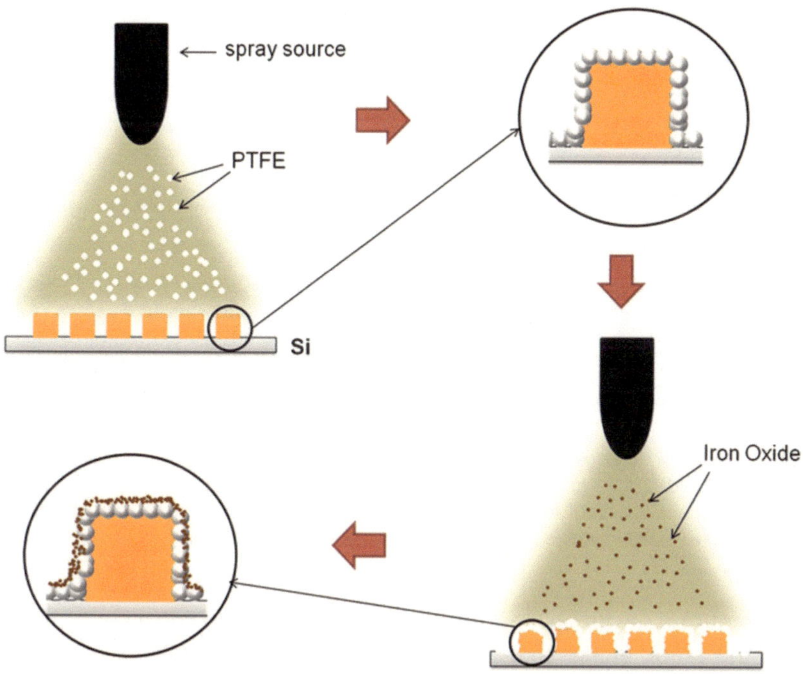

Figure 2. Schematic illustrating the successive coating steps. At first the PTFE solution is sprayed on the substrate with the SU-8 micropillars (top panel). As a second step, the iron oxide NPs are sprayed on top of the SU-8/PTFE micro/submicron rough pattern, thus inducing a three-scale roughness [4].

2.2. Characterization and discussion on the morphology and wetting behavior

In order to characterize the wetting properties of the samples, APCA and tilting angle measurements were taken using a KSVCAM200 contact angle goniometer, Kruss, Germany. The area of the water drop (total drop volume: 10 μl) in contact with the surface was sufficient enough to experience the periodicity of the pattern. The tilting angle measurements were performed by recording the images of the droplets while the underlying substrates were inclined by a tilting base. All measurements were taken a few seconds after the placement of the water droplets.

Scanning electron micrographs were collected with a Nova NanoSEM200 scanning electron microscope (SEM) by FEI Instruments. Low-vacuum configuration was used with a chamber pressure of 0.3 mbar. To decrease the charging effects on the surface of the samples, a flux of water vapor was injected in the chamber. In Figure 3 we present the SEM images of the SU-8 micro-pillars with the different coatings, in high and low resolution, together with the corresponding APCAs.

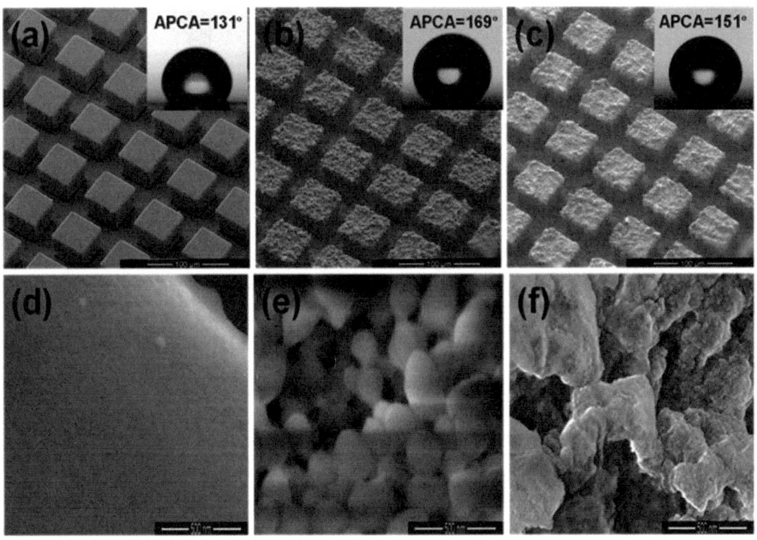

Figure 3. Low-magnification tilted-view SEM images of (a) SU-8 micro-pillars, (b) SU-8 micro-pillars with PTFE sub-micrometer particles sprayed, (c) SU-8 micro-pillars with iron oxide NPs sprayed as a second layer, on top of the already sprayed PTFE sub-micrometer particles. Insets: the APCA of the surface in each case. In (d), (e) and (f) they are presented high-magnification top-view images of the surface of each micro-pillar corresponding to (a), (b) and (c) respectively [4].

As shown in Figure 3, the initial SU-8 uncoated pattern displays an APCA of 131°±7° (Figure 3a and d), significantly higher compared to the APCA measured on the corresponding flat SU-8 surface, (82°±2°). Moreover, this pattern is highly adhesive to water since the droplets stay on the substrate even for 180° tilt. When the PTFE sub-micrometer particles are sprayed on the pillars, the morphology of their surface changes dramatically (Figure 3b and e). In fact, the pillars retain their micro-scale geometrical characteristics, but on top they exhibit sub-

http://dx.doi.org/10.5772/56173

micrometer roughness due to the presence of the PTFE particles. The combination of the micro/submicron roughness with the well-known water-repellent properties of the PTFE makes the patterned surfaces superhydrophobic and self-cleaning with an APCA of 169°±2° and sliding angle ≤1°. In fact, in the absence of the micro-roughness, e.g. at the flat SU-8 surface coated with PTFE particles, the APCA is lower (159°±8°) while the sliding angle ≤ 10°, demonstrating the importance of the micro-patterning.

By using the PTFE-sprayed micro-pillars as a substrate for the successive spraying of a coating of iron oxide NPs, surfaces with micro-, submicron- and nano-roughness can be obtained (SEM images of Figure 3c and f). The underlying PTFE particles seem to be fully covered by the iron oxide NPs. The latter do not appear well separated but rather as a homogeneous coating. After this coating the APCA remains superhydrophobic (151°±6°) (APCA of the flat surface 140° ±10°) but on the other hand the water adhesion is dramatically increased. Indeed the substrates become "sticky" even for 180° substrate tilt angles (Figure 4). Such type of surfaces, that have a very high APCA while at the same time exhibit very high adhesion to water, can be designated as "sticky superhydrophobic" surfaces [23,24]. This state has been also reported as "petal effect" in some part of the literature [25,26].

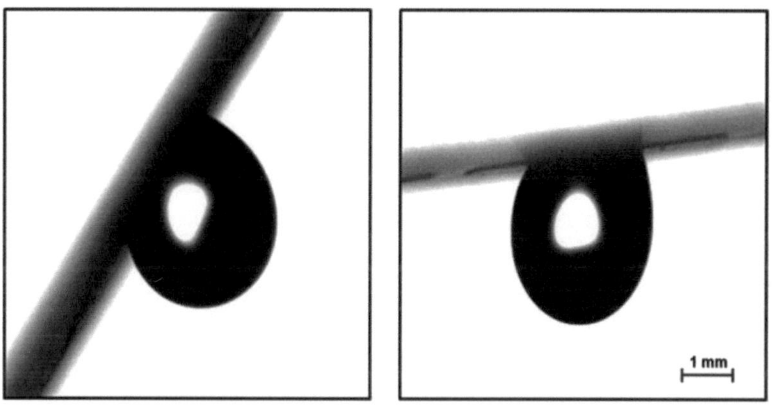

Figure 4. Water drops that remain adhered in an iron oxide/PTFE/SU-8 patterned surface under very high substrate inclinations (tilt angle: 122° on the left image and 173° on the right) [4].

To explain the abovementioned results we will compare them with the well-known Cassie-Baxters' theoretical model [27]. The Cassie-Baxter model is generally applied to predict the APCA on a rough hydrophobic surface and it particularly assumes that a water droplet can wet a surface only partially, due to the trapping of air pockets underneath the droplet at the recessed regions of the surfaces. For such hierarchical surfaces, like the ones fabricated, the water contact angle (WCA) predicted from the Cassie-Baxters' model can be extracted by the following equations [28]:

$$\cos(\theta_{CB}) = f_s[\cos(\theta_Y) + 1] - 1 \qquad (1)$$

$$f_S = \frac{1}{\left(\frac{b}{a} + 1\right)^2} \quad (2)$$

where f_S is the fraction of the solid surface in contact with the droplet and θ_Y is the Young's angle, i.e., the contact angle of the droplet on the corresponding flat surface with the same chemical characteristics. α is the width of the pillars (α=42 μm) and b is the spacing (b=28 μm) between the pillars. The predicted WCA values from Equation (1) for the surface of the coated and uncoated SU-8 micropillars are presented in the Table 1.

Material	APCA/° (flat surface)	APCA/° (patterned surface)	Cassie-Baxter/°
SU-8	82±2	131±7	126
SU-8+PTFE	159±8	169±2	167
SU-8+PTFE+Iron oxide	140±10	151±5	156

Table 1. APCA values of the flat and patterned surfaces and their extracted theoretical values from the Cassie-Baxter model for the three surfaces. The theoretical values were extracted from the flat SU-8 surfaces coated with the same particles as the patterned ones [4].

As shown in Table 1, the Cassie-Baxter model predictions match with the values obtained experimentally. This means that the droplets rest on a surface containing air pockets. However, in terms of water adhesion, the highly sticky uncoated patterns are converted to non adhesive after the application of the PTFE coating. After the subsequent iron oxide coating the superhydrophobic patterns become again highly adhesive. Since the experimental values of all the three cases are in accordance with the Cassie-Baxter model, we must exclude the possibility that there is a complete wetting state (Wenzel) that would increase the water adhesion due to the pinning effects arising from the penetration of the droplet in the inter-pillar spacing. Instead, the capillary effects seem to occur in the upper part of the pillars and are provoked by the final nano-rough coating layer of iron oxide, while the inter-pillar spacing continues to be filled with air. In fact, similar studies that describe such a pinning state occurring in nano-rough hierarchical surfaces have already been published [29-31]. These works present composite micro- and nano-rough systems where the pinning effects take place only in the nano-rough surface while in the micro-scale the wetting state obeys the Cassie-Baxter model.

In summary, after the coating of iron oxide NPs it is possible to fabricate triple-scale rough superhydrophobic surfaces with high water adhesion, and on the other hand it is possible to obtain dual-scale rough SU-8/PTFE coated superhydrophobic surfaces with ultralow adhesion. If an additional layer of PTFE particles is applied on top of the iron oxide coating, the low adhesion state can be recovered again. In such a manner, with successive spray coatings, of PTFE and iron oxide particles, one can obtain surfaces with alternating water adhesion properties on superhydrophobic layers.

3. Laser micromachining and triboelectric deposition

Laser micromachining is a term that includes a variety of processes including hole drilling, ablation, milling, and cutting of soft or hard materials using laser light sources. Specifically laser ablation, is a technique where a focused pulsed laser beam is used in order to remove material from a solid target [32] and thus to modify the morphological features of its surface. In fact, the high energy flux of the pulsed laser beam focused on the target results in the photon absorption by the electrons of the system and in the electrostatic instability of the lattice ions. As a consequence the ejection of the interacting material is induced in the form of plasma and hence the ablation leaves a void in the laser focus region. This lithographic technique is nowadays used widely in the industry for the fabrication of MEMS/NEMS, CMOS, 3D-microstructures, micro-trenches, micro-channels, micro-holes, sub-micrometer gratings, nanophotonics and surfaces dedicated to bacterial activity [33,34]. Laser ablation, apart from the microstructuring of surfaces, has also been used alternatively as a technique to produce NPs by irradiating solid targets [35]. Its main advantages when compared with other lithography techniques are the fast material processing speed, large scan area and single-step capability. Compared to the previous lithography technique described before (UV lithography), this one requires less processing steps and also the use of chemicals is much more limited. However it cannot form ordered and well controlled micron sized features. Nevertheless the laser ablation technique is able to induce roughness on a surface, and subsequently modify its wetting characteristics.

Triboelectric particle deposition is a coating technique based on the physical phenomenon of frictional (or contact) charging of dissimilar materials [36,37]. In other words, during contact, certain materials become electrically charged when they come in contact with another material through friction. It is possible to deposit small polymer particles on a surface by sandwiching them between two surfaces and rub them against one another. In this way, a uniform coating can be obtained on one of the surfaces due to the self-assembly of the particles that they are attracted with electrical forces. The main advantage of this coating method is that it is a dry technique, based on physical concepts and does not require the use of hazardous chemicals.

The laser micromachining can be combined with the triboelectric deposition in order to fabricate surfaces with different roughness scales and different wetting properties [5]. The roughness parameters of the fabricated surfaces can be controlled by tuning the laser parameters such as laser power and focal length, and also by changing the amount of particles deposited during the coating. Both of these techniques are efficient and rapid, and for these reasons their combination can be a potential industrial method for fabricating coatings with special wetting properties.

3.1. Fabrication process and materials

The substrates used in this study are commercial standard silicon wafers (one side polished) having natural oxide surface layers. Both polished and unpolished sides of the wafers were used for laser micromachining. In particular a UV nanosecond laser beam was focused using a cylindrical lens with a focal length of 75 mm. The laser fluencies tested were ranging from

0.5 up to 2 J/cm^2. During the ablation process the silicon wafers were stored in liquid bath of methanol or distilled water with 5 mm of liquid covering their surface. The wafers were moved in precise small steps with an x-y translation stage until a textured surface region of approximately 1 cm^2 was obtained. In Figure 5 there is a schematic representation of the laser micromachining setup. After the ablation, the wafers were removed from the liquid baths and left to dry under ambient conditions.

Apart from the textured silicon wafers, as substrates, we also used various micro-textured silicon SiC surfaces. Typical sandpapers of 600, 800 and 1200 grit sized were used.

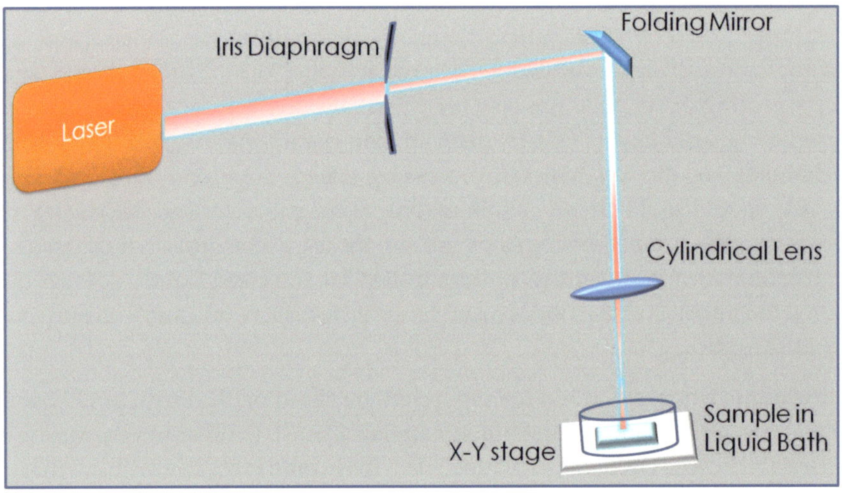

Figure 5. Schematic representation of the laser micromachining setup.

For the coating of the textured silicon wafers, as discussed in Chapter 2, sub-micrometer PTFE particles were used. Certain amount of PTFE powder was spread over commercial Scotch Brite (3M) medium duty cellulosic foam with dimensions 15 cm × 9 cm × 2 cm. A metal rod was used to rub the PTFE powder on the foam surface for a few minutes, in order to spread it uniformly. Subsequently, the foam was continuously rubbed against the textured silicon surfaces in circular motion for a few minutes. At the end of this process, almost all the PTFE particles were transferred to the textured silicon surface forming a thin PTFE layer. It was found that the PTFE particles were strongly attached onto the laser-formed structures, since the post treatment (immersion in different solvents such as acetone, toluene, methanol or chloroform) did not cause particle removal from the surfaces. Figure 6 shows a sketch illustrating the basic steps followed for the triboelectric deposition of the PTFE particles.

A slightly different coating approach was used to apply the PTFE coating on the SiC sandpapers. The PTFE powder was directly deposited on the sandpapers and a cellulosic foam was rubbed over the surfaces directly, for a few minutes, to spread the PTFE particles over the rough surface.

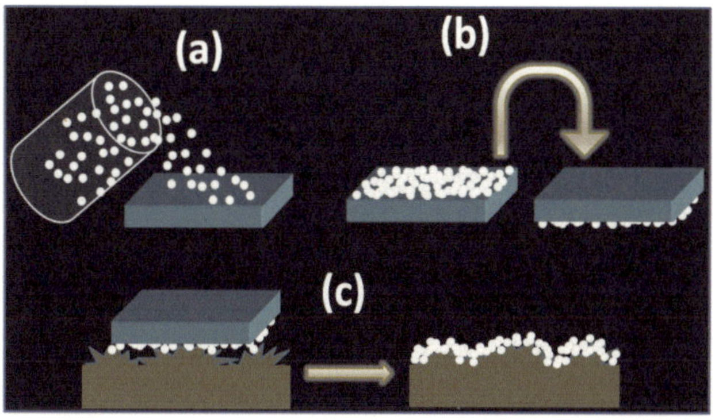

Figure 6. Schematic illustrating the triboelectric deposition of the PTFE particles on textured silicon. (a) The PTFE powder is spread over the foam. (b) The PTFE powder is distributed uniformly with the use of a metal rod. (c) The PTFE particles are rubbed against the textured silicon surface and finally they are transferred to the silicon forming a thin layer.

3.2. Characterization, results and discussion

The morphology of the laser-textured surfaces was inspected with SEM. Figure 7 shows the SEM images of the laser patterned surfaces obtained on the polished surface of a silicon wafer. Densely packed pyramid-like roughness features (pillars with density of 0,8 pillar/μm) of the order of 2 μm in size were produced on the surface.

The topography of the textured surfaces was further examined by means of AFM. In general, scan sizes ranging between 5 and 20 μm, were used with simultaneous acquisition of topography, height, error and phase images. A Park Systems XE-100 AFM was used in non-contact mode with a silicon cantilever. An adaptive scan rate between 0.15 Hz and 0.25 Hz was utilized for all samples. In Figure 8 topographical and error signal images of the textured surfaces described above are shown. To acquire a high quality AFM image of a surface with micrometer features, the deflection (or amplitude) signals (which represent the error signals) have to be minimized. If these signals are minimum, the error images obtained during scanning should look identical to the topographical features. 'Topography' images are generated by displaying the changes in the Z direction of the piezoelectric scanner required to restore the deflection of the cantilever to its predefined set point at each image point. Displaying the transient deviations of the cantilever away from the set point as the tip encounters features during scanning generates the 'error signal' image. It effectively represents a differential of the topography image, since it accentuates sharp turning points in the sample topography (high frequency information) at the expense of smooth slowly undulating areas (low frequency information). Indeed, both the topographical and the error signal images in Figure 8 match perfectly, indicating that there exist no additional phases or impurities on the laser textured surface as a result of the laser exposure. Moreover, from these images, it is seen that the surface topography is made up of collective pyramidal structures which probably were carved out of larger surface bumps during laser ablation. In Figure 9 it is presented a high magnification AFM topography signal of the patterned surface corresponding to Figure 8. In this image the

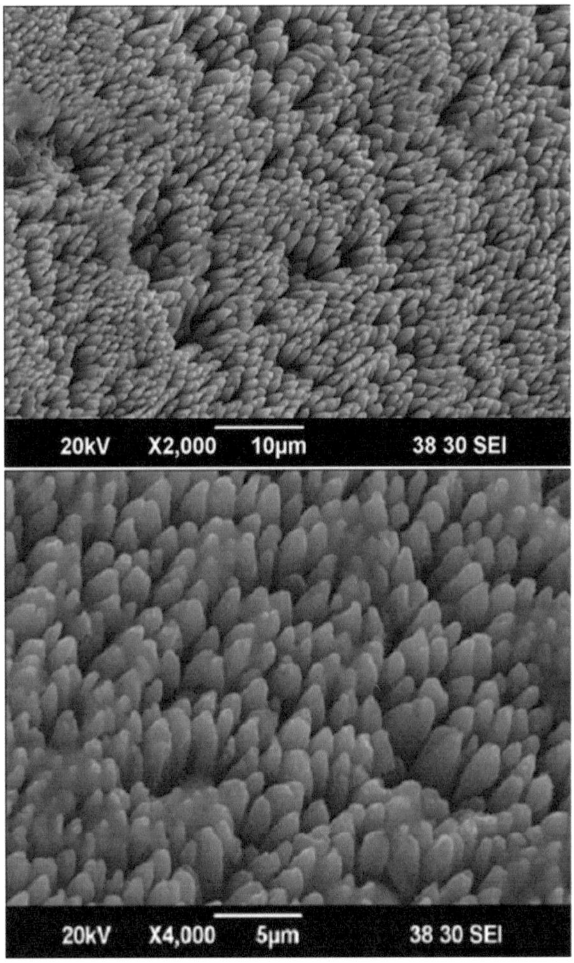

Figure 7. SEM images of low magnification (top) and high magnification (bottom) from a laser-ablated area on a silicon wafer. Closely packed pyramid-like roughness features of average size of 2 μm can be distinguished [5].

morphology of the pyramid-like roughness can be seen in detail. These textured surfaces were realized with a UV nanosecond laser working with a laser fluence of 1 J/cm^2.

From the abovementioned images it can be concluded that four pyramid-like roughness features are packed into a 25 μm^2 silicon surface. Also the base of the pyramids is made up of a layered structure, most probably originating from the laser induced melting-solidification cycles [34]. Roughness analysis indicates that the base of the pyramids is about 2 μm wide and the top is 0.5 μm wide. The size, depth and the distribution of the resultant protruding features, strongly depends on the laser parameters (beam energy and focus spot area). In this case, the samples were prepared by irradiating the wafers with a line focused beam of about 20 mm in length and 0.2 mm in width while rotating the wafer. The line focus is set radially; starting

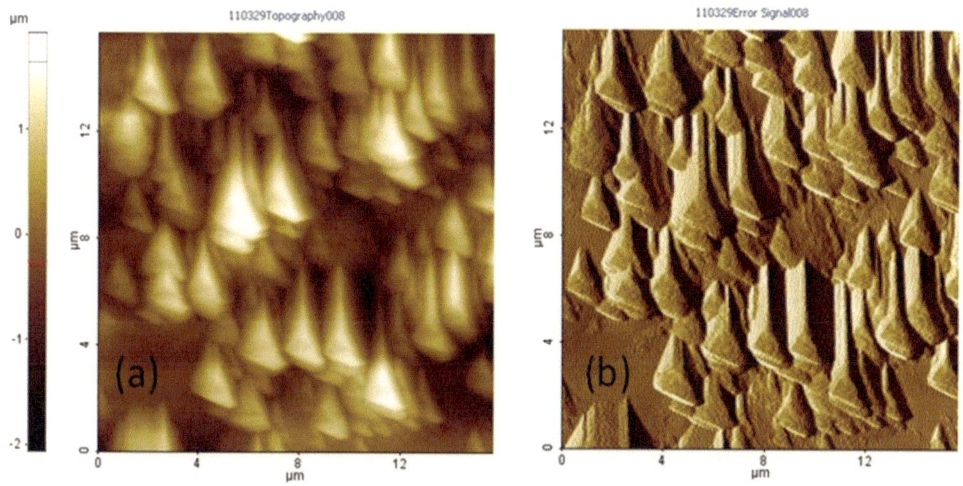

Figure 8. AFM topographical (a) and error signal images (b) corresponding to a laser-textured surface using the UV nanosecond laser at a fluence of 1 J/cm^2 [5].

slightly from the center of the wafer, thus the actual shot dose (i.e. number of laser shots per unit surface area) is changing along the radius.

For the wetting characterization of the surfaces we used the same systems as described in Chapter 2. For water droplet adhesion measurements we used the following method: a water droplet is placed on the surface examined and then, the surface is slowly tilted while a CCD camera captures the changes in the droplet's shape. In the last recorded image, before the droplet's base starts to move, the advancing θ_A and receding θ_B APCA are measured. The difference of the advancing and receding APCA is the CAH. The standard deviation in all the measurements is ±3°. The surfaces of the ablated silicon shown in the Figures 7, 8 and 9 are superhydrophilic. Finite APCAs could not be measured on these surfaces since the water droplets that are placed are completely spread on them after a few seconds from the deposition. For practical purposes an APCA of 0° is assigned.

When PTFE particles are adhered to the laser-formed asperities of the surface by the tribo-electric deposition method, the surface morphology and the wetting characteristics of the original surface change drastically. Two main factors were found to influence the final wetting properties of the composite surface. These are the substrate micro-roughness and the amount of the PTFE sub-micrometer particles that are adhered on it. In Figure 10, two 3D AFM topographic images of an untreated laser textured surface and a triboelectrically deposited PTFE surface on top of the textured surface are presented. The composite hydrophobic surface in Figure 10b displayed an APCA of 135° with CAH of 25°.

The contact angle of a smooth PTFE surface is around 110°. In this case, due to the underlying roughness the APCA is exceeding this value. Detailed APCA and CAH measurements showed that laser-textured silicon surfaces having an average roughness of ~ 5 µm do not become superhydrophobic (APCA > 150°) when coated with triboelectrically charged PTFE particles.

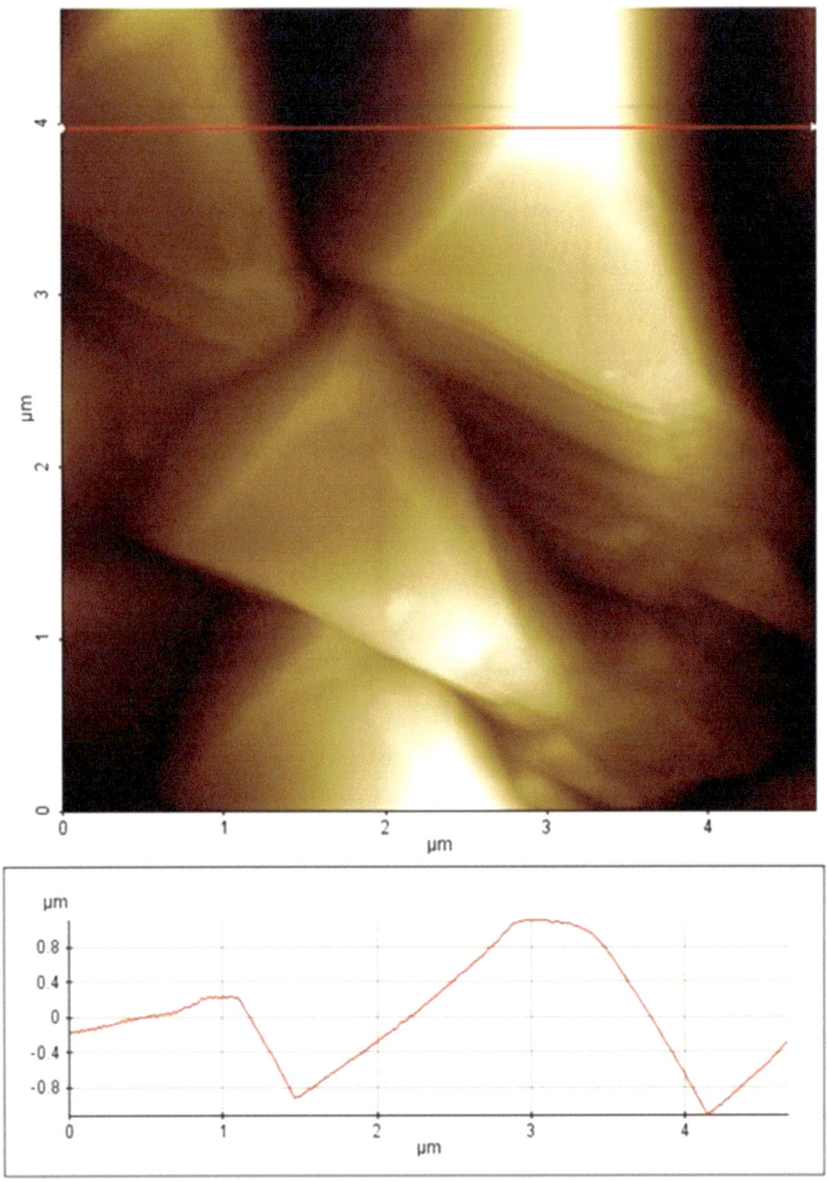

Figure 9. Top: AFM topography of a 5 µm × 5 µm surface area of the laser-microtextured silicon wafer surface. Bottom: Typical roughness profile of the micro-textured surface [5].

The composite hydrophobic surfaces obtained in this way showed relatively high CAH compared to a self-cleaning superhydrophobic surface (CAH < 10°). In average, the CAH obtained was 35°. In Figure 11, detailed APCA and CAH are presented for textured surfaces coated with PTFE. Four different textured surfaces were used as substrates, as a result of the four different laser fluencies that were used for their irradiation (0.5, 1.0, 1.5, and 2.0 J/cm^2).

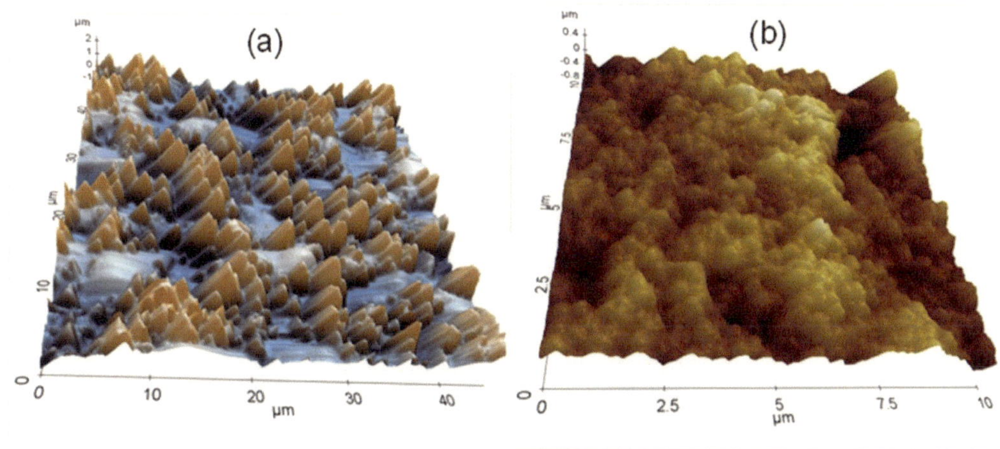

Figure 10. (a) A 3D AFM topography large area scan of textured silicon wafer surface before tribolelectric PTFE adhesion and (b) 3D topographical detail of the resultant surface roughness after the PTFE deposition [5].

Within the range of the laser fluencies that were studied, no major changes in the roughness of the silicon wafer were observed. However, the shape of the individual pyramid like microstructures could be modified without this having an effect on the final wetting properties of the PTFE coated surfaces. As seen in Figure 11, APCAs center around 130° and CAH around 35° for all the surfaces prepared. Although these surfaces cannot be considered superhydrophobic self-cleaning surfaces, water droplets exhibit quite low water adhesion on them and relatively high hydrophobicity. These wetting characteristics outperform certain hydrophobic polymers such as smooth PTFE or poly(dimethyl siloxane) (PDMS).

The four different laser fluencies that were used showed almost identical hydrophobicity. The main reason for this is attributed to the fact that the triboelectric coating creates 1.5 μm thick PTFE films. Therefore, the realization of a conformal superhydrophobic coating was not possible with the resultant silicon wafer surface roughness which was ranging from 2 to 5 μm. In order to circumvent this problem, SiC sandpapers were used of various grit sizes to analyze the effect of surface-microtexture on the degree of final hydrophobicity as a result of tribo-electric PTFE deposition. Figure 12 shows that by tailoring the amount of PTFE adhered per unit area, surface hydrophobicity can be tuned from hydrophobic to self-cleaning superhydrophobic as a function of the surface micro-roughness. In general, for micro-rough surfaces up to 60 μm surface roughness, less than 1.5 mg/cm^2 PTFE attachment is enough to render them superhydrophobic. The rectangular region in Figure 12 indicates the roughness range which could be created by the UV laser micromachining of the silicon wafers. Figure 13 shows measured APCA and CAH as a function of the PTFE quantity deposited on a sandpaper of an average roughness of 16 μm. The best results (APCA > 155° and CAH < 20°) are obtained when the mass of PTFE deposited per surface area is approximately 0.5-0.7 mg/cm^2. A similar analysis on the laser-microtextured silicon wafer shows that much less amount of PTFE per unit area (0.05-0.15 mg/cm^2) is necessary to render the surface hydrophobic (Figure 14). In the case of the laser-textured silicon wafers, the amount of PTFE deposited is not affecting the

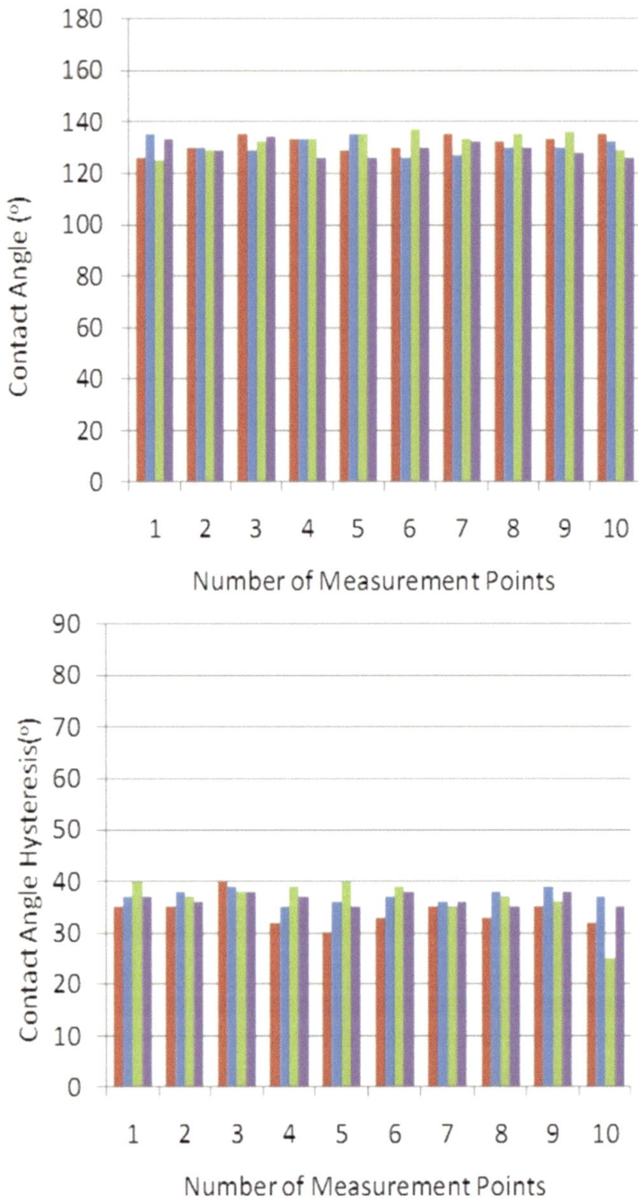

Figure 11. The results from 10 individual measurement points of the APCA and the CAH on PTFE coated laser-textured silicon wafers. The different colors in the graphs indicate the four different laser fluencies used ranging from 0.5 up to 2.0 J/cm^2. The measurements were performed randomly on different parts of the sample to investigate their homogeneity [5].

APCA and the CAH values in such a nice and controllable manner as in the case of the SiC sandpapers.

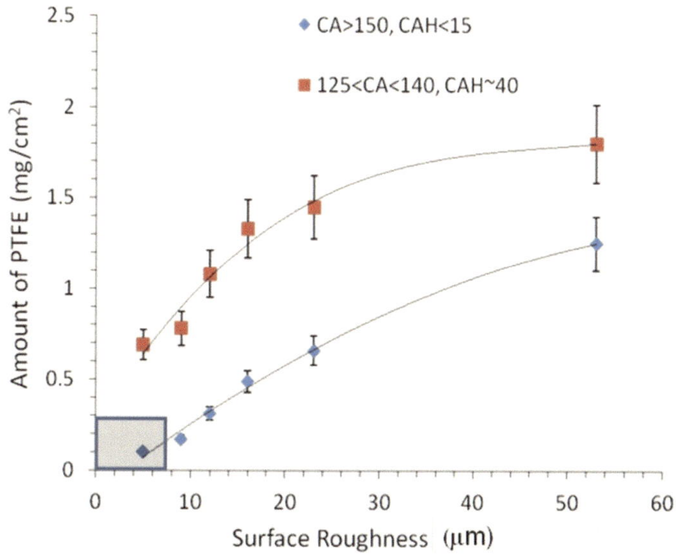

Figure 12. Effect of the amount of PTFE deposited as a function of the surface roughness on the degree of hydrophobicity and water adhesion. The grey box indicates the roughness regions obtained with the UV laser micromachining of the silicon wafers. The surface roughness out of this region is due to the use of SiC sandpapers with different grit sizes. The data set given (CA > 150°, CAH < 15°) indicates self-cleaning superhydrophobicity [5].

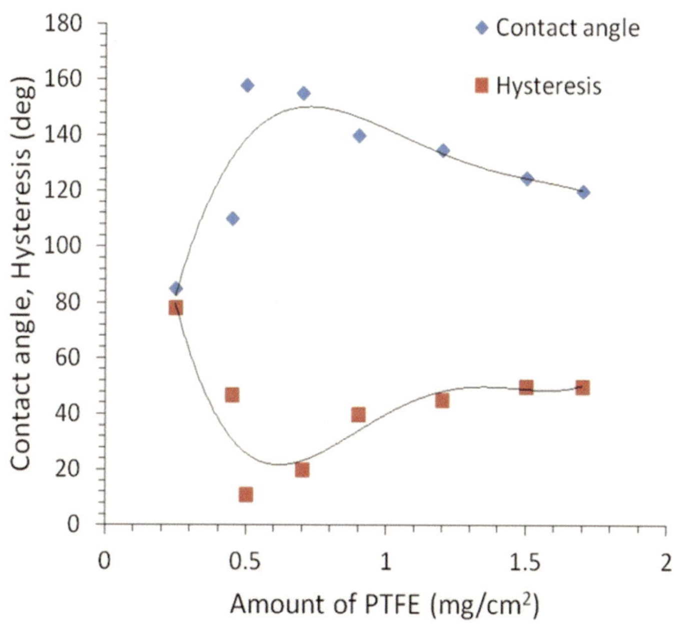

Figure 13. Effect of the amount of PTFE deposited on sandpaper with an average roughness of 16 μm, on the degree of the final hydrophobicity and water adhesion. PTFE measurements were accurate to 0.1 mg/cm^2 [5].

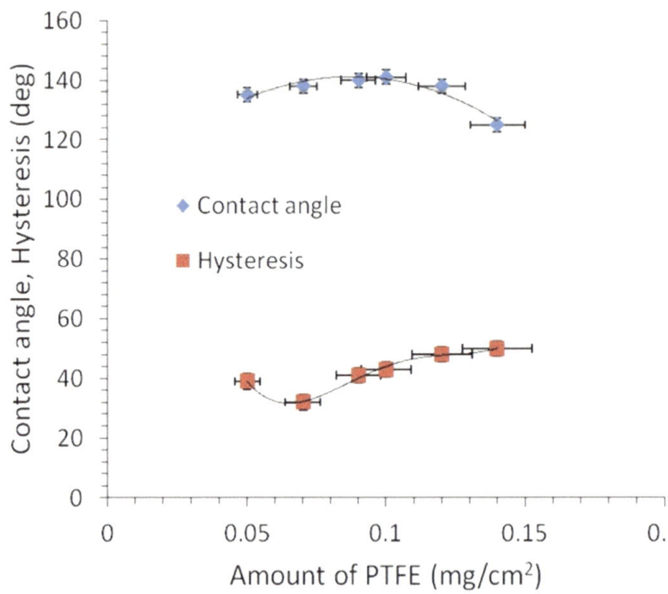

Figure 14. Effect of the amount of PTFE deposited on a laser-textured surface with average roughness of 2 μm, on the degree of the final hydrophobicity and water adhesion [5].

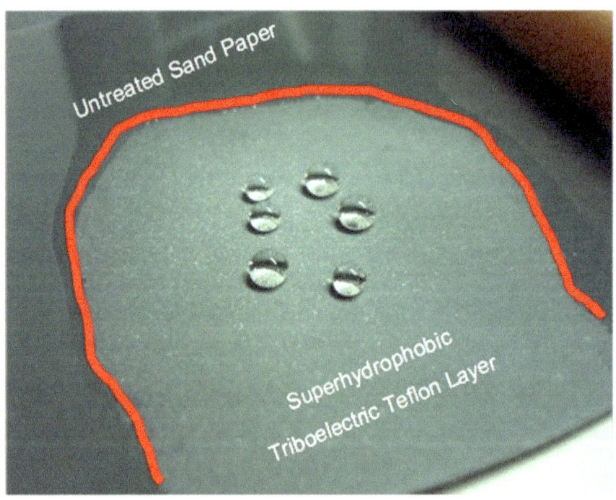

Figure 15. Photograph of a 600 grit SiC sandpaper with a treated with PTFE region and an untreated region. It is clear the difference in the wetting properties of the two surface areas [5].

Finally, in Figure 15, it is shown a photograph of a sandpaper (600 grit, average roughness: 26 μm). A part of which was coated with the triboelectric PTFE deposition and the other was left untreated. The red line in the photo shows the boarders between the treated and untreated region. As it can be observed, the untreated region is completely wet whereas in the treated region the droplets are in a superhydrophobic state.

http://dx.doi.org/10.5772/56173

4. Conclusions

In this work, we demonstrated two different lithography techniques, each one in combination with a different coating method. The purpose of these two fabrication processes is to obtain surfaces with large areas exhibiting special wetting characteristics with a simple, efficient, repeatable and economical strategy. The first process consists of the fabrication of SU-8 micropillar patterns on silicon by means of UV lithography. The patterns are subsequently coated by spraying particles which induce different roughness scales (submicrometer roughness with PTFE and nano-roughness with iron oxide). The substrates, starting from "sticky" hydrophobic (SU-8 uncoated patterned surfaces), by the application of the proper particles, can be converted to superhydrophobic surfaces with ultrahigh or ultralow water adhesion. The second process combines the laser micromachining of silicon wafers with the coating method of triboelectric deposition of charged PTFE submicrometer particles. This green solvent-free fabrication method results in surfaces with dual scale roughness (in micro- and nano-scale). These surfaces, upon the coating application are converted from superhydrophilic to hydrophobic (APCA = 130°) with low water adhesion (CAH ~ 35°). If instead of the silicon wafers are used SiC sandpapers with various grit sizes, then after the triboelectric deposition of PTFE the surfaces become superhydrophobic. The methods described here can find application in the development of microfluidic devices, smart surfaces, biotechnological materials and generally in all kinds of applications that special wetting properties and controllable roughness is required.

Acknowledgements

The authors would like to thank Dr Luigi Martiradonna, Dr George C. Anyfantis, Dr P. Davide Cozzoli and Professor Roberto Cingolani for their contribution in this work, as well as Mr Diego Mangiulo, Mr Gianmichele Epifani and Mr Paolo Cazzato for their technical assistance during the experimental process.

Author details

Athanasios Milionis[1], Ilker S. Bayer[1], Despina Fragouli[1], Fernando Brandi[1] and Athanassia Athanassiou[1*]

*Address all correspondence to: athanassia.athanassiou@iit.it

1 Nanophysics, Istituto Italiano di Tecnologia (IIT), Genoa, Italy

Smart Materials, Center for Biomolecular Nanotechnologies@UNILE, Arnesano (LE), Italy

References

[1] Xia, D, Johnson, L. M, & Lopez, G. P. Anisotropic wetting surfaces with one-dimensional and directional structures: fabrication approaches, wetting properties and potential applications. Advanced Materials (2012). , 24(10), 1287-1302.

[2] Liu, M, & Jiang, L. Switchable adhesion on liquid/solid interfaces. Advanced Functional Materials (2010). , 20(21), 3753-3764.

[3] Lee, J. B, Gwon, H. R, Lee, S. H, & Cho, M. Wetting transition characteristics on microstructured hydrophobic surfaces. Materials transactions (2010). , 51(9), 1709-1711.

[4] Milionis, A, Martiradonna, L, Anyfantis, G. C, Cozzoli, P. D, Bayer, I. S, Fragouli, D, & Athanassiou, A. Control of the water adhesion on hydrophobic micropillars by spray coating technique. Colloid and Polymer Science (2013). , 291(2), 401-407.

[5] Bayer, I. S, Brandi, F, Cingolani, R, & Athanassiou, A. Modification of wetting properties of laser-textured surfaces by triboelectrically charged Teflon particles. Colloid and Polymer Science (2013). , 291(2), 367-373.

[6] Tsougeni, K, Papageorgiou, D, Tserepi, A, & Gogolides, E. Smart" polymeric microfluidics fabricated by plasma processing: controlled wetting, capillary filling and hydrophobic valving. Lab on a Chip (2010). , 10(4), 462-469.

[7] Villafiorita-monteleone, F, Mele, E, Caputo, G, Spano, F, Girardo, S, Cozzoli, P. D, Pisignano, D, Cingolani, R, Fragouli, D, & Athanassiou, A. Optically controlled liquid flow in initially prohibited elastomeric nanocomposite micro-paths. RSC Advances (2012). , 2(25), 9543-9550.

[8] Stratakis, E, Ranella, A, & Fotakis, C. Biomimetic micro/nanostructured functional surfaces for microfluidic and tissue engineering applications. Biomicrofluidics (2011).

[9] Kannarpady, G. K, Lefrileux, Y, Laurent, J, Woo, J, Khedir, K. R, Ishihara, H, Trigwell, S, Ryerson, C, & Biris, A. S. Nanotech (2010). conference proceedings, June Anaheim Convention Center, Anaheim, CA, USA., 21-24.

[10] Huang, X. J, Kim, D. K, Im, M, Lee, J. H, Yoon, J. B, & Choi, Y. K. Lock-and-key" geometry effect of patterned surfaces: wettability and switching of adhesive force. Small (2009). , 5(1), 90-94.

[11] Furstner, R, Barthlott, W, Neinhuis, C, & Walzel, P. Wetting and self-cleaning properties of artificial superhydrophobic surfaces. Langmuir (2005). , 21(3), 956-961.

[12] Feng, J, Tuominen, M. T, & Rothstein, J. P. Hierarchical superhydrophobic surfaces fabricated by dual-scale electron-beam lithography with well-ordered secondary nanostructures. Advanced Functional Materials (2011). , 21(19), 3715-3722.

[13] Shirtcliffe, N. J, Aqil, S, Evans, C, Mchale, G, Newton, M. I, Perry, C. C, & Roach, P. The use of high aspect ratio photoresist (SU-8) for super-hydrophobic pattern prototyping. Journal of Micromechanics and Microengineering (2004)., 14(10), 1384-1389.

[14] Marquez-velasco, J, Vlachopoulou, M. E, Tserepi, A. D, & Gogolides, E. Stable superhydrophobic surfaces induced by dual-scale topography on SU-8. Microelectronic Engineering (2010).

[15] Wagterveld, R. M, Berendsen, C. W. J, Bouaidat, S, & Jonsmann, J. Ultralow hysteresis superhydrophobic surfaces by excimer laser modification of SU-8. Langmuir (2006)., 22(26), 10904-10908.

[16] He, Y, Jiang, C, Yin, H, Chen, J, & Yuan, W. Superhydrophobic silicon surfaces with micro-nano hierarchical structures via deep reactive ion etching and galvanic etching. Journal of Colloid and Interface Science (2011)., 364(1), 219-229.

[17] Athanassiou, A, Fragouli, D, Villafiorita-monteleone, F, Milionis, A, Spano, F, Bayer, I. S, & Cingolani, R. Laser-based lithography for polymeric nanocomposite structures. In: Cui B. (ed.) Recent advances in nanofabrication techniques and applications. Rijeka: Intech; (2012)., 289-314.

[18] Andrzejewska, E. Photopolymerization kinetics of multifunctional monomers. Progress in Polymer Science (2001)., 26(4), 605-665.

[19] Hong, L, & Pan, T. Photopatternable superhydrophobic nanocomposites for microfabrication. Journal of Microelectromechanical Systems (2010)., 19(2), 246-253.

[20] Caputo, G, Cortese, B, Nobile, C, Salerno, M, Cingolani, R, Gigli, G, Cozzoli, P. D, & Athanassiou, A. Reversibly light-switchable wettability of hybrid organic/inorganic surfaces with dual micro-/nanoscale roughness. Advanced Functional Materials (2009)., 19(8), 1149-1157.

[21] Debuisson, D, Senez, V, & Arscott, S. Tunable contact angle hysteresis on micropatterned surfaces. Applied Physics Letters (2011).

[22] Calcagnile, P, Fragouli, D, Bayer, I. S, Anyfantis, G. C, Martiradonna, L, Cozzoli, P. D, Cingolani, R, & Athanassiou, A. Magnetically driven floating foams for the removal of oil contaminants from water. ACS Nano (2012)., 6(6), 5413-5419.

[23] Balu, B, Breedveld, V, & Hess, D. W. Fabrication of "roll-off" and "sticky" superhydrophobic cellulose surfaces via plasma processing. Langmuir (2008). , 24(9), 4785-4790.

[24] Quere, D, Lafuma, A, & Bico, J. Slippy and sticky microtextured solids. Nanotechnology (2003)., 14(10), 1109-1112.

[25] Feng, L, Zhang, Y, Xi, J, Zhu, Y, Wang, N, Xia, F, & Jiang, L. Petal effect: a superhydrophobic state with high adhesive force. Langmuir (2008)., 24(8), 4114-4119.

[26] Bormashenko, E, Stein, T, Pogreb, R, & Aurbach, D. Petal effect" on surfaces based on lycopodium: high-stick surfaces demonstrating high apparent contact angles. Journal of Physical Chemistry C (2009). , 113(14), 5568-5572.

[27] Cassie, A. B. D, & Baxter, S. Wettability of porous surfaces. Transactions of the Faraday Society (1944). , 40(0), 546-551.

[28] Zhu, L, Feng, Y, Ye, X, & Zhou, Z. Tuning wettability and getting superhydrophobic surface by controlling surface roughness with well-designed microstructures. Sensors and Actuators A (2006). SI) , 595-600.

[29] He, Y, Jiang, C, Yin, H, Chen, J, & Yuan, W. Superhydrophobic silicon surfaces with micro-nano hierarchical structures via deep reactive ion etching and galvanic etching. Journal of Colloid and Interface Science (2011). , 364(1), 219-229.

[30] Feng, J, Tuominen, M. T, & Rothstein, J. P. Hierarchical superhydrophobic surfaces fabricated by dual-scale electron-beam-lithography with well-ordered secondary nanostructures. Advanced Functional Materials (2011). , 21(19), 3715-3722.

[31] Teisala, H, Tuominen, M, Aromaa, M, Stepien, M, Makela, J. M, Saarinen, J. J, Toivakka, M, & Kuusipalo, J. Nanostructures increase water droplet adhesion on hierarchically rough superhydrophobic surfaces. Langmuir (2012). , 28(6), 3138-3145.

[32] Wang, X, Mak, G. Y, & Choi, H. W. Laser micromachining and micro-patterning with a nanosecond UV laser. In: Kahrizi M. (ed.) Micromachining techniques for fabrication of micro and nano structures. Rijeka: Intech; (2012). , 85-108.

[33] Fadeeva, E, Truong, V. K, Stiesch, M, Chichkov, B. N, Crawford, R. J, Wang, J, & Ivanova, E. P. Bacterial retention on superhydrophobic titanium surfaces fabricated by femtosecond laser ablation. Langmuir. (2011). , 27(6), 3012-3019.

[34] Leitz, K. H, Redlingshöfer, B, Reg, Y, Otto, A, & Schmidt, M. Metal ablation with short and ultrashort laser pulses. Physics Procedia (2011). , 12-230.

[35] Kalyva, M, Bertoni, G, Milionis, A, Cingolani, R, & Athanassiou, A. Tuning of the characteristics of Au nanoparticles produced by solid target laser ablation into water by changing the irradiation parameters. Microscopy research and technique (2010). , 73(10), 937-943.

[36] Bayer, I. S, Caramia, V, Biswas, A, Cingolani, R, & Athanassiou, A. Metal-like conductivity exhibited by triboelectrically deposited polyaniline (emeraldine base) particles on microtextured SiC surfaces. Applied Physics Letters (2012). , 100-201604.

[37] Liu, C, & Bard, A. J. Electrostatic electrochemistry at insulators. Nature Materials (2008). , 7-505.

Section 2

Resist

Chapter 6

Resist Homogeneity

Nima Arjmandi

Additional information is available at the end of the chapter

http://dx.doi.org/10.5772/56618

1. Introduction

A key element in all lithography processes is resist material; unfortunately its inhomogeneity is the main source of most lithography problems. Resist inhomogeneity can be any kind of variation in its thickness, arrangement of its molecules, chemical composition, and concentration of materials in it. Homogeneity of the resist layer before exposure and also after development is required to achieve low product variation, high production yield, high resolution and small patterns. If the resist layer does not have enough homogeneity and so its thickness, chemical composition or the molecular structure changes on a wafer or from a wafer to another wafer; dimensions of the patterns will change from one device to the other. This change will cause increased variation of the product's specifications. Dramatic variations may even mar some of the products and will reduce production yield. Resist inhomogeneity in relatively small dimensions increases the so called line edge roughness. This is generally defined as the root mean square of the line width variation along a patterned line. In the other words, if the resist properties randomly changing from place to place; dimensions of the final pattern will randomly change from place to place. For example, if one tries to pattern a line with a constant width, while the resist properties are changing in distances smaller than the length of the line, the line width will fluctuate along its length. If one tries to pattern a line thinner than these fluctuations; at some parts the width of the fabricated line will be zero. So, such a variation in the resist properties can limit both the ultimate patterning resolution and the smallest achievable pattern. Even if the resist inhomogeneity does not considerably alter the pattern geometry, it can cause variation in the etch resistance of the resist. Consequently, device dimensions and properties may vary randomly.

This chapter will cover different types of inhomogeneities in different resists and lithography processes, their origins, effects; and methods to increase the resist homogeneity. At first the basic issues in wafer level homogeneity in different step of lithography will be discussed; and effect of each lithography parameter in each step will be describe. Then, last section will focus on the most important problems in nano lithography. These problems include, resist's molecular agglomeration and line edge roughness, which are the main limiting factors of nano lithography's ultimate resolution and yield. Latest developments in solving these problems and improving the resist homogeneity to nanometer scale will be described. Here by an industrial lithography process capable of making patterns smaller than 5nm will be described in detail. These are one of the smallest patterns ever made in an industrial lithography and an important key factor in successful fabrication of these patterns is making a homogeneous layer of resist.

2. Resist material

First criteria for choosing an appropriate resist material is dictated by the exposure technology. Exposure technology determines the wavelength of the photon or electron beam the resist should be sensitive to. Resist's tune to be negative or positive, is generally choosen to minimize the exposed area. The reason is, exposing a large area and keeping small dimentions unexposed, usually increases inhomogeneity and so called proximity effects. To trasfer the resist's pattern to other layer(s) two different types of processes can be used, etching and lift-off. Etching transfers the resist's pattern to the underlying layer; while lift-off transfers the negative of the resist's pattern to the layers which are deposited after lithography on top of the resist. Thus, etching processes are transfering the positive of the resist's image while lift-off processes are transfering its negative (reversed) image. So, choosing resist's tune to be negative or positive, also depend on the pattern transfer process to be etching or lift-off. For instance, if it is desired to pattern a hole by electron beam lithography in a wide blank film of aluminum, and it is designed to transfer the pattern by reactive ion etching (RIE). Then, a positive tune resist is needed. Because, if a negative tune resist be used, it will be necessary to expose a very large area, and such an exposure will introduce proximity problems, will take a very long time and will increase the exposure noise. So, the edge roughness and other inhomogeneities and deformities will increase. On the other hand if the pattern will be transferred by lift-off, a negative tune resist should be used. If one needs to make a pillar, the choice of negative or positive tune resist will be the opposite.

Next constrain for choosing a resist is its underlying layer and/or layer that may be coated on top of it. Because some resists do not adhere to some materials at all, may degrade near them, or react with them. So, the resist should be compatible with the other materials and processes used in device fabrication. For example it is not possible to use PMMA resists on copper.

An other common constrain to choose a resist is its etch resistance. In the other words, the resist should be able to hold up during the etching process. Most of the resist materials are designed to hold up in a specific etching process and using them with other pattern transfer process will introduce process complications or loss of patterning quality. For instance, SU-8 resists pill off in KOH wet etching solutions and cannot be used to gether with KOH wet etching. PMMA resists cannot hold up in oxigen plasma. So, they cannot be used with reactive ion etching processes which are incorporating oxigen plasma.

With modern exposure tools and established resists, generally the contrast is not a major issue and desired contrast can be easily achieved by tuning the exposure dose and development parameters. Thus, after considering the mentioned constrains, usually there are plenty of different resists applicable to the process. In order to minimize the product variation, maximize the yield and improve the resolution, it is necessary to choose a resist which can give the highest homogeneity.

Most of the commercially available and widely used resists are composed of a photon or an electron sensitive polymer, a solvent and some additives. The ultimate homogeneity of a resist layer is dictated by the structure of the polymer material itself. In most of these resists the Steric compatibilities between the tactic sequences of the polymer molecules [19] and Van der Waals attractions [17] drives the smaller polymer molecules to diffuse and attach to the bigger molecules and make granules of about 20 nm big (figure 1).

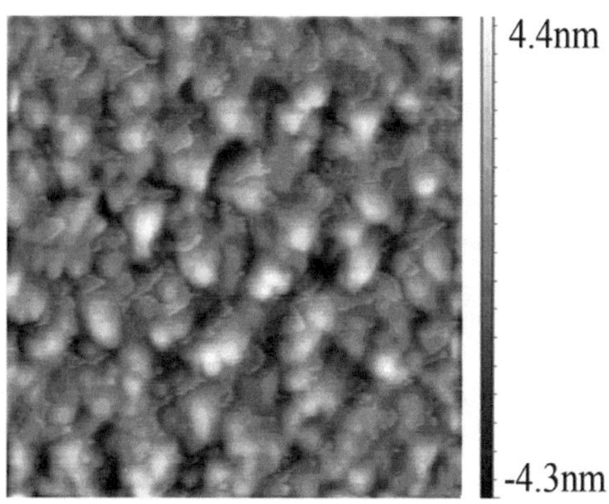

Figure 1. Granular Structure of Resist. 200x200 nm atomic force microscope (AFM) scan of a resist's surface is showing its granular structure resulted from agglomeration of polymer molecules. The resist is a commercial PMMA and coated on a silicon wafer in a standard process.

If the mentioned forces be enough strong, and the polymer chain be enough long and flexible; they can even make the polymer chain to collapse on itself. The diffusion rate of the developer is much faster between the granules and at their boundaries than inside the granules. In addition, these granules are much less soluble in the developer than the rest of the material. So, the resist material in the granules will not dissolve in the developer and these granules will be extracted and released one by one in to the developer.

Consequently, the size and solubility of these granules will limit the lithography resolution and line edge roughness (figure 2) [9, 19]. So, in order to increase the resist homogeneity, it is necessary to reduce the aggregation and formation of these granules; or choose a resist with less aggregation or smaller granules.

Generally, if the final resist layer will be much thicker than the length of the polymer chain; then, more rigid polymers with more 3-dimentional structures will result in less aggregation, e.g. if final resist layer be thicker than about 100 nm; surface roughness of hydrogen silses-

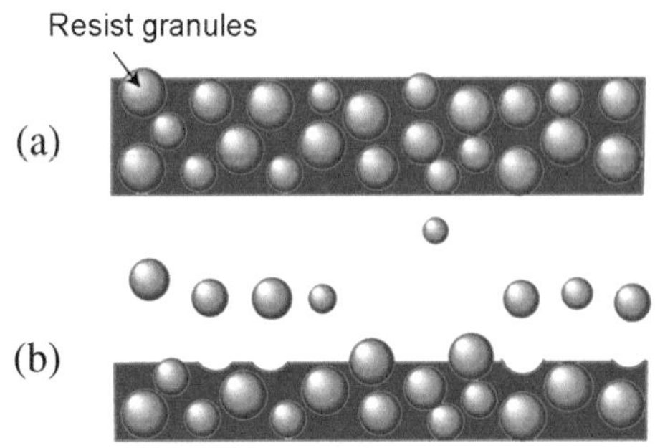

Figure 2. Effect of Resist's Granular Structure in Development. a) Schematic representation of a resist layer with granular structure. b) During the development the granules are being extracted and a rough edge is left behind.

quioxane (HSQ) or Calixarene be less than poly methyl methacrylate (PMMA), SAL ® (Shipley Inc.) or ZEP ® (Nippon Zeon Ltd.) [17]. In such conditions, smaller polymer molecules will results in smaller granules, e.g. Calixarene results in a smoother surface than HSQ if the resist thickness is more than about 100 nm. On the other hand, if the resist thickness is in the order of or less than the polymer length, more flexible and longer polymers will result in a more homogeneous surface, e.g. if the resist thickness be 20 nm, Novolac will result in a rougher surface than polyvinyl phenol [21].

Although, the existence of smaller polymer chains in a solution of longer polymers, prevents the resist layer from shrinking after removal of the solvent; they will increase the granule formation by diffusing into the bigger chains. So, the best polymer to make a very thin and homogeneous layer is a long and heavy polymer with a narrow molecular weight distribution and a one dimensional and flexible structure.

The other ingredient of a resist is its solvent which is a liquid that transfers the resist material to the substrate. This solvent should not degrade and should prevent the polymer molecules to reach each other and from particles. Because, if the resist molecules get very close to each other they will bind by the mentioned forces and make small particles or aggregations. To reduce the polymer aggregation, it is required to reduce the concentration of the polymer in the solvent. Hereby, the probability of two polymer molecules colliding to each other will reduce; but, to prevent partial coverage of the substrate by the resist layer and also to prevent the early aging of the resist and formation of particles, the polymer concentration should not reduce below a certain value. In addition, if a very low viscosity solvent be used and a thicker layer be required, it will be necessary to reduce the spin coating speed too much, which will increase the wafer level resist thickness variation.

To improve the surface coverage of the resist, other materials can be added to the resist solution to improve its adhesion to the surface. In some resists these are enhancing the sensitivity, dark erosion or contrast as well.

Potentilly the most homogeneous resists are self assembled surface mono layers (SAM) [12, 15, 18, 2]. These are a few nanometer long molecules that can bind to the substrate's dangling bonds only at a specific site located at one end of the molecules (figure 3). In ideal case, interaction between the molecules themselves and the substrate, results in a perfect 2-dimentional lattice of an arranged mono layer on the substrate.

Unfortunately, SAMs are very sensitive to the material beneath them. So, not only they are very sensitive to surface contamination, but also, it is not possible to use them on different materials. In addition, they have a very limited etch resistance. Hydrogenated dangling bonds have also been used as an atomic monolayer resist [16]. Although they may have potential of providing atomic resolution lithography, they are even more severely suffer from the lack of etch resistance and high sensitivity to the surface. Monolayer resists are mostly in the research stage and yet they have not found vast industrial applicability.

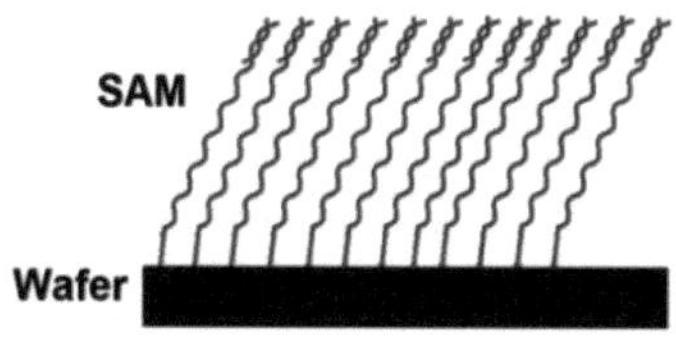

Figure 3. Schematic representation of a self assembled surface mono layer (SAM) resist. In this case linear resist molecules have made an arranged coating on the wafer.

3. Resist storage and aging

The resist material should be stored in a clean low sodium glass, Teflon or HD-PE bottle. There should not be any plastic with a softener near the resist. The bottles can be cleaned in hot acetone and then in hot isopropyl alcohol (IPA) and should completely dry before putting the resist in it. After each refilling a waiting period of several hours is necessary to outgas air bubbles from the resist. The filled bottles should be stored in a dry and dark place, preferably inside the clean room. The storage temperature should be enough high to prevent particle formation and sedimentation, and enough low to prevent evaporation, degradation and decomposition of photo initiator. Usually the best storage temperature is between 5 °C to 10 °C.

Inappropriate storage or aging of the resist will result in formation of nitrogen bubbles, and particles and gradually conglomeration of them to bigger clusters. All of them will result in so called comet like inhomogeneity (figure 5.a). In an aged resist the photo initiator is lost. Therefore, development rate decreases and dark erosion increases. Aging also reduces the adhesion of the resist to the surface. This reduction will result in undesired undercuts, pill off of the resist during the development and will change the pattern dimensions. It is also recommended not to store partially used bottles for a long time. Because, the nitrogen can dissolve in the resist and the dissolved nitrogen can cause bubbles or popping during the baking or exposure. Frequent opening of the resist bottle will also result in evaporation of the solvent which will increase the resist viscosity and thickness.

4. Surface preparations of the substrate

Substrates should be cleaned before the resist coating. This cleaning decrease the surface roughness by removing particles and contamination; and also it will increase the resist homogeneity by removing contaminants and increasing the adhesion of the resist. Loss of adhesion may reduce the homogeneity by resulting in pill-off of the resist, or under cut in some parts. Contamination on the surface will result in pattern variations and homogeneity reduction as well.

In general the cleaning process will depend on the substrate's material. In the case of silicon substrate, an appropriate cleaning process can contain the following steps: first, dipping in piranha (H_2O_2(25%) : H_2SO_4 (97%) 1:3 v:v) for more than 5min; hereby not only most of contaminants will dissolve in the solution, but also, SiO_2 will grow in to the silicon. So, the Si/SiO_2 interface will be kept clean. Second, dipping in diluted HF (1-5%) will remove the oxide with all the remained contaminants attached to it. Third, dipping in SC-1 (H_2O_2 (25%) : NH_4OH (25%) : H_2O 1:1:5 v:v:v) for about 10 min at 75 °C; forth, dipping in diluted HF again. Fifth, dipping in SC-2 (HCl (30%): H_2O_2 (25%) : H_2O 1:1:6) for about 10 min at 80 °C; and finally, dipping in the diluted HF again and then blow drying with nitrogen. After this process the surface dangling bonds of the silicon should be hydrogenated (H-passivated), and consequently the resist will have a very good adhesion on it. If for any reason it does not be possible to completely remove the oxide from the surface, the surface will be hydrophilic to some degree. The resist will not have a good adhesion to a hydrophilic surface. In addition, a hydrophilic surface will easily absorb moisture. So, it is necessary to bake it for a few minutes at more than 120 °C (preferably in an oxygen free environment). This baking process will dehydrate the surface e.g. in the case of silicon it will remove the OH- groups from the silicon dangling bonds.

If for any reason it does not be possible to produce sufficient adhesion between the substrate and the resist material. It will be necessary to use an adhesion promoter. It is a material that coats the surface before resist coating. On one side it binds to the substrate and on the other side adheres to the resist layer which comes on top. For example, Hexamethyldisilazane (HMDS) binds to the SiO_2 surface, and then releases its ammonia. So, a methyl group remains bound to the surface (figure 4). Many resists have a very good adhesion to this methyl group. Possibly the HMDS coating will be thicker than a monolayer. Such a HMDS layer can release considerable amount of ammonia during the baking process. The released ammonia diffuses in to the resist material and crosslinks the resist molecules. In this case the complete development of the resist will be impossible. So, considering the fact that we need just a monolayer of the HMDS, thinner HMDS coatings are better. It is also necessary to dehydrate the substrate before HMDS coating. Another common adhesion promoter is TI PRIME ® (Microchemical Inc.); it works like HMDS; but, it automatically makes a monolayer by baking at 120 °C. So, it does not have the complications of the HMDS.

http://dx.doi.org/10.5772/56618

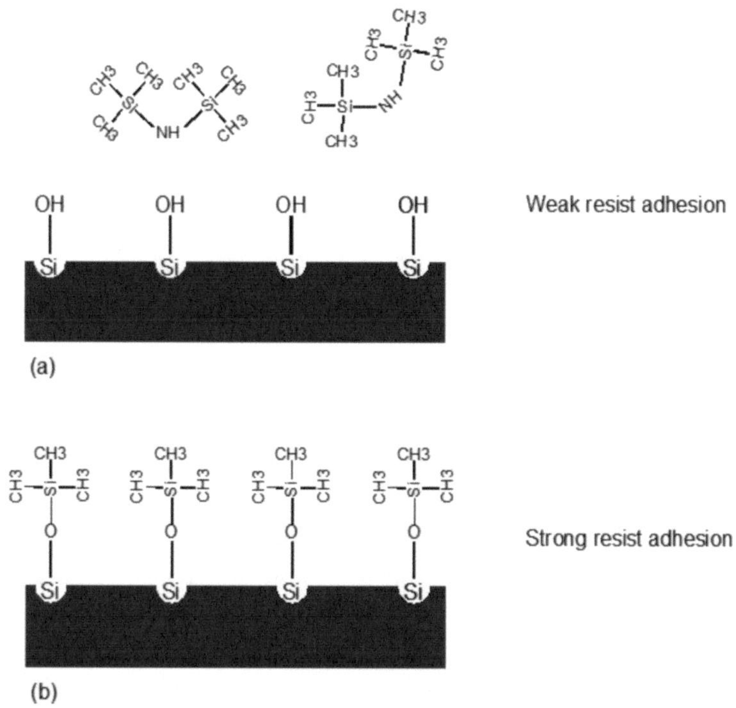

Figure 4. Schematic representation of chemistry of HMDS adhesion promoter working on a silicon substrate. a) Dangling bonds of silicon atoms and native oxide are occcupied with OH groups, leaving a hydrophilic surface that cannot adhere to resists. b) HMDS molecules have left their NH group and bind to the silicon atoms on the surface, leaving a hydrophobic surface that strongly adheres to resists.

5. Resist coating

Some resists are available in rolled sheets of several centimeters by a few meters. These sheets can be laminated on the substrate. In the case of the liquid resists, it is possible to put a layer of the resist solution on the substrate by wetting a cylinder by the resist solution and rolling the cylinder on the substrate; but, the most widely used methods of resist coating are spaying the resist solution on the substrate; or dispensing it on the substrate; and then, rotating the substrate with high speed to obtain a very thin and homogeneous layer.

By rotating the wafer, the centrifugal force tries to spread the resist to the edges of wafer and eventually fling it off the edges of the wafer. On the other hand, the adhesion of the resist to the substrate and viscosity of the resist are trying to keep the resist on the substrate. At the same time, evaporation of the solvent gradually increases the resist viscosity and makes it impossible for the centrifugal force to move the resist material off the wafer. In ideal situation, the end result of these competing forces will be a homogeneous and thin layer of the resist material on the substrate.

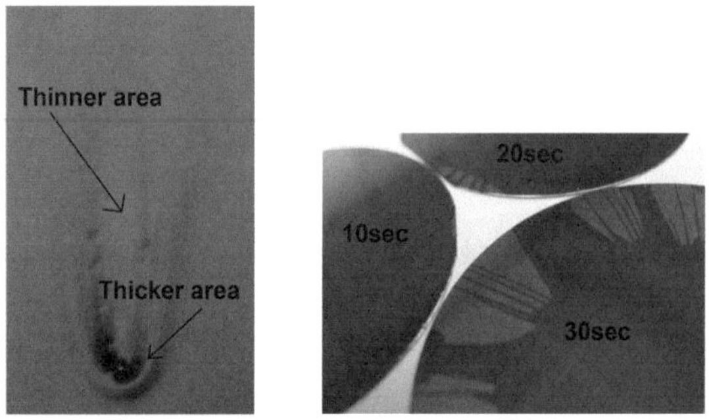

(a) (b)

Figure 5. Resist Inhomogeneities Caused by Spinning. a) A comet inhomogeneity in the resist which is caused by a particle in the resist. b) Inhomogeneities caused by insufficient amount of dispensed resist and too long acceleration time. The lower wafer accelerated in 30 s, the upper one in 20 s and the left one in 10 s.

The spin coating consists of the following stages [14]:

1. At first small puddle of the resist is dispensed on the wafer near its centre. The amount of the dispensed resist usually is a few milliliters or less. For bigger wafers and more viscose solutions, more resist should be dispensed; and for less viscose resists and smaller wafers, less resist is needed. In many cases it is beneficial to dispense the resist through a micrometer sized filter to filter out the particles and bubbles. It is necessary to dispense the resist from close proximity of the wafer and prevent dropping off the resist on the wafer. Otherwise air bubbles may generate and cause comet shape thickness variations (figure 5.a). These so called comets can also be generated by bubbles or particles in the resist or on the wafer. If the amount of the dispensed resist be too small, depending on the acceleration, some triangular shaped areas with no resist will appear near the edges (figure 5.b).

2. Wafer should be accelerated to the final rotation speed and this acceleration should be precisely tuned. When the wafer rotates with a constant speed with no accelleration, the centrifugal force moves the resist outward on a radial direction. So, if there be some topographical features perpendicular to this radial movement, it will not be covered homogeneously by the resist. By the angular acceleration, the resist will make a spiral path instead of a direct radial path. This spiral movement during accelleration can increase the homogeneity of the coating on the topographical features. If the acceleration be too low (long acceleration time), some triangular shaped area with no resist will appear near the edges (figure 5.b). When the rotation speed of the wafer becomes enough high, aggressive expulsion of the resist from the wafer edges can be observed (figur 6.a). Due to the acceleration and the speed difference between the upper and lower layers of the resist, spiral vortices may be briefly visible. It is very important not to stop the spinning at this time. Otherwise, large resist thickness variations will remain on the wafer. By continuation of spinning, eventually the resist will be thin enough that the viscous shear drag exactly balances the rotational acceleration.

3. In this stage, wafer rotates with a constant speed and the resist rotates with the same speed meanwhile the resist thickness gradually reduces. The resist thickness is almost uniform at this stage (figure 6.b), and gradual chang of interference colors of the whole surface of the resist due to reduction of its thickness is often visible. If a relatively thick resist layer is desired, it is possible to stop the spinning at this stage; but, in order to have high reproducibility and lower wafer to wafer thickness variation, usually the spinning is continued for a longer time. In addition, edge effects are more pronounced at this stage. Because, the fluid flows very gradually outward to edges, this flow is not enough strong to instantaneously make the resist to fly off the edges. Infact the resist is gathered on the edges and only when a significantly thick layer forms on the edges, small amounts of resist can fly off the wafer in small droplets or thin fibers. This thicker resist layer on the edges is called edge bead. The edge bead not only reduces the useful area of the wafer but may cause sticking to the mask. The edge beads are inversely proportional to the spinning speed.

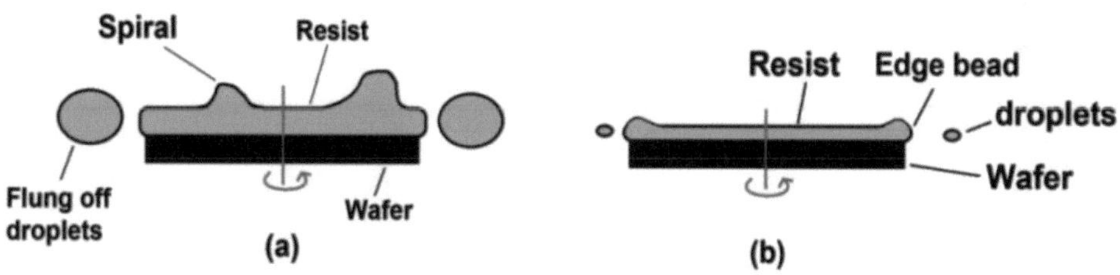

Figure 6. Formation of Edge Bead during the Spinning. a) Schematic representation of the second stage of spin coating when resist explodes from the wafer edges. b) Schematic representation of the third stage of spin coating when resist flys off the wafer in very small amounts and edge bead forms.

4. By continuation of spinig flow of resist on the substrate reduces to zero. At this stage most of the thickness reduction is due to the solvent evaporation. This evaporation is enhanced by the flow of air due to the rotation. This stage happens at the relatively flat tail of the spin-curve (figure 7). The best time to stop the spinning is during this stage when the highest homogeneity and reproducibility is achievable. The overall spinning time required to get to this point depends on the resist viscosity and the spinning speed. For more viscose resists it takes more time to get to this point and the final resist layer will be thicker. By using higher spinning speeds, this time will reduce as well as the resist thickness.

If C be the polymer concentration, η the intrinsic viscosity, ω the number or rotation per minute and K, γ, β and α fitting constants; the final resist thickness obtained by spin coating will be:

$$T = \frac{KC^{\beta}\eta^{\gamma}}{\omega^{\alpha}} \tag{1}$$

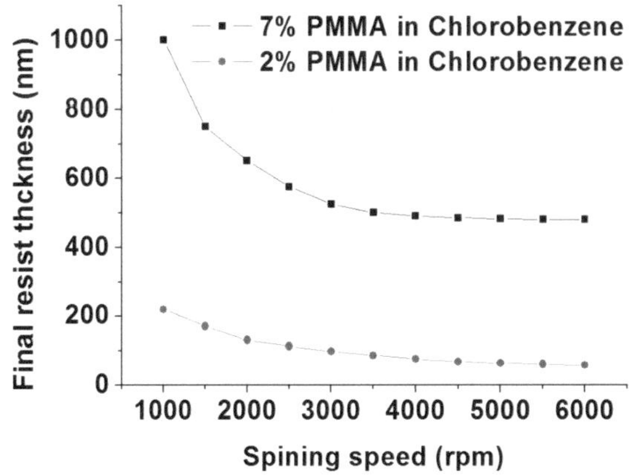

Figure 7. Spin speed curves for 2 and 7% PMMA in chlorobenzene.

6. Baking

After spin coating of thick resist layers, it is usually beneficial to wait for few minutes. This waiting time allows the resist to relax on the surface. Thus, the edge bead and other thickness variations will reduce. Then the resist should be baked. The baking is a process in which the wafer is heated to 50 °C to 250 °C before exposure. This process is necessary to evaporate the solvent content and harden the reisit to prevent formation of nitrogen bubbles, sticking to the mask, improving the resist adhesion and minimizing the dark erosion. After spin coating the solvent content of the resist is usually about 20 to 40%. During the baking process this solvent content reduces by diffusion from depth to surface and evaporation of the solvent at the surface. So, the solvent content of the resist at the resist top surface is less than areas deep in the resist. By increasing the baking time the solvent content decreases (figure 8).

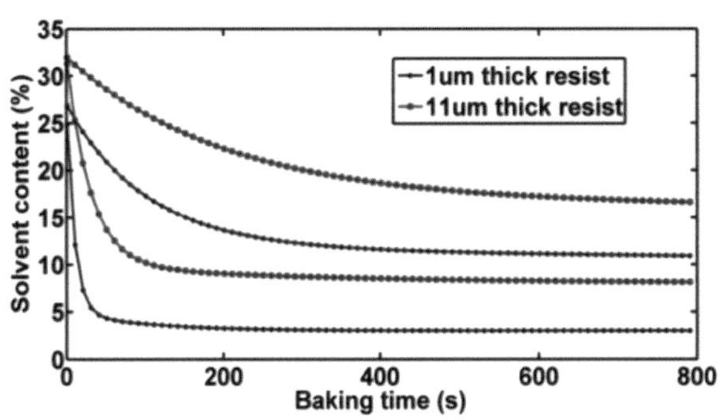

Figure 8. Solvent content of a PMMA based resist as function of time for two different resist thicknesses. Baking was done on a hot plate at 90 °C.

http://dx.doi.org/10.5772/56618

In order to have a more reproducible process, the baking time should be in the flat region of the figure 8. To have a homogeneous resist layer the solvent content of the resist should be less than a few percents. To increase the throughput it is desired to reduce the baking time by increasing the baking temperature; but, too high baking temperature may introduce cracks in thicker resists and eventually degrade the resist material. Too high temperature or too long soft bake will also decrease the development rate and contrast.

Baking can be done on hot plates or in itrogen oven (in a state of the art process that will be described in the last section a rapid thermal processing system should be used). Usually baking in nitrogen oven takes more time than hot plate, but, it can result in a better resist homogeneity. Regardless of equipment used for baking, temperature gradients on the wafer and in different parts of the equipment should be minimized.

During the baking process, water content of the resist layer reduces. While a certain amount of water in the resist layer is necessary to have a reasonable development rate. Considering the fact that the water is absorbed from ambient on the surface, some time after baking is needed to let the water to penetrate deep in to the resist layer. So, if the exposure done immediately after the baking, development rate will be different in different depths in the resist. Such an inhomogeneity will change the pattern's cross section profile (figure 9). The time needed to have a homogeneous rehydration, is varying between a few seconds to tens of minutes depending on the resist thickness and other parameters. If the humidity of the yellow room be less than 45% the resist will not be able to absorb enough water, and if the humidity of the room be more than 55% the resist adhesion will decrease by absorption of water on the substrates surface. The best condition for the yellow room is 22 °C and 45% ralative humidity.

Since most of the resist materials are sensitive to ultraviolet, UV free yellow light is used in areas of the clean room where lithography in done; so, lithography rooms are also called yellow rooms. Although the resist coated substrates can be stored for months in yellow room conditions without chemical degradation of resist material, any unnecessary elongation of the lithography process is not recomended and can reduce the homogeneity.

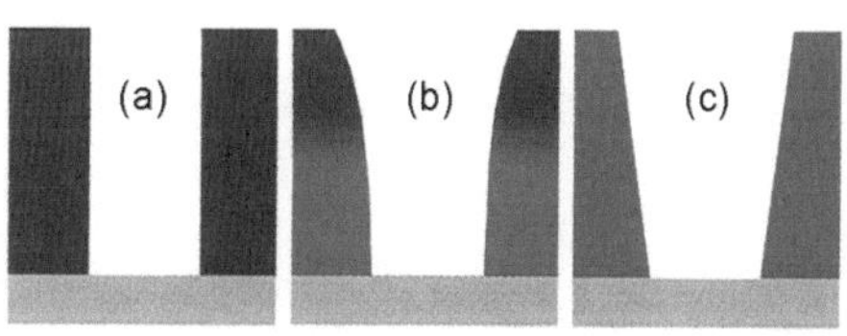

Figure 9. Effect of resist rehydration on the pattern cross section after development. a) Rehydration was complete. b) Rehydration was happened in the upper half of the resist layer. c) There was no rehydration.

7. Exposure

In order to have a homogeneous pattern in ultraviolet (UV) lithography, the exposure dose variation should be less than 10% along the wafer. In addition, the exposure should be done on an anti vibration table. The wafer and the mask should be precisely aligned to the lenses or mirrors. The exposure tool and especially the mask should be clean and free of particles. If the chromium mask has defects in the form of small holes or cracks, using lower exposure dose will reduce their effect. In other words, although increasing the exposure dose increases the contrast, it can reduce the homogeneity. Reducing the gap between the mask and the wafer will enhance the resolution as well as the homogeneity. However, this reduction usually increases the probability of mask contamination and may damage it, and shortens mask's longivity. Variation of the reflection coefficient of the substrate beneath the resist can also cause inhomogeneity in the UV exposure.

In the case of electron beam lithography it is recommended to put the samples at the centre of the sample stage directly under the column. Because, for exposing the areas away from the centre, the beam should bend more; and this bending increases the beam noise by coupling the electrical noise of the deflector; enlarges the spot size, and increases the effects of aberations. The electrical and acoustic noise should be minimized during the electron beam exposure. Turning off the lights of the lithography room, especially florescent lights during the exposure decrease the noise level. Any electrical inhomogeneity, edge or thick insulator layers on the substrate can introduce randomized beam fluctuation. This fluctuation is caused by charging, and inhomogeneous or fluctuating electrical fields near the wafer.

8. Development

Frequent opening or exposure of the developer to air leads to carbon dioxide absorption by the developer and after exhaustion of its buffer; it will result in lower development rate. In addition, by using a developer for lot of wafers, concentration of the developed resist in it will increase. Consequently, the development time will increase too. As a rule of thumb, if the resist content of a developer reaches one tenth of a percent, the development rate will decrease about ten percent. Very low development rate usually is accompanied by reduction of contrast and increase of dark erosion. Consequently it will pronounce the effect of resist thickness variations and inhomogeneities. On the other hand, if the development time be too short; reproducibility and homogeneity will reduce. Because, it will be difficult to produce a repeatable and homogeneous flow of the developer on the substrate, in a very short time compared to the development time. Also strong developers tend to roughen the resist's surface and decrease the resolution by making a thick gel like layer between the developer and the resist (these effects will be further discussed in the next section). Usually preferred development time is between 30 s to 3 min. Development time is a function of temperature, concentration and exposure dose. So, it is necessary to precisely control the temperature of the developer and keep it homogeneous by constant stirring. Development rate increases with the exposure dose to some point and then it remains constant (figure 10). Thus, to have a better repeatability and homogeneity, the exposure dose should be in the flat region of figure 10's curve. Substrate compatibility of the developer should be considered too. For example alkaline developers attack aluminum and its alloys.

http://dx.doi.org/10.5772/56618

Some times after the development the wafers are baked again. This application of high temperature after development is called hard bake. Although hard baking can increases the stability of the resist layer and its adhesion to the substrate; but, it can also result in cracks, rounding of the edges, reflow of the resist and consequently reducing the resolution.

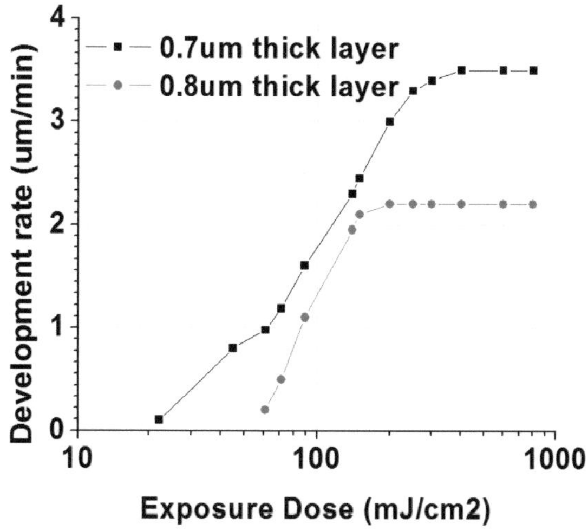

Figure 10. Development rate versus exposure dose for a UV resist (AZ 9260) that baked for 10 min at 100 °C.

9. Resist homogeneity in nano lithography

As it mentioned earlier in this chapter, the most important problem that limits the ultimate resolution of industrial lithography can be granular structure of the resist material (figure 1). The formation mechanism of these granules is:

After spin coating and during backing process, bigger polymer molecules fold on themselves; due to the attraction between the monomers on the same chain. Hence, end-to-end distance of the polymer reduces to much shorter than its length. Moreover, solvent evaporation reduces the radius of gyration. During the solvent evaporation, the big polymer molecules immobilize sooner, while the smaller molecules are still highly mobile and because of the intermolecular attractions, these small molecules penetrate to the long molecules. This aggregation makes about 20 nm to 60 nm big granules.

Because the development is much faster at the granule boundaries, granules release one by one into the developer instead of dissolving (figure 2). This effect increases the RMS of surface roughness to more than 250 pm and the RMS of line edge roughness to more than 2 nm, and determines the smallest patternable structure. An approved method to reduce this problem is a very short time thermal processing, which does not give enough time to the short polymer molecules to penetrate into the longer molecules, while using higher temperature to evaporate all of the solvent in the limited time. A detailed study of using this method to obtain a very homogeneous layer of PMMA and making patterns smaller than 5 nm by electron beam lithography (EBL) follows.

As a typical substrate, p-type <100> silicon is used in this study. Usually it is needed to pattern a thin film on the substrate. This thin film may be a part of the final device or maybe used as a mask to make patterns in other layers. Here, a thin film of ruthenium is used on the substrate. Ruthenium has high ion milling etch rate, so, it does not need a thick layer of resist to hold up the etching process; ruthenium has high conductivity, so, the charging effect during the electron beam exposure will be minimum; it has small grain size, so, it can make a thin, continuous and smooth film which will not increase the resist inhomogeneity; ruthenium has compatibility with different surfaces [24], and the possibility to be used as a mask to etch other material beneath the ruthenium. In addition, ruthenium activates C–H and C–C bonds and benzene decompose on ruthenium at 87 °C, thus it helps the solvent to outgas during baking, while there is no dehydrogenations bellow 277 °C. Thus, it will not damage the PMMA in the backing process, while PMMA has good adhesion on it. To make the ruthenium layer; after RCA cleaning of the substrate, 10 nm ruthenium sputtered on the samples. In order to improve PMMA adhesion and uniformity, samples annealed at 400 °C. Hereby, the contact angle of PMMA solution on the substrate's surface reduced to less than 5 °C.

Samples were coated with 2% 950K PMMA in chlorobenzene by spinning at 6000 rpm for 50 s with 1s acceleration and 10 s deceleration time. This long deceleration time increases the resist homogeneity. The best thermal processing for this process found to be: increasing the sample temperature from 21 to 250 °C in 1 s and keeping it in that temperature for about 15 s and cooling it down to 21 °C in 1 s (figure 11).

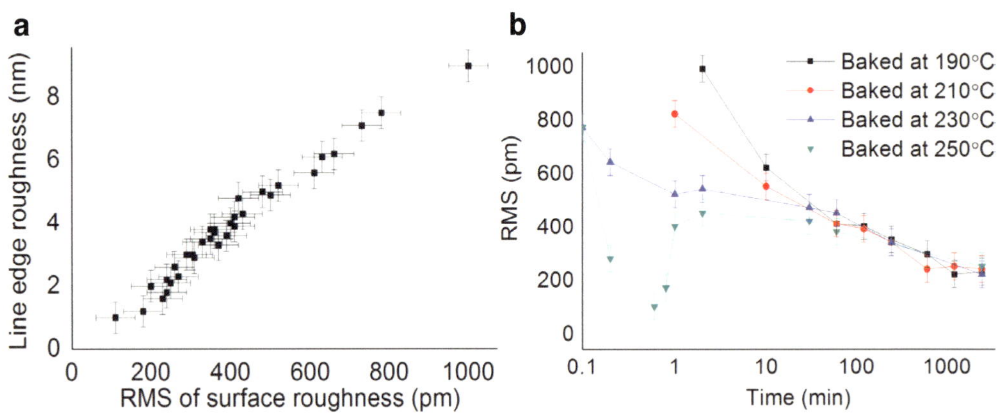

Figure 11. Effects of Different Baking Processes on Line Edge Roughness. a) RMS of surface roughness versus line edge roughness; obtained by baking the resist in different thermal processes. b) RMS of surface roughness versus baking time for different baking temperatures.

Highest resolutions and smallest patterns in industrial lithography have been achieved by using electron beam to pattern PMMA. By using PMMA as the electron beam resist; lines as thin as 5 nm [23, 3, 22] and 10 nm gaps [7, 13] have been reported. 5 nm pillars have been achieved by etching thicker pillars [5, 10]; but, without using the described method to solve the problem of resist aggregation, holes smaller than 10 nm had not achieved [8, 1, 6].

Unlike patterning lines, in patterning holes, there is a very confined space for developer to penetrate through the exposed area. In addition, there is a very small space for the dissolved resist to diffuse in to the bulk developer. So, to overcome this problem, and to overcome the intermolecular force between the unexposed walls and the exposed PMMA, ultrasonic agitation should be used.

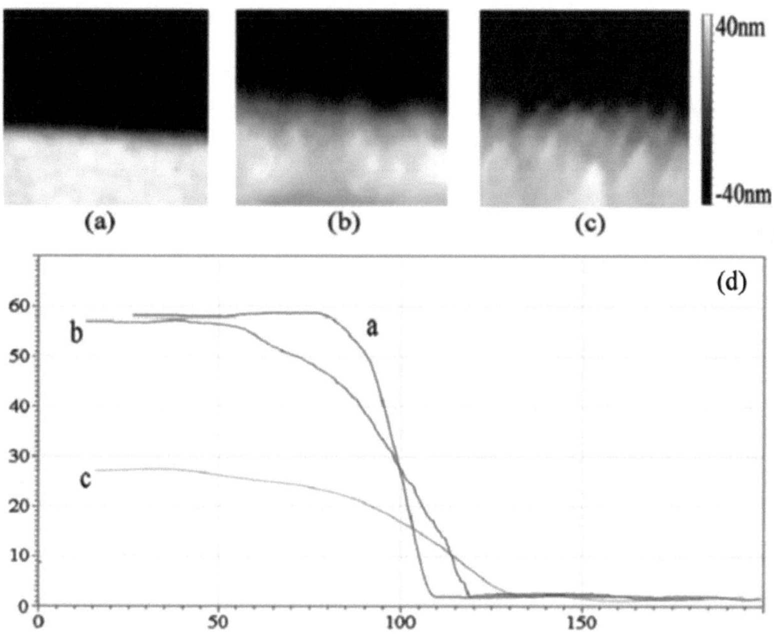

Figure 12. AFM Investigation of Effect of baking process on Line Edge Roughness. a) Line edge of a PMMA that baked for 15 s at 250 °C. b) Line edge of a PMMA that baked for 3h at 165 °C. c) Line edge of a PMMA that baked for 10 s at 300 °C. d) Cross section profile of a, b and c; axes are scaled in nanometers. Lowest line edge roughness and highest contrast are obtained in sample a, while in sample c the resist is degraded.

Other problems in patterning small holes are the gel thickness at the interface between the exposed resist and the developer (related to the radius of gyration) and developer induced swelling. To overcome these problems a weak developer, and cold development, should be used. This process is successfully used to make sub 5 nm patterns. For making such a small pattern, the only available industrial lithography process is electron beam lithography (EBL). In this study, electron beam exposure performed by Vistec VB_6HR with 50 keV acceleration voltage and 150 pA current (by increasing the acceleration voltage and decreasing the current, spot size reduces). Optimized development condition is 7:3 v:v isopropyl alcohol (IPA) in water in ultrasonic bath at 12 °C. After the development, ion milling for about 65 s at 80 mA beam current with 375 V acceleration voltage used to transfer the PMMA pattern to the ruthenium layer beneath. Then, the resist residue striped by 1, 2-dichloroethane:acetone 1:1 v:v to completely remove the PMMA without affecting the underlying surface [11].

The almost linear relation between the surface roughness and line edge roughness (Figure 11.a) substantiates a common source for both of them and the mentioned theory of PMMA granules. In samples which have been baked at temperatures lower than 230 °C, surface roughness and line edge roughness are improving by increasing the baking time (Figure 11.b). This improvement is due to reduction of solvent content in the resist by evaporation. By baking in higher temperatures for a very short time, a significant improvement in surface roughness and line edge roughness appears. By baking at 250 °C for 15 s the lowest surface roughness and line edge roughness are obtained which are about 100 pm (figure 11.b) and 1 nm (figure 12.a), respectively. By reducing the baking time, PMMA molecules do not have enough time to diffuse and form the granules, while due to the high temperature all the solvent evaporates. By further increasing the baking temperature, resist thinning and line edge rounding increases considerably, which is an indication of losing the contrast and degradation of the resist material (figure 11.b).

If ultrasonic agitation do not be used for development, small holes will not completely develope, no matter for how long they were developed (a partially undeveloped hole could not result in a highly visible object in SEM image after ion milling). Doing the development in higher temperatures takes much shorter time but in higher or lower temperatures or stronger developers (such as methyl isobutyl ketone:IPA 3:1 v:v), contrast will be much lower and pattern edge roughness will be much higher that patterning small holes will not be possible. However, by using the mentioned optimized development process, holes smaller than 5 nm were obtained on the PMMA (figure 13).

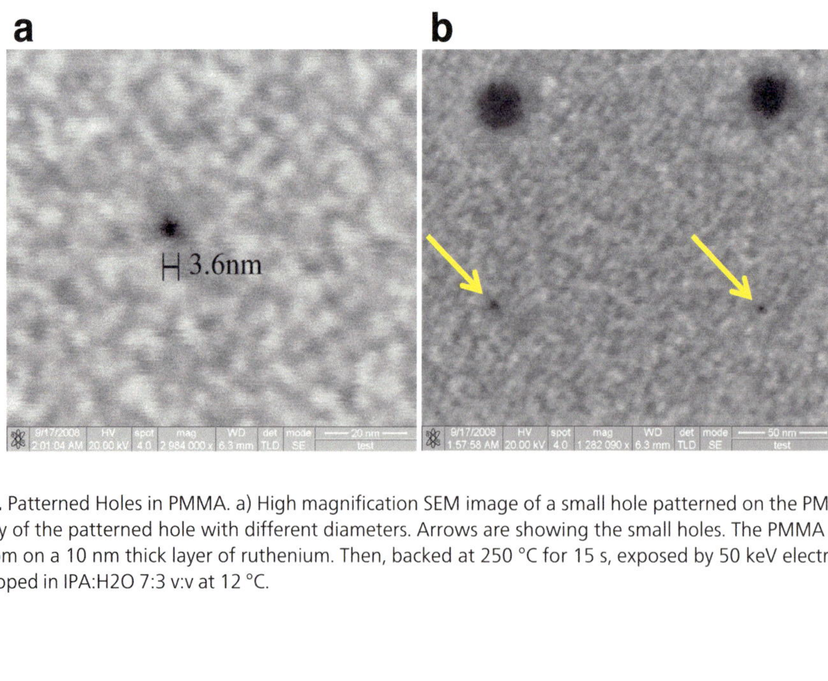

Figure 13. Patterned Holes in PMMA. a) High magnification SEM image of a small hole patterned on the PMMA layer. b) An array of the patterned hole with different diameters. Arrows are showing the small holes. The PMMA was spun at 6000 rpm on a 10 nm thick layer of ruthenium. Then, backed at 250 °C for 15 s, exposed by 50 keV electron beam, and developed in IPA:H2O 7:3 v:v at 12 °C.

http://dx.doi.org/10.5772/56618

By baking the samples at 250 °C for 15 s, the development time was more than 2 min, but for the samples baked in lower temperature for longer time, it is less than 1 min. The reason is, if the PMMA has granular structure, the development rate at the boundaries is much higher than in the granules. So, the granules release during the development. However, by using the fast baking process, resist is homogeneous. So, there is no granule boundary and it takes a longer time for the PMMA to gradually dissolve in the developer.

The resist thickness in this process is about 58 nm and before and after the development, there is no measurable change in the resist thickness. The resist can hold up for about 70 s in the mentioned ion milling process. So, it can be used as a mask for etching 20 nm of copper, 15 nm of aluminum, ruthenium, tantalum, or some other materials in ion milling. Hereby, open holes smaller than 5 nm obtained on the ruthenium without significant pattern enlargement during the etching process (figure 14). It worth to mention that the etch rate of the ruthenium layer in SF_6/O_2 plasma in typical reactive ion etching (RIE) processes is almost zero. Therefore, the ruthenium thin film can be used as a mask for RIE of silicon, silicon dioxide or nitride.

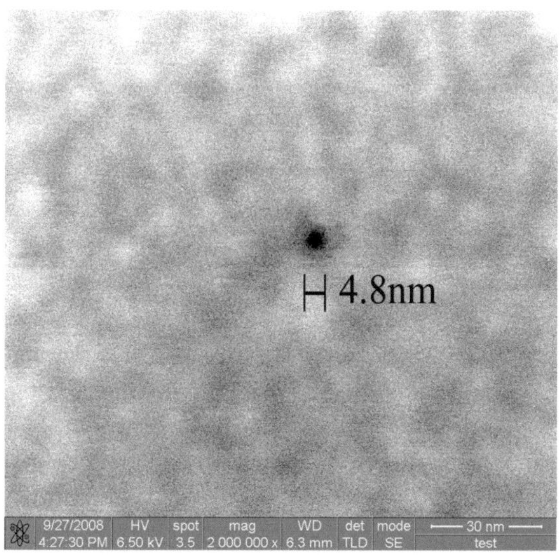

Figure 14. High magnification SEM image of a hole patterned in the ruthenium layer by transferring the PMMA pattern to it by ion milling.

10. Conclusion

Wafer level homogeneity of resist layer is crucial to minimizing product variation and it is also one of the most important factors in having a high production yield. This chapter provided the guidelines for increasing the wafer homogeneity. In order to precisely optimize the process parameters, it is necessary to perform a range of experiments in the framework of a design of experiment (DOE). The guidelines and trends provided here will be useful to find the optimum condition in a shorter time. These guidelines can also be used as a troubleshooting tool for solving common lithography problems. In the last section, nanometer scale inhomogeneity of the resist which is a consequence of polymer agglomerations, described. Overcoming the problem of these nanometer scale inhomogeneities is important in nanolithography, especially to improve the resolution and making small patterns. A state of the art process to solve this problem described in details. This process has resulted in some of the smallest patterns ever made in an industrial lithography. However, polymer resists have their own intrinsic homogeneity limitations. For atomic resolution lithography, further research in using surface monolayers or crystalline resists for making patterns; and application of atomic layer deposition instead of etching, will be valuable.

Author details

Nima Arjmandi[1,2*]

Address all correspondence to: Nima@Arjmandi.Org

1 Laser and Plasma Research Institute, Shahid Beheshti University, Evin, Tehran, Iran

2 IMEC Kapeldreef, 3001 Leuven, Belgium

References

[1] Awad, Y.; Lavallee, E.; Lau, M.N.; Beauvais, J. & Drouin, D. Arrays of Holes Fabricated by Electron Beam Lithography Combined With Image Reversal Process Using Nickel Plus Reversal Plating. *J. Vac. Sci. Technol. A* 22-3, 1040-1043 (2004)

[2] Carr, D.W.; Lercel, M.J.; Whelan, C.S. & Craighead, H.G. High-Selectivity Pattern Transfer Process for Self-Assembled Monolayer Electron Beam Resists. *J. Vac. Sci. Technol. A* 15-3, 1446-1450 (1997)

[3] Chen, W. & Ahmed, H. Fabrication of 5-7nm Wide Etched Lines in Silicon Using 100kev Electron Beam Lithography and Poly(Methylmethacrylate), *App. Phys. Lett.* 62-13, 1499-1501 (1992)

[4] Chen, W. & Ahmed, H. Fabrication of Sub-10nm Structures by Lift-Off and by Etching After Electron Beam Exposure of Poly (Methylmethacrylate) Resist on Solid Substrates.. *J. Vac. Sci. Technol. B* 11-6, 2519-2523 (1993)

[5] Chen, W. & Ahmed, H. Fabrication of High Aspect Ratio Silicon Pillars of <10nm Diameter, *App. Phys. Lett.* 63-8, 1116-1118 (1993)

[6] Craighead, H.G. 10-nm Resolution Electron Beam Lithography, *J. App. Phys.* 55-12, 4430-4434 (1984)

[7] Cumming, D.R.S.; Thomas, S.; Beaumont, S.P. & Weaver, J.M.R. Fabrication of 3nm Wires Using 100kev Electron Beam Lithography and Poly (Methylmethacrylate) Resist, *App. Phys. Lett.*.68-3, 322-324 (1995)

[8] Dial, O.; Cheng, C. & Scherer, A. Fabrication of High-Density Nanostructures by Electron Beam Lithography, *J. Vac. Sci. Technol. B* 16-6, 3887-3890 (1998)

[9] Dobisz, E.A.; Brandow, S.L.; Snow, E. & Bass, R. Atomic Force Microscope Studies of Nanolithography Exposure and Development of Polymethylmethacrylate, *J. Vac. Sci. Technol. B* 15-6, 2318-2322 (1997)

[10] Fischer, P.B.; Dai, K.; Chen, E. & Chou, S. 10nm Si Pillars Fabricated Using Electron-Beam Lithography, Reactive Ion Etching, and HF Etching. *J. Vac. Sci. Technol. B* 11-6, 2524-2527 (1993)

[11] Hang, Q.; Hill, D. & Bernstein, G.H. Efficient Removers for Poly(Methylmethacrylate). *J. Vac. Sci. Technol. B* 21-1, 91-97 (2002)

[12] Lercel, M.J.; Tiberio, R.C.; Chapman, P.F.; Craighead, H.G. Self-Assembled Monolayer Electron-Beam Resists on GaAs and SiO2. *J. Vac. Sci. Technol. B* 11-6, 2823-2828 (1993).

[13] Liu, K.; Avouris, P.H.; Bucchignano, J.; Martel, R. & Sun, S. Simple Fabrication Scheme for Sub-10nm Electrode Gaps Using Electron-Beam Lithography, *App. Phys. Lett.* 80-5, 865-867 (2001)

[14] Luurtsema, G.A. Spin Coating for Rectangular Substrates, Thesis submitted to the department of electrical engineering and computer science, *University of Berkeley* (1997)

[15] Marrian, C.R.K.; Perkins, F.K.; Brandow, S.L.; Koloski, T.S.; Dobisz, E.A. & Calvert, J.M. Low Voltage Electron Beam Lithography in Self-Assembled Ultrathin Film with the Scanning Tunnelling Microscope, *App. Phys. Lett.* 64-3, 390-392 (1993)

[16] Masu, K. & Tsubouchi, K. Atomic Hydrogen Resist Process With Electron Beam Lithography for Selective Al Patterning. *J. Vac. Sci. Technol. B.* 12-6, 3270-3274 (1994)

[17] Namatsu, H.; Nagase, M.; Yamaguchi, T.; Yamazaki, K. & Kurihara, K. Influence of Edge Roughness in Resist Pattern on Etched Patterns, *J. Vac. Sci. Technol. B* 16-6, 3315-3321(1998)

[18] Whelan, C.S; Larcel, M.J. & Craighead, H.G. Improved Electron-Beam Patterning of Si With Self-Assembled Monolayers, *App. Phys. Lett.* 69-27, 4245-4247 (1996)

[19] Yamaguchi, T.; Namatsu, H.; Nagase, M.; Yamazaki, K. & Kurihara, K. Line With Fluctuations Caused b Polymer Aggregates in Resist, *J. Photopolymer Sci. Technol.* 10-4, 635-639 (1997)

[20] Yamaguchi, T.; Namatsu, H.; Nagase, M.; Yamazaki, K. & Kurihara, K. Nanometer-Scale Line With Flactuations Caused by Polymer Aggregates in Resist Films. *App. Phys. Lett.* 71-16, 2388-2390 (1997)

[21] Youshimura, T.; Shiraishi, H.; Yamamato, J. & Okazaki, S. Nano Edge Roughness in Polymer Resist Patterns, *App. Phys. Lett.* 63-6, 764-766 (1993)

[22] Yamazaki, K. & Namatsu, H. (5-nm-Order Electron-Beam Lithography for Nanodevice Fabrication, *Jap. J. App. Phys.* 43-6B, 3767-3771 (2003)

[23] Yasin, S.; Hasko, D.G. & Ahmed, H. Fabrication of <5nm With Lines in Poly(Methylmethacrylate), *Applied Physics Letter* 78-18, 2760-2762 (2001)

[24] Arjmandi, N.; Lagae, L.; Gorghs, G. Enhanced resolution of poly (methylmethacrilate) electron beam lithography by thermal processing. *J. Vac. Sci. Technol. B* 27, 1915-1919 (2009)

[25] Arjmandi, N.; Stakenborg, T.; Lagae, L.; Borghs, G. Sub 5nm Electron Beam Lithography, *Micro Nano Engineering conference*, 2009, Ghent

Section 3

Nanoimprint

Chapter 7

Soft UV Nanoimprint Lithography and Its Applications

Hongbo Lan

Additional information is available at the end of the chapter

http://dx.doi.org/10.5772/56186

1. Introduction

Large-area nanopatterning technology has demonstrated high potential which can significantly enhance the performance of many devices and products, such as LEDs, solar cells, hard disk drives, laser diodes, display, etc [1]. For example, nano-patterned sapphire substrates (NPSS) and photonic crystals (PhC) have been considered as the most effective approaches to improve the light output efficiency (internal quantum efficiency and external quantum efficiency) of LEDs and beam shaping [2,3]. The solar cells with sub-micro anti-reflective coating exhibited higher photocurrent and higher power conversion efficiency compared to those without nanostructures [4]. Moreover, the ability to produce large-area micro- and nanostructures on non-planar surfaces is of importance for many applications such as optics, optoelectronics, nanophotonics, imaging technology, NEMS, and microfluidics [5]. However, creating large-area nanostructures on curved or non-planar surfaces are extremely difficult using existing patterning approaches. Furthermore, a variety of existing nanopatterning technologies such as electron beam lithography (ELB), optical lithography, interference lithography (IL), etc., cannot cope with all the practical demands of industrial applications with respect to high resolution, high throughput, low cost, large area, and patterning on non-flat and curved surface. Therefore, new high volume nanomanufacturing technology strongly needs to be exploited and developed so as to meet the tremendous requires of rapid growing markets.

Nanoimprint lithography (NIL) has now been considered as a promising nanopatterning method with low cost, high throughput and high resolution, especially for producing the large-area micro/nano scale patterns and complex 3-D structures and as well as high-aspect-ratio features. Due to these outstanding advantages, it was accepted by International Technology Roadmap for Semiconductors (ITRS) in 2009 for the 16 and 11 nm nodes, scheduled for industrial manufacturing in 2013. Toshiba has validated NIL for 22 nm and beyond. NIL has also been listed as one of 10 emerging technologies that will strongly impact the world by

MIT's Technology Review. The resolution potential has been demonstrated by the replication of 2.4-nm features. It is expected to play a critical role in the commercialization of nanostructure applications [6-8].

Compared to other NIL processes (thermal NIL or hot embossing, UV-NIL with rigid mold) and nanopatterning methods, soft UV-NIL using a flexible (or soft) mold has been proven to be a very promising approach for making large-area patterns up to wafer-level in the micrometer and nanometer scale, fabricating 3-D micro/nano structures and high-aspect-ratio features, especially producing large-area patterns on the non-planar surfaces even curved substrates at low-cost and with high throughput. Since the soft mold (stamp, template) is adopted, the soft UV-NIL process has some unique advantages compared with the traditional UV-NIL with rigid mold. These strengths include: (1) Cost reduction. Cheap soft molds can be easily replicated from one expensive master, significantly reducing cost of the master template fabrication. (2) Conformal contact. Conformal contact between the undulated (or curved, waviness, bow, warp) substrate and the mold can be achieved over large areas without applying high external pressure. (3) Insensitive to particle contaminants. Particle contaminants are less problematic as the soft mold can locally deform around a particle avoiding damage to the mold or substrate which lead to improve the yield of the process and to extend the application fields. (4) Avoiding anti-adhesive layer due to the low surface energy of flexible mold materials. (5) Low imprinting and demolding force. (6) Utilizing gradually sequential micro-contact and "peel-off" separation method for thin film type molds. However, soft UV-NIL process has also some inherent drawbacks. (1) Deformation and distortion of soft molds. Due to the relatively low Young's modulus, the deformation of soft molds under pressure remains a major issue which limits the resolution, uniformity and reproducibility of imprinted patterns. High aspect ratios structures and dense patterns are not stable and tend to collapse. (2) Poor dimensional stability. Due to the poor solvent resistance as well as deformation of pressure and thermal expansion, the dimensional stability of imprinted patterns is determined and degraded. (3) Short mold lifetime. Since the hardness and resistance to solvent are poor, soft molds have relatively low mold lifetime. These limitations must be solved and overcame for extensive applications [8-13].

Currently, full wafer imprint up to 300mm, and 12.5nm resolution patterns have been achieved by using the soft UV-NIL. Soft UV-NIL has been considered as one of the most promising solution implementing mass production of micro/nanostructures over large areas at low cost for the applications in compound semiconductor optoelectronics and nanophotonic devices, especially for LED patterning [2, 14].

As an emerging cost-effective nanopatterning technique, soft UN-NIL involves two basic aspects: fundamental investigation and application research. The fundamental investigation comprises of theoretical basis and key enabling techniques including process, mold (material, fabrication of working stamp and master template), material (resist, functional material, etc.), tool. The application research mainly covers a variety of practical applications suitable for soft UV-NIL, such as LED patterning, optical components, nanophotonics, biological applications, etc. Author proposed an infrastructure of soft UV-NIL, as shown in Figure 1.

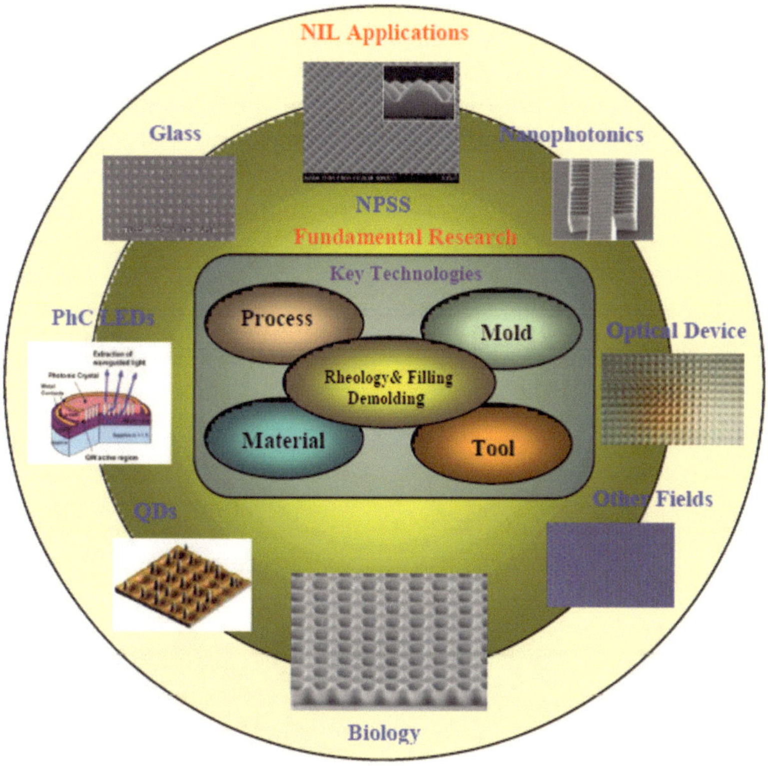

Figure 1. An infrastructure of soft UV-NIL process

2. Principle and process of Soft UV-NIL

2.1. Principle and process flow

The whole process flow of forming micro/nanostructures by soft UV-NIL is composed of four steps: the fabrication of a master template, the replication of a soft mold (or working mold) by this master template, the imprinting in the UV curable resist using the replicated soft mold, and the replicated patterns transfer form UV curable resist to the substrate or functional materials by etching or lift-off process. Together, these steps affect the quality of the final replica in terms of resolution, uniformity, fidelity, patterning area, and line edge roughness.

A master template is firstly fabricated by EBL, IL or other patterning technologies (e.g. block copolymers, AAO, FIB, photolithography, etc.). Then, the surface of the master is treated forming an anti-adhesive layer. The liquid mold material is spin coated or casted the master template to duplicate a patterning layer. Subsequently, a backplane or a flexible layer is bonded to the patterned layer. After cured thermally or UV curing, the soft composite mold is peeled off form the master template. The soft mold obtained is a negative copy of the master template.

The fabrication process of soft UV-NIL includes four steps (as shown in Figure.2): (a) Firstly, a UV-curable resist, which is liquid at room temperature, is spin-coated or dispensed on the substrate. (b) Subsequently, the soft mold is pressed into the resist on the substrate with a low pressure, and adjusts to the waviness or curvature of the substrate until completely conformal contact is achieved. Due to high flexibility, the soft mold can well adapt its shape to the waviness of the substrate obtaining good conformal contact between the non-flat substrate and the mold. (c) After filling all cavities or trenches of the mold, UV curing that solidifies the liquid resist due to cross-linking is carried out by a UV light source. (d) Finally, the soft mold is released, leaving the UV-curable resist patterned. Low viscosity UV-curable resist, commonly comprised of a low molecular weight polymer and photoinitiator, is significantly essential for easily filling the nanocavities of the mold [8-9, 15].

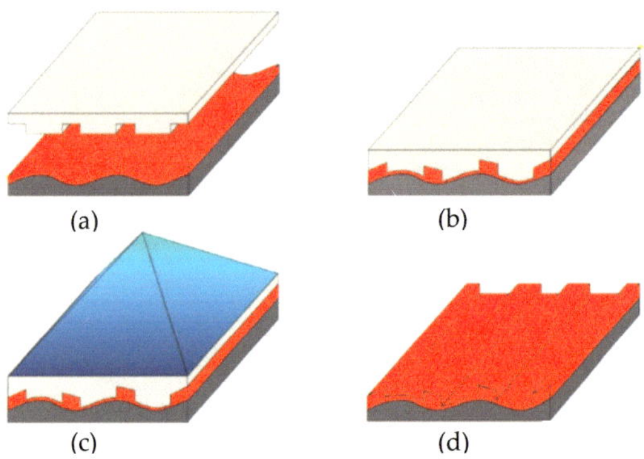

Figure 2. Process flow of soft UV-NIL [15]

The removal of the residual layer is typically performed by using a reactive ion etching (RIE) process with an oxygen plasma. The molded resist can then serve as a mask for further processing steps or be used as a functional layer itself. The most common methods for further processing are using the structured resist as an etch mask for patterning the substrate or as a mask for a lift off process for structuring of functional materials like metals. Besides using the structured resist as a mask for further processing, the resist pattern can also be used directly as a functional layer [16, 17].

2.2. Variations of soft UV-NIL

Some variations of soft UV-NIL have been proposed and developed, e.g., SCIL (Substrate conformal imprint lithography) and UV-enhanced SCIL developed by Philips and SUSS, Soft Molecular Scale Nanoimprint Lithography (SMS-NIL) developed by EVG, and full wafer soft UV-NIL, etc. Verschuuren *et al.* [1, 14, 18-19] proposed substrate conformal imprint lithography (SCIL), which combines the advantages of a soft composite mold for large area patterning with the advantages of a rigid glass carrier for low pattern deformation and high resolution. SCIL is based on a combination of the sequential imprinting method and the sol-gel resist. Figure 3 presented the schematic illustration of the SCIL imprint and separation sequences. To achieve a substrate conformal contact between a working mold and a substrate, SCIL process relies on a sequential imprinting process. The approaching of the flexible mold starts from one side and spreads to the whole mold subsequently by releasing the vacuum in the grooves step by step and applying a small over pressure of 20 mBar on the mold (step 1 to 3). After conformal contact over the entire substrate is carried out, the imprint resist is cured by UV exposure or in case of using imprint sol-gel based resists diffusion of the sol-gel solvent into the PDMS mold (step 4). The automatic separation of the mold from the substrate is performed by switching on the vacuum in the grooves consequently, which is opposite to the imprint process (step 5 to 7). This results in a low force peeling action which removes the stamp from the patterned resist layer and avoids damage to mold or transferred patterns. SCIL has demonstrated sub-10nm resolution over 150mm-diameter substrates. The SCIL technology can well cope with non-ideal substrates and implement full wafer imprinting in a single step. For the SCIL process, the curing of sol-gel resist relies on the diffusion of solvents into the PDMS mold. Depending on the operation and preparation conditions, the curing time varies from 5 to 15 minutes. In order to improve throughput (reducing curing time) and repeatability of the process, a UV enhanced SCIL process using UV curable resist has been developing. Fader *et al.* recently introduced UV-SCIL with purely organic UV-curing materials showing curing times of 17s [20]. The excellent performance of SCIL in respect to substrate conformity and pattern fidelity over large areas makes this imprint technology a powerful tool, especially for applications like LED/VCSEL, optical elements or patterned media [21].

Author developed a full wafer soft UV-NIL with a tri-layer composite mold. The composite mold includes a thin layer of fluoropolymer-based material as the patterning layer, a thick layer of *s*-PMDS as intermediate flexible or cushion layer, and a thin glass sheet as the support layer. Figure 4 illustrated the schematic diagram of the proposed full wafer soft UV-NIL process. The imprinting process is performed by a sequential and micro-contacting solution starting from the center to two sides of the mold. The separation process employs a continuous 'peel-off' demolding mode starting from two sides to the center of the mold. Compared to the SCIL, the distinct advantages of the process include: (1) The imprinting and demolding procedure take the mold center as axis of symmetry, are carried out at the same time on two sides with higher throughput and easier eliminating trapped air bubble. (2) An enhanced demolding approach is adopted. (3) Since the imprinting procedure is performed under a low vacuum pressure environment, it can better remove the trapped air bubbles and provide completely conformal contact [22].

There is a big difference for the imprinting and demolding mode between soft UV-NIL using flexible mold and NIL with rigid mold. Using gas-assisted imprinting method can easily achieve uniform distribution of applied pressure over the entire substrate, gradually and exactly load imprint force. The sequent contact mechanism prevents the flexible mold from trapping air bubbles and therefore ensures that the mold follows exactly the undulating topography over whole substrate surface. A combination of the sequent micro-contact and gas-assistedmprinting can ensure to achieve uniform pressure and good conformal contact, avoid the trapped air bubble defects, reduce the deformation of soft mold. The peel-off

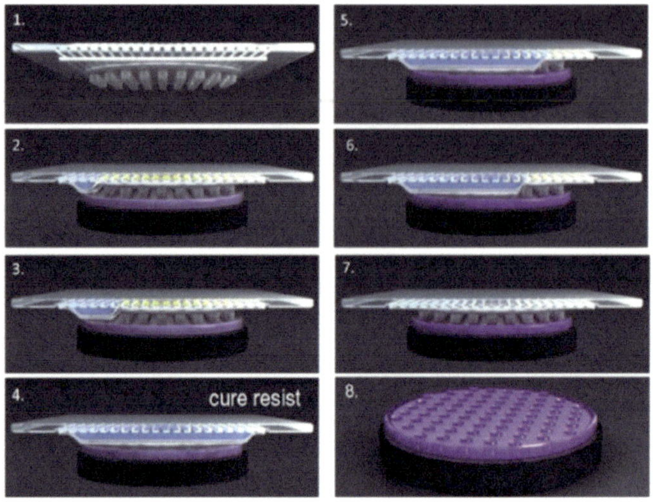

Figure 3. Schematic of substrate conformal imprint lithography (SCIL) [1]

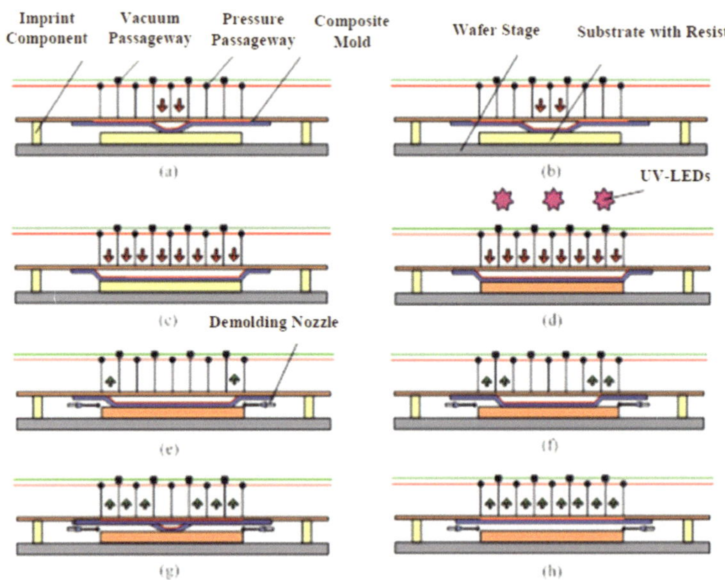

Figure 4. Schematic of a full wafer soft UV-NIL process for non-ideal substrates

demolding approach can be utilized which results in a low separating force that avoids damage to the patterned resist as well as to the mold, and improving the mold lifetime [18, 23].

In order to satisfy the demands of a variety of practical applications, lots of new soft UV-NIL processes have been being proposed and developed. Here only presents some principal and typical variants of soft UV-NIL processes. As the progresses in the soft UV-NIL and continu-

http://dx.doi.org/10.5772/56186

ously growing application needs, much more innovative processes or methods regarding soft UV-NIL will emerge in the future.

3. Various types of soft molds

The flexible mold is the most important elements for soft UV-NIL. The performance of flexible mold has a decisive effect on the soft UV-NIL in term of resolution, patterning area, throughput, uniformity of the imprinted patterns and the residual layer, and reproducibility of imprinted structures.

Since the soft UV-NIL was introduced, various structural types or configurations have been developed and employed. The most common type of the soft mold is a bi-layer structure which is composed of a rigid glass backplane for mechanical stability (optional) and a patterned soft PDMS (the commercially available PDMS brand Sylgard 184, also known as soft-PDMS or *s*-PDMS from Dow Corning Inc.) imprint layer that adapts perfectly to the waviness or bow of a substrate [12, 24-26], illustrated in Figure 5(a). However, due to the low Young's modulus, high viscosity and swell problem of *s*-PDMS, hard-PDMS (*h*-PDMS) which has higher Young's modulus (a Young's modulus of around 9 MPa) and lower viscosity, has been developed as patterned layer material. Figure 5(b) illustrated such a composite mold, in which a thin *h*-PDMS layer with relief structure is supported by a thick layer of *s*-PDMS. The thin layer of *h*-PDMS is able to ensure a good replication of the nanostructures due to his higher Young's modulus, the thick commercial *s*-PDMS top layer maintains a global flexibility of the whole mold allowing perfect conformal contact even for a non-flat substrate at low imprint pressure [27-28]. Figure 5(c) showed a tri-layer composite flexible mold which includes a thin glass backplane, a soft PDMS flexible layer and a *h*-PDMS patterned layer. Although the *h*-PDMS stamp worked well for replication, there were still limitations in using *h*-PDMS as a patterning layer, such as cracking during the mold fabrication step and degraded conformal contact with the substrate compared with the PDMS material normally used. *X*-PDMS, which has a higher Young's modulus up to 80 MPa, the highest cross linking density and lower diffusion, has been developed by Philips and SUSS as a patterned layer for the tri-layer mold for the SCIL and UV enhanced SCIL applied. The composite mold includes a *X*-PDMS patterned layer, a low modulus PDMS intermediary layer, a thin glass sheet, as shown in Figure 5 (d). The in-plane stiffness of the mold avoids pattern deformation over large areas, while out-of-plane flexibility allows conformal contact to underlying surface features [18, 29]. In Figure 5(e) offers the fifth structure of the flexible mold. The first layer is a 2-mm thick cushion layer cast from PDMS. The second layer is a thin flexible plate attached to the PDMS cushion using plasma-activation, which results in an irreversible strong bond between the glass and the PDMS. The third layer is a thin layer of *h*-PDMS spin-coated onto the master. One of the advantages of this composite mold is the reduction of the lateral deformation since the patterned layer is closely anchored to the rigid glass plate [12].

In order to improve Young's modulus of a patterned layer for enhancing the resolution of replicated pattern, PMMS is also used to be a structured layer material. A tri-layer mold configuration was proposed which consists of a rigid carrier, a PDMS buffer layer (Young Modulus $E = 0.75$ MPa) and a PMMA patterned layer ($E = 5.2$ GPa), as shown in Figure 5 (f). The introduction of a thin layer of hard material over the PDMS buffer ensures a good hardness for stamping all types of patterns and a reduced pattern deformation over a large area because of the flexibility of PDMS buffer layer. In order to overcome the cracks and fractures during the imprinting and detaching process for high modulus elastomers, a UV cured rigid polymer has been used as the patterning layer. The hybrid mold composed of a thin (100-200 nm) photocured feature layer and the thick (~2 mm) elastic PDMS support. An interpenetrating polymer network was formed between the interface of the UV-cured rigid patterning layer and the flexible PDMS substrate to provide excellent adhesion of the two distinct materials [30].

Due to the excellent properties of fluorinated materials for an flexible mold used in soft UV-NIL such as low surface energy, excellent mechanical property, high UV-transparency, low viscosity, high chemical durability, thermal stability, etc., fluorinated materials (e.g., PTFE, ETFE, Teflon,) have been considered as the most promising candidate material of patterning layer. Figure 5 (g) and (h) demonstrates a composite mold including a thin layer of PFPE (*a*-PFPE) as the patterning layer and a thin, flexible, polyethylene terephthalate (PET) sheet as the flexible backing layer. The mold has the ability to fabricate high dense and sub-20 nm feature patterns without cracking or tear-out defects that typically occur with high modulus elastomers [31-33].

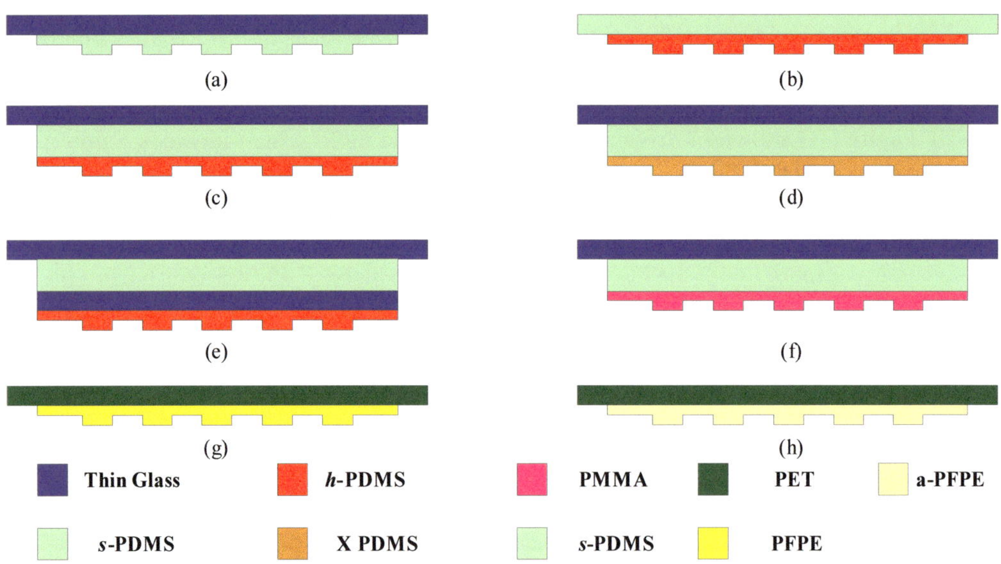

Figure 5. Various types of soft molds used by soft UV-NIL

Based on the presentations and analysis above, we can generalize a structural model of a common soft mold, as shown in Figure 6. It commonly comprised of three layers: a patterning layer, a buffer layer, and a support layer. The fundamental properties of the patterning layer involve the low surface energy, excellent mechanical property, high UV-transparency, low

http://dx.doi.org/10.5772/56186

viscosity, and thermal and chemical stability. The buffer layer is an intermediate cushion layer which has high flexibility to ensure the intimate conformal contact between the non-flat substrate and the mold. The support layer has mainly twofold: avoiding the lateral dimension of the mold and attaching or fixing the stamp to holder (or chuck). High Young's modulus materials with proper flexibility commonly need be adopted, such as a thin glass backplane or a PET sheet. Of course, there are more variations following the basic configuration. The composite mold should have global flexibility and local rigidity for achieving high resolution patterns over large areas.

Figure 6. Structural model of a common soft mold

A flexible thin-film (or membrane) can provide better conformal contact with the non-flat substrate (or large area surface) to be patterned without applying large pressure during the imprinting. A further benefit is that demolding can be achieved by peeling the mold from the substrate with an effectively smaller demolding area and low release force. This is much easier than the detaching of a rigid mold, where the hard mold needs to be separated from the substrate as a whole in a parallel way. Therefore, apart from the structure and material used of a flexible composite mold, the thickness of each layer of the mold is also be optimized.

4. Materials used and fabrication methods for a variety of soft molds

Apart from the structural style discussed in Section 3, the performances and capabilities of soft molds are largely dependent on the properties of the materials used. Nowadays, a wide variety of materials have been utilized to fabricate flexible molds [7, 8, 10]. This section primarily discusses the various material used for flexible molds.

4.1. PDMS-based materials

4.1.1. s-PDMS

Up to now, PDMS is undoubtedly the most widely used soft mold material. There are several reasons that PDMS emerged as a kind of standard material for the soft mold. It has a low Young Modulus and low surface energy that allows for conformal contact and easy release from both a master template and imprinted patterns. It is a durable material with fair chemical resistance and has good optical transparency down to a light wavelength of approximately 256 nm. It is a relatively tough material with a high elongation at break (> 150%) that allows for significant deformation before failure during patterning conditions. It can be easy handled and has high gas permeability. Most importantly, *s*-PDMS (Sylgard™184 from Dow Corning) is commercially available in kits that allow for inexpensive fabrication of flexible molds from polymer precursors. However, there are some inherent disadvantages to *s*-PDMS which severely limit its capabilities in soft UV-NIL. The low Young Modulus below to 2.0 MPa limits the replication of both the high-density and high resolution features and is also a detriment for forming high-aspect ratio structures as fabricating such features will be apt to collapse, merge, or buckle. Moreover, *s*-PDMS tends to absorb organic solvents and monomers. This leads to fluctuations in the resist composition and swelling issue of the mold. This becomes a serious problem when trying to pattern certain biological materials or for the fabrication of functional nanostructures with controlled surfaces. Its poor solvent resistance has a serious effect on reproducibility due to degradation in the course of patterning repeatedly. Its high elasticity and thermal expansion can lead to deformation and distortions during the fabrication can result in loss of critical dimensions. The surface energy of *s*-PDMS is not low enough to duplicate profiles with high fidelity. In addition, the *s*-PDMS remains its high viscosity (for Sylgard 184, this is 3900 mPa s). As a consequence, 3D nanostructures and fine features of a master mold are difficult to be fully filled which results in a loss of feature height and some defects and uniformity of replicated patterns. Besides, the dimensional change (i.e., thermal shrinkage) due to thermal expansion after thermal curing makes it difficult to apply it for large-area mass production, especially multilevel pattern registration over a large area [10, 12, 13, 26, 34-37].

Simple PDMS molds are typically replicated by first mixing two commercial PDMS components: 10:1 PDMS RTV 615 (part A) siloxane oligomer and RTV 615 (part B) cross-linking oligomers (General Electric). The mixture is then casted on a master template and degassed in a low pressure vacuum chamber. A curing time of 24 h and a curing temperature of 60 °C are usually recommended in order to reduce roughness and to avoid a build up of tension because of the thermal shrinkage. The replicated mold is left to cool to room temperature, carefully peeled off from the master template. In the subsequent fabrication step the unstructured backside of the PDMS layer was treated with oxygen plasma for 30 s in order to temporarily increase the surface energy and thereby ensure a stable bonding of the PDMS layer to the quartz or glass backplane. Sometimes, the mold is also treated with silane based anti-sticking treatment to further reduce the low PDMS surface energy. During the fabrication, curing times and temperatures, which have large effect on the elastic modulus and hardness of the replicated soft mold, must be exactly controlled. The vacuum degree in degassing phase is also a key process parameter. The previous results have indicated that a higher vacuum degree and a lower temperature are the favorable conditions to get better comprehensive physical properties of a PDMS-based soft mold [13, 26].

4.1.2. h-PDMS

An obvious way to overcome unwanted deformations is to use a mold with a higher Young's modulus. Therefore, there have been attempts to utilize materials of high Young's modulus for flexible molds. *h*-PDMS was developed at IBM as early as 2000 [36]. They tried to formulate a better imprint material by trying different combinations of vinyl and hydrosilane end-linked polymers and vinyl and hydrosilane copolymers, with varying mass between cross-links and

http://dx.doi.org/10.5772/56186

junction functionality. A nanoimprint resolution record of 2 nm has been achieved by using soft molds based on *h*-PDMS [37]. Viheriälä *et al.* described a formulation of *h*-PDMS [12].

The procedure for producing a bi-layer *h*-PDMS/*s*-PDMS mold is illustrated in Figure 7. The *h*-PDMS prepolymer, prepared by referring to the formulation in [12], is spin coated onto a Si master template fabricate by EBL at 5000 rpm for 30 sec and then degassed in vacuum for 10 min (the thickness of the *h*-PDMS is about 5-8μm). A mixture of conventional *s*-PDMS (1:10) is then casted the spin coated *h*-PDMS layer curing at 60 °C for 24 hours, the bi-layer soft mold can then be peeled off from the master template and it is treated with trichloromethylsilane (TMCS) [13].

In addition, Pei *et al.* presented a *h*-PDMS-based tri-layer composite mold and its fabrication process for nanostructured amorphous silicon photovoltaics [38]. Although the *h*-PDMS-based mold worked well for replication, there are still some limitations for its applications: (1) releasing it from the master caused cracking across the face of the mold; (2) external pressure is required to achieve conformal contact with a substrate, which created long-range, nonuniform distortions over the large areas of contact [39, 40].

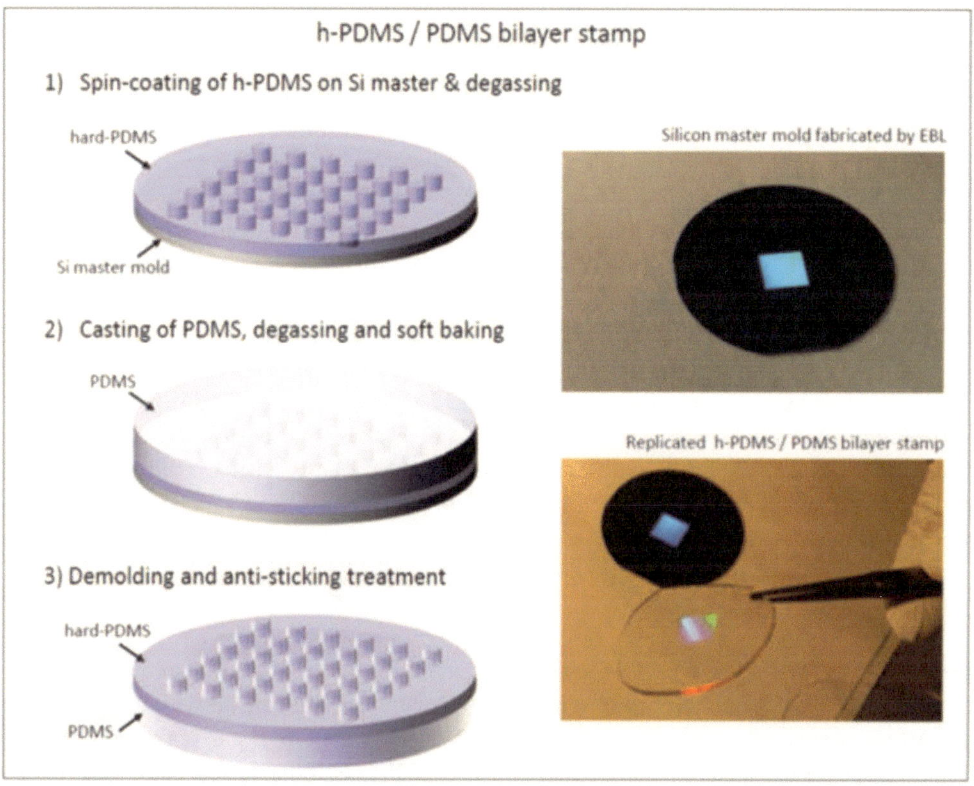

Figure 7. Schematic of the fabrication process of a *h*-PDMS/*s*-PDMS bilayer composite mold and a pictures of a Silicon master template and of a replicated soft mold after peeling off [13]

4.1.3. X-PDMS

Phlips and SUSS developed a new high modulus silicone rubber (*X*-PDMS) which is made from combination of vinyl-modified linear di-methyl-siloxanes and vinyl-modified quaternary siloxanes. The latter component increases the intrinsic crosslink density in the rubber and thereby the Young's modulus. The mixture is cross linked with hydride modified linear siloxanes using a platinum catalyzed vinyl-hydride addition reaction. By changing the linear to quaternary siloxane ratio, they synthesized rubbers with Young's Modulus up to 80 MPa. The rubber material with the highest attained stiffness allows the faithful replication of dense sub-10 nm features while still providing conformal contact over a full wafer [18, 29].

4.1.4. hv-PDMS

Conventional PDMS requires a long, thermal cure under pressure which is very time consuming and can take many hours. Photocurable PDMS (*hv*-PDMS) can overcome deformations associated with thermal curing of conventional PDMS. *hv*-PDMS also provides a tensile modulus higher than that of *s*-PDMS and an elongation at break that is much higher than that of *h*-PDMS, which makes it easier to handle than *h*-PDMS and to be less susceptible to mechanical deformation. Choi and Rogers developed a photocurable PDMS system [41]. *hv*-PDMS has been used as a mold material to successfully replicate 300 nm width by 300 nm spacing by 600 nm height lines which could not be replicated in either *s*-PDMS(feature collapse failure) or *h*-PDMS (fracture failure).

4.1.5. Diluted PDMS

The main drawback of conventional *s*-PDMS materials is the high viscosity, which is for Sylgard 184 (Dow Corning) 3900 mPa.s. It is extremely difficult for high viscous s-PDMS to fill the nanocavities of the master template for fabricating high resolution soft molds. Thus, the resolution of the soft mold is limited by an inappropriate material flow for pattern geometries within the sub-100 nm regime. To overcome this problem, using triethylamine, toluene and hexane as solvent to low the viscosity of *h*-PDMS has been reported and demonstrated the imprinting of 75 nm lines with a pitch of 150 nm. koo *et al.* have reported an improved mold fabrication process using Sylgard 184 diluted with toluene. Dots with a resolution of 50 nm are well replicated and an excellent imprint homogeneity across a 4 inch wafer with one imprint step only has been demonstrated [42].

4.2. Fluorinated polymer materials

Fluorinated polymer offers an ideal material for soft molds. Compared to other materials used by soft molds, Fluorinated polymer materials have many outstanding advantages: extremely low surface energy, suitable Young's modulus (10Mpa~2GPa), high gas permeability, good mechanical strength, solvent resistance, chemical stability, visible transparency, and tunable modulus. These characteristics open up the possibility of fabricating high performance flexible molds for soft UV-NIL to pattern a wide variety of nanostructures and materials for real applications [43,44].

4.2.1. Perfluoropolyether (PFPE)

Recently perfluoropolyether (PFPE) and its derivatives have gained popularity as flexible mold materials. PFPE is the main component of fluoropolymer that is composed of only carbon, fluorine and oxygen [44]. The pre-polymer mixture is a liquid at room temperature, and can be cross-linked under UV exposure or thermal annealing. The PFPE-based materials contain several distinctive properties such as high chemical resistance, extremely low surface energy, high gas permeability, high solvent resistance, high elastic recovery and good mechanical strength. In particular, the modulus of the elastomer can be easily tuned by changing the molecular weight between crosslinks of the precursor allowing for higher fidelity molds, and are formed via UV curing in several minutes [45-49].

4.2.1.1. HPFPE

HPFPE is a copolymer resin of a perfluoropolyether (PFPE) and a hyperbranched polymer (HP). The HPFPE exhibited low viscosity, high Young's modulus, and lower surface energy, and enhanced stability. The low surface energy of the HPFPE resist resulting from the acrylic acid functional groups on the backbone of the fluorinated polyether precludes any adhesion of the polymer to the mold. By adjusting the weight percent of the multifunctional HP and diluter, 1, 6-hexanediol diacrylate (HDDA), it is possible to modify the viscosity of the obtained HPFPE copolymer resist and yield a HPFPE copolymer resist with a high Young's modulus. By optimizing the soft UV-NIL process with this flexible HPFPE mold, the imprinting results exhibited near zero residues at the bottom of the resist grooves, and no sticking over a large area, and patterning a 50 nm linewidth and a 200 nm period [50].

4.2.1.2. a-PFPE

Unlike certain PFPE formulations described earlier, which have a Young's modulus of around 4 MPa, the Young's modulus of acryloxy PFPE (denoted as *a*-PFPE) is 10.5 MPa. The low surface energy (~18.5 mN m^{-1}) as well as the chemical inertness of *a*-PFPE eliminates the necessity of treating the surfaces of the masters with fluorinated silanes to avoid sticking during casting and curing of the mold. Figure 8 illustrated the fabrication of a composite *a*-PFPE mold using backing layers of *s*-PDMS or polyethylene terephthalate (PET). The photocurable fluorinated acrylate oligomer CN 4000 is mixed with 0.5 wt % of a photoinitiator (Darocurr 4265, Ciba Specialty Chemicals) and filtered through a 0.22 μm syringe filter. A thin layer (~2 μm) of the *a*-PFPE resin is spin coated (4000 rpm, 30 s) on the master and cured in UV light (350–380 nm, 4 mW cm^{-2}) under nitrogen purge for 2 *h*. Then a layer (~4 mm) of Sylgard 184 PDMS (*s*-PDMS) is poured onto the *a*-PFPE layer and cured at room temperature for ~48 *h* or at 65°C for 2 *h*. The *a*-PFPE/*s*-PDMS composite mold is peeled off from the master [51]. The high resolution capabilities of *a*-PFPE, together with other properties such as resistance to swelling, chemical inertness, and photocurability make it a promising alternative to PDMS for soft UV-NIL.

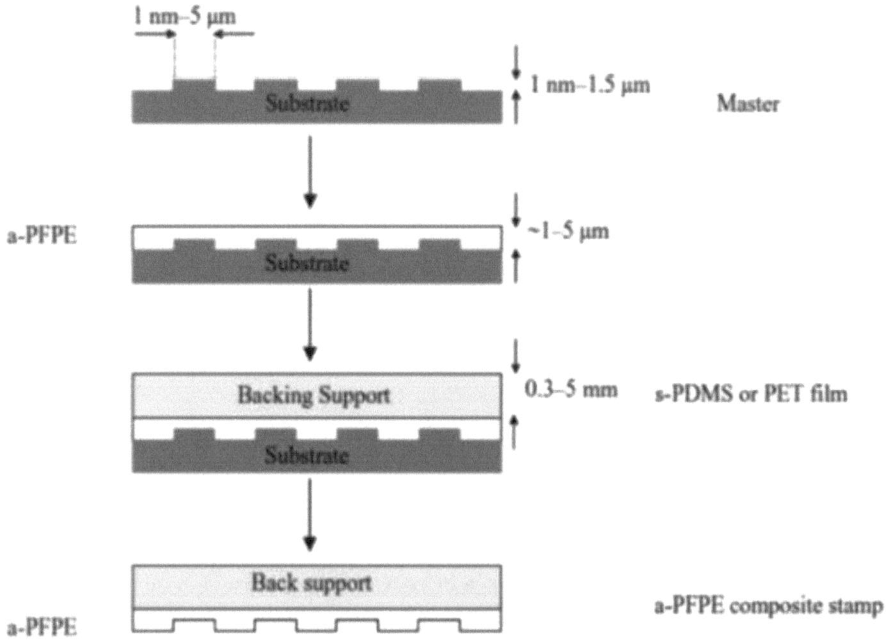

Figure 8. Fabrication of a composite a-PFPE mold with either s-PDMS or PET films as backing supports [51]

4.2.1.3. PFPE-DMA (Photochemically curable fluoropolymer)

The a-,w-methacryloxy functionalized PFPE (PFPE-DMA) is a photochemically curable fluoropolymer. A commercially available hydroxy-terminated PFPE (Solvay Solexis, *Mn*=3800 $gmol^{-1}$) with isocyanato ethyl methacrylate (Aldrich) is used to yield a methacryloxy-functionalized polymer which is a pourable liquid of low viscosity (0.36 Pas) at room temperature. A photoinitiator, 1-hydroxycyclohexyl phenyl ketone (1 wt%, Aldrich), is dissolved into the PFPE-DMA to make a photocurable liquid resin. The PFPE-DMA based mold involves the following features: easy fabrication, the ability to make conformal contact, remarkably low surface energy, resistance to swelling by small organic molecules, and enduring multiple printing procedures. It is able to replicate sub-100-nm sized features with no indications of limits to going to even smaller in size [52].

4.2.2. Ethylene(Tetrafluoroethylene) (ETFE)

The ETFE is a copolymer of ethylene and tetrafluoroethylene. It has some outstanding properties: an exceptional toughness and flexibility, a relatively high stiffness (elastic modulus ~ 1.2 GPa), a high thermal stability (a high melting point in the range of 255–280 °C). ETFE has superior mechanical properties compared to Teflon AF in pressure assisted imprinting at high temperatures. The flexibility and low surface energy of ETFE provide a clean mold release without fracture or deformation of the embossed structures. Patterning over large areas is made possible because of the good conformal contact originating from the high flexibility and low-adhesion of the ETFE mold. ETFE-based molds also provide a high fidelity of reproduction. ETFE is transparent in the visible and in the UV spectrum (91–95% transmittance in the 200–800 nm range for a 25μm sheet), it can be used to structure a photocurable polymer and to solidify it by irradiation with UV light at room temperature, providing a cheap alternative to quartz or other transparent inorganic mold materials.

Combined with the advantage of their low adhesion and flexibility, ETFE molds can therefore be used to replace brittle and expensive inorganic materials, as well as deformable elastomers and other plastic molds which lack strength when pressure is applied. The strength and robustness of ETFE molds is essential for embossing large microstructures as well as small nanostructures with lateral dimensions approaching 10 nm. Some commercial ETFE products including Fluon ETFE, Tefzel® ETFE resin, have been directly utilized [53].

4.2.3. Teflon

Teflon AF 2400 form Dupont has been adopted as a typical rigiflex mold material. Teflon AF 2400 is a copolymer of 2,2-bistrifluoromethyl- 4,5-difluoro-1,3-dioxole and tetrafluoroethylene. The polymer has a tensile modulus of ca. 1.6 GPa (almost a thousand times harder than the elastomeric PDMS material), which is stiff enough for patterning small features without mold deformation. It can be used for UV-NIL in place of a quartz mold. This material has a low surface energy of ~16 mN m^{-1} which can be used as a mold material without any surface treatment. It is inert to all solvents except for perfluorinated solvents (3M, FC-77) such that there is no swelling problem. It is also inert to all chemicals. Therefore, the inert nature with a low surface energy makes it easy to demold after the imprinting process without any surface treatment and without deterioration in surface properties over many imprinting cycles. In addition, it is transparent to light in the region between deep UV and near infrared. It has high gas permeability. The property can eliminate the trapped air bubble during full wafer imprinting. Teflon® AF (amorphous fluoropolymer by Dupont) can have a life greater than 1000 impressions in a very clean environment. There are two ways of preparing the Teflon mold: solvent-casting method and compression molding [24, 54-56].

4.2.4. Others fluoropolymer materials

A typical concern for high modulus materials is their tendency to crack and break due to their brittle nature. Choi *et al.* fabricated a fluorinated organic-inorganic hybrid mold with high modulus of 115 MPa by using a nonhydrolytic *sol-gel* process which can produce a crack-free mold without leaving any trace of solvent. Various nanometer scale patterns including sub-100 nm patterns have been obtained using the fluorinated hybrid mold [56]. A new commercially fluorinated mold material has been developed by AGC Chemicals Co. The fluoropolymer material has good durability (>250 J/cm^2) against UV irradiation, good transparency at UV region (>95% at 375 nm wavelength), high tensile modulus (1.2GPa), CTE (70–80 ppm/K), and low surface energy. T_g of this fluororesin is about 110°C. The thickness of the fluororesin could be controlled from 60 nm to 15 μm by changing concentration and spin coat conditions. Furthermore, it has high chemical durability [57-58]. Haatainen *et al.* demonstrated the fabrication of the F-template by combining the fluorinated mold material and thermal step and stamp nanoimprint lithography (SSIL) method, as well as presented imprinted results using the F-template. The patterns including gratings with 50 nm features and dot arrays of 350 nm diameter features have been achieved [59].

4.3. UV curable materials

4.3.1. Poly(Urethaneacrylate) (PUA)

The PUA based materials exhibit higher rigidity and better impact strength compared with PDMS without significant compromises in flexibility enabling micro-and nanopatterning when the PUA mold thickness is below 50 mm. It is almost impermeable to gases, inert to chemicals and solvents such that there is no swelling problem, and it is transparent to light in the UV and visible regions. However, the rigidity of PUA material is not enough high. It is necessary to further improve the Young's modulus of PUA for fulfilling higher resolution pattern replication. The mechanical properties of PUA based materials largely depend on their cross-linking density, UV-exposure time, UV light wavelength/intensity, as well as the PUA precursor composition. According to the optimizing results of mechanical properties (hardness and effective modulus) form Kim, the PUA replica mold demonstrated very high mechanical properties of hardness (0.15 GPa) and elastic modulus (2.7 GPa) due to the increased cross-linking density of the PUA precursor at an optimized UV-exposure time of 600 s. The PUA replica mold demonstrated potential for the fabrication of multi-scale line-and-space patterns with sizes of 350 nm or less with good uniformity and reproducibility over large areas [48, 60-61].

4.3.2. Modulus-tunable UV-curable materials

A UV-curable mold material which consists of a functionalized prepolymer with acrylate group for cross-linking, a monomeric modulator, a photoinitiator, and a radiation-curable releasing agent for surface activity, has been developed. The mechanical properties of the mold can be tailored by the chain length of an acrylate modulator in the cross-linking reaction. This tunability can be utilized to obtain a proper balance that is needed for a given patterning technique between the rigidity requirement (tensile modulus of 320 MPa) of a mold for patterning a fine structure and the flexibility requirement (tensile modulus of 19.8 MPa) for a conformal contact [62].

4.3.3. UV-curable inorganic–organic hybrid polymer — Ormostamp

Both the transparency and the thermal stability are principal material properties for flexible molds used in soft UV-NIL. A facing challenge in the UV-NIL process is the high transparency of the mold material in the UV-range, which is the characteristic wavelength range for the majority of photo initiators used in photoresist materials. The novel mold material system based on a – Ormostamp – offers high UV-transparency even after thermal exposure at 270 °C. In this case 90% transparency remains at 350 nm. The elasticity and hardness of the mold material are also critical factors for the transfer of nanostructures and dense patterns as well

http://dx.doi.org/10.5772/56186

as high-aspect-ratio features. Physical properties of Ormostamp are modulus of elasticity of 0.650 GPa, hardness of 0.036 GPa, liquid viscosity of 0.75 Pa s. The values of Ormocomp modulus of elasticity and hardness are sufficient for patterning both micro- and nanostructures without any cracks and fractures (too brittle material) or deformations (too soft material). In addition, the relatively low viscosity of Ormostamp is important for efficient filling of the master template cavities and allows various deposition techniques, which make the material handling and further processing easier [12, 63-65]. The Ormostamp working mold possesses high transparency and thermal and UV stability which are essential for soft UV-NIL. Up to now, only quartz stamps exhibited all these features. Excellent pattern transfer fidelity has demonstrated for 100 nm structures.

4.4. Summary and discussion

In order to meet various differently functional demands of the soft UV-NIL process, applied materials of soft molds should possess a number of desirable properties such as physical rigidity for high resolution, flexibility for intimate conformal contact, low surface energy for high quality demolding, high UV transparency for fast curing resist, low viscosity for easy and fast filling into the nano-cavities or features of a master at low pressure to achieve a high resolution mold, small curing-induced shrinkage for dimension accuracy and stability, and chemical inertness for mold durability, as well as thermal stability, easy material processing and high pattern transfer fidelity, etc. A proper balance between the mold rigidity required for patterning a very fine and dense structure and the flexibility needed for a conformal contact with the substrate is at the core of the successful applications. Table 1 summaries these mold materials.

Various fluorinated polymer materials with ultra-low surface energy and suitable rigidity have received enough attention to become ideal materials of flexible molds for soft UV-NIL. The crack and too friable issues for some materials (such as Teflon) with high Young's modulus should be overcome. A composite mold combining a rigid patterning layer using fluoropolymer-based materials and a flexible support layer shows better performance with higher resolution, easy demolding and intimate confocal contact capability as well as longer mold lifetime. Furthermore, the discovery of newly suitable materials (mold and resist) and control their properties will play a critical role for enhanced soft UV-NIL process.

5. Applications of soft UV-NIL

Soft UV-NIL has been employed to fabricate various micro/nanostructures and devices for nanoelectronics, optoelectronics, nanophotonics, optical components, glass, biological applications, etc. It has become a perfect match for some emerging application fields that are in great need of large area patterning of submicro and nano scale features at a low cost, such as patterned magnetic media, light emitting diodes, optical metamaterials and plasmonic devices for chemical and bio-sensing applications, etc. In particular, this technique has demonstrated great commercial prospects in several market segments, LEDs, laser diodes,

Item	Sub-class		Young's Modulus (MPa)	Surface Energy (mN/m)	Viscosity (mPa.s)	Curing mode
PDMS-based materials	*s*-PDMS		< 2	21-24	~3900	Thermal
	h-PDMS		8-12	~20	Tunable	Thermal
	X-PDMS		~80			Thermal
	hv-PDMS		~3-4	~20		UV-Light
Fluorinated polymer materials	PFPE-based materials	PFPE	4	12		UV-Light
		HPFPE	4-5.4	17-22	300-900	UV-Light
		a-PFPE	10.5	~18.5	60 cps at 25 °C	UV light
		PFPE-DMA	4	16.3	360	UV light
	ETFE		~1.2GPa	15.6		Thermal
	Teflon AF 2400		1.6 GPa	~16		Thermal
UV curable materials	PUA		2.7 GPa	23		UV light
	Modulus-tunable UV-curable materials		Tunable 19.8-320			UV light
	Ormostamp		650		750	UV-curable

Table 1. Summary of various soft mold materials

solar cells, optical elements, patterned media, flat panel displays, micro-lens, and functional polymer devices [7-12, 15-17].

5.1. LED patterning

Light efficiency enhancement and manufacturing cost reduction have always been regarding as the two most crucial issues in LED industry, particularly for the large-scale realization of solid state lighting. Compared to other technologies improving the LED performance, two emerging techniques, photonic crystals (PhC) and nanopatterned sapphire substrate (NPSS), have shown higher potential in output efficiency enhancement and beam shaping. NPSS and Photonic Crystal based LEDs have been considered as the most promising solutions for high brightness LEDs. The typical characteristics of LED epitaxial wafers and sapphire substrates are with large variation in wafer topography (varying TTD), high bow and warp, surface roughness with surface protrusions with micron size, and particle contaminations, etc. And these materials tend to be fragile or brittle. Due to the non-planar and rough nature of the LED epitaxial wafer and substrate, existing nanopatterning technologies cannot well meet the requirements of producing these nanostructures in both technology and cost level which mainly originated form the new challenging issues form LED patterning. Due to a very small depth of focus, optical lithography techniques have insufficiently fidelity for LED patterning. As warpage increases with larger wafer sizes, the ability of the photolithography tool to compensate for substrate warpage becomes even more critical. Interference lithography is another method of generating periodic patterns over large areas at low cost. Although the patterns made by IL are highly uniform and have superior long-range order, these patterns are usually in very simple geometric forms of grating lines and 2-D dots, and their dimensions are difficult to reduce to sub-100 nm due to light diffraction. Furthermore, it is unsuitable for high volume production processes because the optical configuration has to be modified to realize different patterns. In addition, this approach requires a strict control of the environment to maintain stable fringe patterns. Soft UV-NIL with flexible mold has the capability of nanopatterning on non-flat surface over large areas and is less-sensitive to the production atmosphere. Compared to ICs industry, the LED application is much more relaxed than IC's for overlay and defect density. Therefore, soft UV-NIL has been considered as one of the most suitable solution for LED patterning. Due to its cost-effectiveness combined with superior processing performance, soft UV-NIL will play a crucial role in moving the LED industry into a new realm of nanopattered LEDs with ultra-high efficiency [2, 14, 17, 65]. Figure 9 showed some cases related to LED patterning using soft UV-NIL. In addition, some commercial companies such as SUUS, Obducat, EVG, Toshiba, Aurotek, Luminus, etc. have been developing the process and equipment of soft UV-NIL for high volume producing PhC LEDs and NPSS.

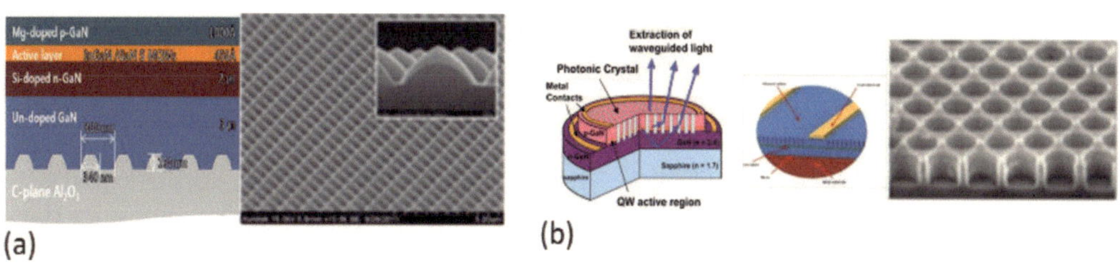

Figure 9. LED patterning using soft UV-NIL [2, 14, 66, 67] (a): NPSS-based LEDs; (b): PhC LEDs)

5.2. Nanophotonic devices

Nanophotonics is a rapidly growing field with great commercial potential. It is a wide field covering many interesting applications branching from cutting edge science including plasmonics, metamaterials, cavity quantum electrodynamics in high-Q cavities all the way to applied sciences like silicon nanophotonics for on chip optical interconnections and single frequency semiconductor light sources. Most of the practical device demonstrations in these fields utilize nanopatterned surfaces. Applications require patterning of nanoscopic gratings, photonic crystals, waveguides and metal structures. Viheriälä *et al.* demonstrated soft UV-NIL nanophotonics applications including distributed feedback laser diodes, plasmonic nanostructures, and patterned facets of optical fibres. Soft UV-NIL will play a critical important role in the commercialization of many nanophotonic applications since it offers excellent cost effectiveness and requires relatively low capital investment. High-density plasmonic nanostructures have been realized on a large area (1 mm^2) using the soft UV-NIL technique. The obtained dimensions of the nanodisks are 65 nm in diameter, 180 nm in periodicity and 25 nm in height with the soft *h*-PDMS/*s*-PDMS mold. Cattoni *et al.*, have successfully realized a Localized Surface Plasmon Resonance (LSPR) biosensor based on $\lambda^3/1000$ plasmonic nanocavities fabricated by Soft UV-NIL on large surfaces (0.5-1 cm^2). These structures present nearly perfect omnidirectional absorption in the infra-red regime independently of the incident angle and light polarization and outstanding biochemical sensing performances with high refractive index sensitivity and figure of merit 10 times higher than conventional LSPR based biosensor [12, 13, 18, 68-71]. Figure 10 presented some nanophotonic devices made using soft UV-NIL.

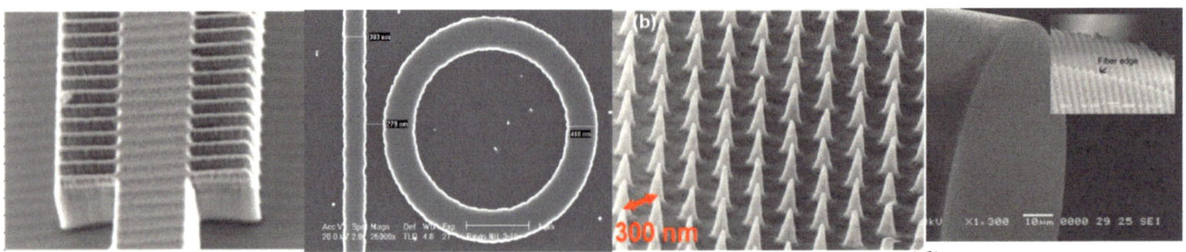

(a) (b) (c) (d)

Figure 10. Nanophotonic devices made using soft UV-NIL [12, 13, 72] (a: DFB-laser diodes; b: Silicon microring resonators; c: Metal nanostructures; d: Optical fibre with the imprinted blazed grating)

5.3. QDs-based devices

Quantum dot (QD) arrays have now been attracting tremendous attention due to the potential applications in various high performance devices (e.g., QD lasers, 3rd generation solar cells, single photon emitters, QD memories, etc.), the fundamental investigation of quantum computing and quantum communication, and in the exploration or observation of novel physical phenomena. Currently, the major challenging issues in commercialized application of QD arrays include fabrication of large-area, defect-free, highly uniform and ordering QDs, accurate positioning for individual QD nucleation site, and reproducibility in size and spatial distribution, which all crucially determines optoelectronic performance and consistency for these QDs-based functional devices and the investigation of fundamental physical properties for QDs. In order to accurately control the size, position, density and composition of epitaxially grown self-assembled quantum dots, a variety of strategies including buried stressor dislocation networks, multiple-layer heteroepitaxy structures to control the stress distribution, quantum dots growth on patterned substrates using various nanofabrication techniques have been proposed in the past decade. Among these, the QDs growth on patterned substrates has been considered as the most straightforward approach to control the size, density and position of QDs so as to achieve highly uniform and ordering QD arrays. Furthermore, growing QD arrays on the patterned substrates has the ability to control the absolute lateral position of quantum dots on a long-range scale. Compared to other nanopatterning approaches, soft UV-NIL technique has high potential to create large-area, low defects patterned substrates with low cost and high throughput. Tommila *et al.* reported on the development of UV-NIL process for patterning GaAs substrates, which are used as templates in seeded S–K growth of QDs.

Soft UV-NIL has also been used to pattern GaAs (1 0 0) substrate into periodic nucleation sites for the growth of InAs site-controlled quantum dots. The incorporation of soft UV-NIL and MOCVD may be a promising method of forming large-area, site-controlled, highly uniform and ordered arrays of quantum dots with low-cost and high throughput to satisfy the requirements of mass production for QDs and QD arrays [73-76].

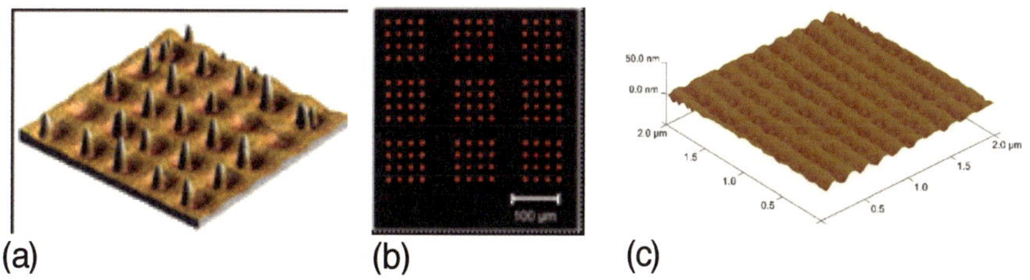

Figure 11. QDs-based devices fabricated using soft UV-NIL [73-76] (a: Site-controlled InAs QDs on UV-NIL patterned surface; b: Nanoimprinted QD arrays for sensing and detecting biomolecules; c: Quantum dot optoelectronics fabrication using soft UV-NIL)

6. Conclusions

Soft UV-NIL by using a flexible mold has been proven to be a cost-effective high volume nanopatterning method for large-area structure replication up to wafer-level (300mm) in the micrometer and nanometer scale, fabricating complex 3-D micro/nano structures, especially making large-area patterns on the non-planar surfaces even curved substrates at low-cost and with high throughput. In particular, it provides an ideal solution and a powerful tool for mass producing micro/nanostructures over large areas at low cost for the applications in compound semiconductor optoelectronics and nanophotonic devices, especially for LED patterning. That opens the way for many applications not previously conceptualized or economically feasible.

Soft UV-NIL has been regarded as the closest process for the industrial application of NIL. In particular, the applications in LED patterning and wafer level micro optics have demonstrated significantly commercial prospect. Soft UV-NIL and its variations (e.g., Roll-type nanoimprint using soft molds) will become more and more important for these applications in large area patterning, fabricating 3-D micor/nanostructures and forming patterns on the non-planar or curved surface. There is a plenty of room to enhance the resolution, patterning area, mold lifetime, yield for the promising patterning method.

Acknowledgements

This work was financially supported by National Science Foundation of China - Major Research Plan "Fundamental Study on Nanomanufacturing" (Grant No. 91023023) and Program for New Century Excellent Talents in University (Grant No.NCET-11-1029).

Author details

Hongbo Lan[1,2*]

Address all correspondence to: hblan99@gmail.com

1 Nanomanufacturing and Nano-Optoelectronics Lab, Qingdao Technological University, Qingdao, China

2 Qingdao Bona Optoelectronics Equipment Co., Ltd., Qingdao National High-Tech Industrial Development Zone, China

References

[1] Verschuuren, M A, & Sprang, H A. Large-area Nanopatterns: Improving LEDs, Lasers, and Photovoltaics. SPIE. http://spie.org/x87355.xml.(2012).

[2] Lee, Y C, & Tu, S H. Improving the Light-emitting Efficiency of GaN LEDs Using Nanoimprint lithography. In: Cui B. (ed.) Recent Advances in Nanofabrication Techniques and Applications. Rijeka: InTech; (2011). , 173-195.

[3] Baek, J H, Kim, S M, & Lee, I H. Control of Characteristic Performance by Patterned Structure in Light-emitting Diodes. Proc. of SPIE (2011). B.

[4] Yang, Y, Mielczarek, K, & Aryal, M. Nanoimprinted Polymer Solar Cell. ACS NANO (2012). , 6(4), 2877-2892.

[5] Farshchian, B, Amirsadeghi, A, & Hurst, S M. Soft UV-nanoimprint Lithography on Non-planar Surfaces. Microelectronic Engineering (2011). , 88-3287.

[6] Chou, Y, Krauss, P, & Renstrom, P. Imprint Lithography with 25-nanometer Resolution. Science (1996). , 272(5258), 85-87.

[7] Schift, H. Nanoimprint lithography: An Old Story in Modern Times? A Review, J. Vac. Sci. Technol. B. (2008). , 26(2), 458-480.

[8] Lan, H, Ding, Y, & Liu, H. Nanoimprint Lithography Principles, Processes and Materials. Nova Science Pub Inc.; (2011).

[9] Guo, J. Recent Progress in Nanoimprint Technology and its Applications. Journal of Physics D: Applied Physics (2004). RR141., 123.

[10] Guo, J. Nanoimprint Lithography: Methods and Material Requirements. Advanced Materials (2007). , 19(4), 495-513.

[11] Bender, M, Fuchs, A, & Plachetka, U. Status and Prospects of UV-Nanoimprint Technology. Microelectronic Engineering (2006). , 83-827.

[12] Viheriälä, J, Niemi, T, & Kontio, J. Nanoimprint Lithography- Next Generation Nanopatterning Methods for Nanophotonics Fabrication. In: Kim K. (ed.) Recent Optical and Photonic Technologies (2010). , 275-298.

[13] Cattoni, A, Chen, J, & Decanini, D. Soft UV Nanoimprint Lithography: A Versatile Tool for Nanostructuration at the 20nm Scale. In: Cui B. (ed.) Recent Advances in Nanofabrication Techniques and Applications Rijeka: InTech; (2011). , 139-156.

[14] Ji, R, Hornung, M, & Verschuuren, M. UV Enhanced Substrate Conformal Imprint Lithography (UV-SCIL) Technique for Photonic Crystals Patterning in LED Manufacturing. Microelectronic Engineering (2010).

[15] Glinsner, T, Plachetka, U, & Matthias, T. Soft UV-based Nanoimprint Lithography for Large Area Imprinting Applications. Proc. of SPIE (2007).

[16] Hiroshima, H. Nanoimprint with Thin and Uniform Residual Layer for Various Pattern Densities, Microelectronic Engineering (2008).

[17] Lan, H, & Ding, Y. Nanoimprint lithography. In: Wang M. (ed.) Lithography Rijeka: InTech; (2010). , 457-494.

[18] Verschuuren, M. A. Substrate Conformal Imprint Lithography for Nanophotonics. PhD thesis. Utrecht University; (2011).

[19] Hornung, M, Ji, R, & Verschuuren, M. Inch Full Field Wafer Size Nanoimprint Lithography for Photonic Crystals Patterning. (2010). th IEEE Conference on Nanotechnology. , 2010, 339-342.

[20] Schmitt, H, Duempelmann, P, & Fader, R. el at. Life Time Evaluation of PDMS Stamps for UV-enhanced Substrate Conformal Imprint Lithography. Microelectronic Engineering. (2012). S., 275-278.

[21] SUSShttp://www.suss.com/index.php,(2012).

[22] Lan, H. A full wafer-scale UV Nanoimprint Lithography Process and Tool for Patterning Photonic Crystal for HB-LEDs. BIT's 1st Annual world congress of Nano-S&T-2011. Oct. Dalian; (2011).

[23] Hiroshima, H. Release Force Reduction in UV Nanoimprint by Mold Orientation Control and by Gas Environment. J. Vac. Sci. Technol. B. (2009). , 27(6), 2862-2865.

[24] Rogers, J A, & Lee, H H. Unconventional Nanopatterning Techniques and Applications. John Wiley & Sons, Inc.; (2008).

[25] Plachetka, U, Bender, M, & Fuchs, A. Comparison of Multilayer Stamp Concepts in UV-NIL. Microelectronic Engineering (2006). , 83-944.

[26] Bender, M, Plachetka, U, & Ran, J. High Resolution Lithography with PDMS Molds. J. Vac. Sci. Technol. B. (2004). , 22(6), 3229-3232.

[27] . Bilayer Transparent Molds for High Resolution Soft UV Nanoimprint Lithography. http://jnte08.trans-gdr.lpn.cnrs.fr/FILES/p12.pdf.

[28] Barbillon, G. Plasmonic Nanostructures Prepared by Soft UV Nanoimprint Lithography and Their Application in Biological Sensing. Micromachines (2012). , 3-21.

[29] Verschuuren, M A, & Brakel, R. van de Laar H W J J., *et al*. VCSEL, LED, Thin-film PV production by Substrate Conformal Imprint Lithography. International Conference on Micro and Nano Engineering Berlin; (2011).

[30] Roy, E, Kanamori, Y, & Belotti, M. Enhanced UV Imprint Ability with a Tri-layer Stamp Configuration. Microelectronic Engineering (2005).

[31] Williams, S S, Retterer, S, & Lopez, R. High-resolution PFPE-based Molding Techniques for Nanofabrication of High-pattern Density, Sub-20 nm Features: a Fundamental Materials Approach. Nano Lett. (2010). , 10-1421.

[32] Gilles, S, Meier, M, & Prömpers, M. UV Nanoimprint Lithography with Rigid Polymer Molds. Microelectronic Engineering (2009). , 86-661.

[33] Choi, D G, Jeong, J H, & Sim, Y. Fluorinated Organic-inorganic Hybrid Mold as a New Stamp for Nanoimprint and Soft Lithography. Langmuir (2005). , 21-9390.

[34] Choi, W M, & Park, O O. Soft-imprint Technique for Submicron-scale Patterns Using a PDMS Mold. Microelectronic Engineering (2004).

[35] Koo, N, Plachetka, U, & Otto, M. The Fabrication of a Flexible Mold for High Resolution Soft Ultraviolet Nanoimprint Lithography. Nanotechnology (2008).

[36] Schmid, H, & Michel, B. Siloxane Polymers for High-resolution, High-accuracy Soft Lithography. Macromolecules (2000). , 33-3042.

[37] Hua, F, Sun, Y, & Gaur, A. Polymer Imprint Lithography with Molecular-scale Resolution. Nano Letters (2004). , 4(12), 2467-2471.

[38] Pei, L, Balls, A, & Tippets, C. Polymer Molded Templates for Nanostructured Amorphous Silicon Photovoltaics. J. Vac. Sci. Technol. A. (2011).

[39] Li, Z, Gu, Y, & Wang, L. Hybrid Nanoimprint-soft Lithography With Sub-15 nm Resolution. Nano Lett. (2009). , 9(6), 2306-2310.

[40] Odom, T W. Christopher Love J., Wolfe D.B., *et al*. Improved Pattern Transfer in Soft Lithography Using Composite Stamps. Langmuir (2002). , 18-5314.

[41] Choi, K M, & Rogers, J A. A Photocurable Poly(dimethylsiloxane) Chemistry Designed for Soft Lithographic Molding and Printing in the Nanometer regime. J. AM. CHEM. SOC. (2003). , 125-4060.

[42] Koo, N, Bender, M, & Plachetka, U. Improved Mold Fabrication for the Definition of High Quality Nanopatterns by Soft UV-Nanoimprint Lithography Using Diluted PDMS Material. Microelectronic Engineering (2007). , 84-904.

[43] Boday, D. J. The State of Fluoropolymers. In Smith D (ed.) Advances in Fluorine-Containing Polymers. ACS; (2012). , 1-7.

[44] Con, C, Zhang, J, & Jahed, Z. Thermal Nanoimprint Lithography Using Fluoropolymer Mold. Microelectronic Engineering. (2012). In press.

[45] Williams, S, Retterer, S, & Lopez, R. High-resolution PFPE-based Molding Techniques for Nanofabrication of High-pattern Density, Sub-20 nm Features: a Fundamental Materials Approach. Nano Lett. (2010). , 10-1421.

[46] Gilles, S, Diez, M, & Offenhausser, A. Deformation of Nanostructures on Polymer Molds During Soft UV Nanoimprint Lithography. Nanotechnology (2010).

[47] Mühlberger, M, Bergmair, I, & Klukowska, A. UV-NIL with Working Stamps Made From Ormostamp. Microelectronic Engineering (2009).

[48] Kim, J K, Cho, H S, & Jung, H S. Effect of Surface Tension and Coefficient of Thermal Expansion in 30 nm Scale Nanoimprinting with Two Flexible Polymer Molds. Nanotechnology (2012).

[49] Jayakumar, P, Ho, Y T, & Soo, H. Adhesion Force Measurement Between the Stamp and the Resin in Ultraviolet Nanoimprint Lithography-an Investigative Approach. Nanotechnology (2009).

[50] Zhu, Z, Li, Q, & Zhang, L. UV-based Nanoimprinting Lithography with a Fluorinated Flexible Stamp. J. Vac. Sci. Technol. B. (2011).

[51] Truong, T T, Lin, R, & Jeon, S. Soft Lithography Using Acryloxy Perfluoropolyether Composite Stamps. Langmuir (2007). , 23(5), 2898-2905.

[52] Rolland, J P, Hagberg, E C, & Denison, G M. High-resolution Soft Lithography: Enabling Materials for Nanotechnologies. Angew. Chem. Int. Ed. (2004).

[53] Barbero, D R, Saifullah, M, & Hoffmann, M. P., *et al.* High-resolution Nanoimprinting with a Robust and Reusable Polymer Mold. Advanced Functional Materials (2007). , 17(14), 2419-2425.

[54] Kim, M J, Park, J E, & Song, S. Simple "Solutal" Method for Preparing Teflon Nanostructures and Molds. J. Vac. Sci. Technol. B. (2007). , 25-1412.

[55] Khang, D Y, & Lee, H H. Sub-100 nm Patterning with an Amorphous Fluoropolymer Mold. Langmuir (2004). , 20-2445.

[56] Khang, D Y, Kang, H, & Kim, T. Low Pressure Nanoimprint Lithography. Nano Lett. (2004). , 4-633.

[57] Choi, D G, Jeong, J J, & Sim, Y S. Fluorinated Organic-inorganic Hybrid Mold as a New Stamp for Nanoimprint and Soft Lithography. Langmuir (2005). , 21-9390.

[58] Kawaguchi, Y, Nonaka, F, & Sanada, Y. Fluorinated Materials for UV Nanoimprint Lithography. Microelectronic Engineering (2007). , 84-973.

[59] Haatainen, T, Mäkelä, T, & Ahopelto, J. Imprinted Polymer Stamps for UV-NIL. Microelectronic Engineering (2009). , 86-2293.

[60] Kim, J Y, Park, H S, & Kim, Z S. Fabrication of Low-cost Submicron Patterned Polymeric Replica Mold With High Elastic Modulus Over a Large Area. Soft Matter. (2012). , 8-1184.

[61] Suh, D, Choi, S J, & Lee, H H. Rigiflex Lithography for Nanostructure Transfer. Advanced Materials (2005). , 17(12), 1554-1560.

[62] Yoo, P J, Choi, S J, & Kim, J H. Unconventional Patterning with a Modulus-tunable Mold: From Imprinting to Microcontact Printing. Chem. Mater. (2004). , 16-5000.

[63] Klukowska, A, Kolander, A, & Bergmair, I. Novel Transparent Hybrid Polymer Working Stamp for UV-imprinting. Microelectronic Engineering (2009). , 86-697.

[64] Mühlberger, M, Bergmair, I, & Klukowska, A. UV-NIL with Working Stamps Made From Ormostamp. Microelectronic Engineering (2009).

[65] Byeon, K J, Hong, E, & Park, H. Full Wafer Scale Nanoimprint Lithography for GaN-based Light-emitting Diodes. Thin Solid Films (2011). , 519-2241.

[66] Hung, R. Luminus Devices, Inc. (2012).

[67] Uhrmann, T. Patterned Sapphire Substrates (PSS): Making LEDs Brighter and Cheaper. (2012).

[68] Jukka, V, Milla-riina, V, & Juha, H. Soft Stamp Ultraviolet-nanoimprint Lithography for Fabrication of Laser Diodes. Proceedings of the SPIE (2009). O.

[69] Shi, J, Chen, J, Decanini, D, et al. Fabrication of Metallic Nanocavities by Soft UV Nanoimprint Lithography. Microelectronic Engineering (2009).

[70] Barbillon, G. Soft UV Nanoimprint Lithography: A Tool to Design Plasmonic Nanobiosensors. In : Kostovski E (ed.) Advances in Unconventional Lithography. Rijeka: InTech; (2011). , 3-14.

[71] Weng, Y J, Weng, Y. C, & Yang, S Y. Fabrication of Optical Waveguide Devices Using Gas-assisted UV micro/nanoimprinting with Soft Mold. Polymers for Advanced Technologies (2007). , 18(11), 876-882.

[72] Spillane, S, Xu, Q, & Fattal, D. Fabrication of Nanophotonic Structures for Information Processing. Proc. of SPIE (2008).

[73] Lan, H, & Ding, Y. Ordering, Positioning and Uniformity of Quantum Dot Arrays. Nano today (2012). , 7(2), 94-123.

[74] Meneou, K. Pathways for Quantum Dot Optoelectronics Fabrication Using Soft Nanoimprint Lithography. Dissertation, University of Illinois at Urbana-Champaign; (2010).

[75] Oh, Y, Lee, K, & Ko, S. Quantum Dot Arrays Patterned by Direct Nanoimprint. The 10th International Conference on NNT, Shilla Jeju, Korea, (2011).

[76] Tommila, J, Tukiainen, A, & Viheriälä, J. Nanoimprint Lithography Patterned GaAs Templates for Site-controlled InAs Quantum Dots. Journal of Crystal Growth (2011). , 232(1), 183-186.

Chapter 8

Sub-30 nm Plasmonic Nanostructures by Soft UV Nanoimprint Lithography

Grégory Barbillon

Additional information is available at the end of the chapter

http://dx.doi.org/10.5772/56119

1. Introduction

The capability to fabricate nanostructures of high density and high resolution over large areas is important point for fundamental and applied research as subwavelength optical nanostructures, optoelectronics and biosensors [1–5]. Various lithographic techniques such as focused ion beam lithography [6] and electron beam lithography [7, 8] are mainly used to pattern sub-100 nm structures on large surfaces. However, these two techniques are slow to obtain these surfaces and their equipments are expensive. Moreover, charge effect on insulating surface can alter the regularity of the pattern shape. Thus, these techniques will not be suitable for a mass production. In addition, alternative methods emerged in the past decades, and are not very expensive and fast to realize high density nanostructures. Moreover, these methods offer a better compatibility for biology and chemical applications [9, 10]. One of these recent techniques is the soft UV Nanoimprint Lithography, which is very promising for the periodic nanostructures fabrication with a high density and high resolution on large surfaces for a reasonable cost [11, 12]. However, a limiting factor of UV-NIL is the resolution of the fabricated molds [13, 14]. In soft UV-NIL, cast molding processes are used for flexible mold realization. An advantage of the soft UV-NIL technique is the obtaining of a great patterns homogeneity on a large zone. This chapter proposes to present in details the soft UV-NIL and its use for the fabrication of sub-30 nm plasmonic structure on large area.

2. Principle & fabrication process of soft UV nanoimprint lithography

2.1. Principle of UV-NIL

Soft UV-NIL has two fundamental steps and is illustrated in figure 1. Firstly, a UV transparent mold with nanostructures on its surface is pressed into a UV sensitive resist. The UV sensitive resist, which is liquid at room temperature, is typically spin coated on the substrate. The UV transparent stamp is deposited on the substrate with a low pressure between 0 and 1 bar [15], at room temperature. Next, the soft stamp is released. The first step duplicates the nanopattern of the transparent mold in the UV sensitive resist. The second step is the removal of residual layer of UV sensitive resist. This step is realized by anisotropic etching such as Reactive Ion Etching (RIE) in order to obtain the desired patterns into the UV resist. During the nanoimprint step, the resist is cured by a UV source (for example, a simple UV lamp or other available systems).

The main advantages of soft UV-NIL are the transparent flexible stamp and a low viscosity UV-curable resist. The replication of the flexible stamps is typically obtained by molding and curing a polymer from a 3D template. The UV transparent flexible stamp fabrication is mainly realized with poly(dimethylsiloxane) PDMS [16–18]. PDMS offers a good chemical stability and a high optical transparency. Moreover, the deformation risk of the soft stamp is minimized with the use of low viscosity UV-curable resists, which permits 3D patterning at low pressure without any heating cycles.

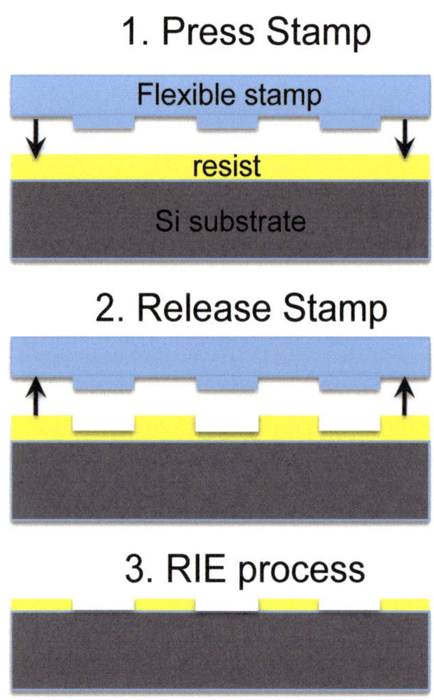

Figure 1. Scheme of UV nanoimprint lithography process: (1) Imprint with the stamp (+ curing with UV light), (2) Stamp withdrawal and (3) Removal of the residual layer with RIE process.

2.2. Master mold fabrication

The master mold used to fabricate flexible stamp is made using the electron beam lithography (EBL). EBL technique allows an excellent accuracy, a very high resolution, and an capability to pattern a large variety of geometries. In the example presented here for the Si master mold fabrication, a PMMA layer of 100 nm (PolyMethylMethAcrylate A2 resist: 950 PMMA A2, MicroChem Corp.) is deposited by spin-coating on Si substrate and baked at 170 °C during 30 min. An EBL system (Leica EBPG5000+) is used to expose PMMA A2, employing an accelerating voltage of 100 kV. Then, the sample is developed in a methylisobutylketone (MIBK)/isopropanol (IPA) solution at room temperature for 35 s followed by a rinsing of 10 s in IPA and thoroughly dried with N_2 gas. Next, the patterns designed in PMMA are transferred into the silicon substrate via a suitable RIE process. The RIE conditions are: 10 sccm for O_2, 45 sccm for SF_6 with P = 30 W, a pressure of 50 mTorr and an autopolarization voltage of 85V [15, 17]. Then, the PMMA mask is removed with a lift-off process in trichloroethylene at 80 °C. Next, the Si master mold surface is treated with HF and H_2O_2 in order to obtain a SiO_2 thin surface layer, then modified with an anti-sticking layer (TMCS: TriMethylChloroSilane) for decreasing the surface energy (Si+TMCS = 28.9 mN/m) in order to easy remove the PDMS molds [17, 19, 20]. In figure 2, SEM images of nanostructures obtained on a Si substrate are presented.

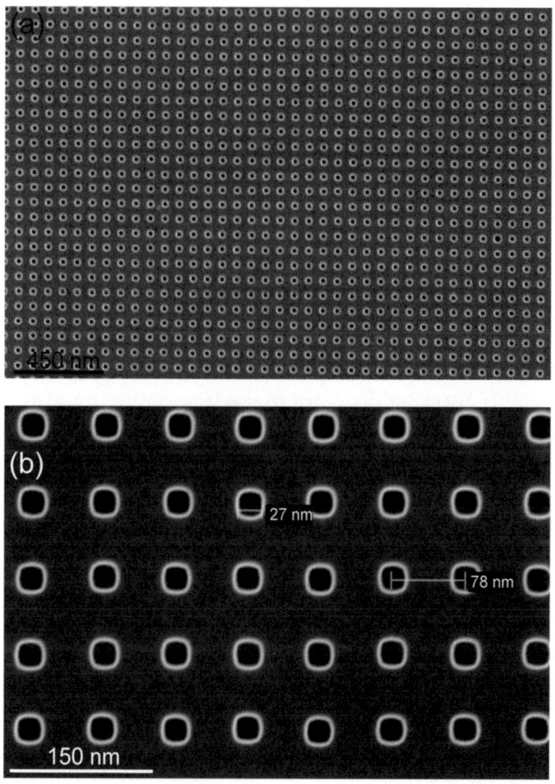

Figure 2. SEM images of Si master mold designed by EBL: nanoholes of diameter ~ 27 nm and ~ 78 nm of periodicity. (a) zone of some μm^2 and (b) zoom of the previous zone.

For the master mold, the dimensions of obtained nanoholes are ~ 27 nm of diameter and ~ 78 nm of periodicity on a zone of 1 cm^2. The programmed dimensions for the electron beam lithography are 25 nm for the diameter and 75 nm for the pitch. Consequently, the dimensions obtained experimentally are in good agreement with those programmed by taking into account the errors of measurements, which could be observed on the nanostructure dimensions with SEM.

2.3. Flexible UV-transparent stamp fabrication

2.3.1. Polymeric materials for stamp fabrication

An advantage of polymeric materials for the replication of original master mold in the nanoimprint process is the low cost of fabrication compared to EBL. With a single and expensive master mold, a large number of polymeric stamps can be replicate and use in the nanoimprint process. Moreover, the excellent flexibility of the elastomeric material offers a good contact between the stamp and the substrate on large areas at low pressures (tens of bars) and on non-planar substrates. Various polymeric materials have been used for stamp fabrication as cross-linked novolak based epoxy resin [21], polycarbonate resins [22], fluoropolymer materials and tetrafluoroethylene (PTFE) [23]. In addition, poly(dimethylsiloxanes) (PDMS) have very interesting properties as a stamp elastomer. The first of these properties is a conformal adhesion of the stamp with the substrate on large areas without any external pressure.

Indeed, PDMS has other attractive properties: (1) its flexibility, which allows a good accuracy of relief shapes replication in the fabrication of the patterning elements, (2) its low Young's modulus (750 KPa) [24] and its low surface energy which allows conformal contact with surface without applied pressure and non-destructive release from designed structures [25], (3) its good optical transparency to a UV light source [26], and (4) its commercial availability in bulk quantities at low cost. The standard PDMS has some advantages, however a number of properties inherent to PDMS limits its performances in the soft UV-NIL. First, the Young's modulus of standard PDMS is low and can limit the fabrication of high density patterns at a sub-100 nm scale due to the collapse of structures. Second, the surface energy ($\sim$ 20 mN/m) of PDMS is not low enough and that does not make it possible the duplication of profiles with a high fidelity. Moreover, the high elasticity and thermal expansion can lead to deformations and distortions during the fabrication process. In general, long range deformations can be avoided by using a thin glass backplane which preserves a global flexibility. In addition, short range deformations can be avoided only by increasing the elastic modulus of PDMS (see paragraph 2.3.2).

2.3.2. Standard PDMS stamp fabrication process

Standard PDMS stamps are mainly realized with a mixture of two commercial PDMS components: (10:1) PDMS RTV 615 siloxane oligomer and RTV 615 cross-linking oligomers (General Electric). This mixture is deposited then casted on the Si master mold and degassed in a dessicator. Standard PDMS is cured at 60 °C for 24 h in order to reduce roughness and to avoid a build up of tension due to thermal shrinkage. If longer curing times and higher temperatures are used, then the elastic modulus and hardness of the polymer are increased (up to x2). However, a higher roughness and deformations can be observed. The stamps are cooled to room temperature, thoroughly peeled off from the master mold and treated with the TriMethylChloroSilane (TMCS) anti-sticking layer in order to reduce the low PDMS surface energy. These stamps are not suitable for the replication of sub-100 nm structures (or with a high aspect ratio) due to the low elastic modulus of PDMS. To solve this problem of low Young's modulus, a modified PDMS called hard-PDMS was already developed.

2.3.3. *Bilayer hard-PDMS/PDMS stamp fabrication process*

To increase the resolution and fidelity of structures, the mechanical properties of the soft stamp need to be improved. Odom *et al.* [28] developed a bilayer stamp of hard-PDMS and standard PDMS, which presents, as advantages, a rigid layer to obtain a high resolution pattern transfer and an elastic support for obtaining a conformal contact even at a low imprint pressure. The hard-PDMS has an attractive property: a lower viscosity of its prepolymer in comparison to standard PDMS. The hard-PDMS prepolymer viscosity is obtained by decreasing of the chain length during its preparation. Thus, the accuracy of replication is improved especially for high-density and small patterns. Another groups [23, 29] also studied the viscosity reduction of the prepolymer for an good replication of the master mold. In this case, the PDMS prepolymer viscosity was decreased with the introduction of a solvent in the mixture. This solvent used with an excessive amount of modulator allows to delay the cross-linking.

The figure 3 represents the fabrication process of the bilayer hard-PDMS/PDMS stamp. The hard-PDMS is a specific thermocured siloxane polymer based on copolymers Vinylmethylsiloxane-Dimethylsiloxane (VDT301) and Methyl-hydrosilane-Dimethylsiloxane (HMS-301) from ABCR, respectively, 34 g and 11 g [27]. In addition, before degassing the mixture with a mixing machine we add 50 μL of platinum catalyst, and 0.5% w/w modulator tetramethyl-tetravinyl cyclotetrasiloxane from FLUKA to the mixture [30]. Then, the hard-PDMS is spin coated on the silicon master mold and the used thickness is mainly 5-8 μm and supported by a standard PDMS layer ($\sim$1.5 mm) (see figure 3). The standard PDMS layer keeps a good flexibility and adaptation on the spin coated wafer during imprint transfer [31]. Then, the bilayer stamp is placed on a glass carrier. The standard PDMS (RTV 615) with its curing agent are mixed before deposit on the thin hard layer PDMS (H-PDMS). Finally, the sample is cured at 60 °C during 24 hours and treated with a TriMethylChloroSilane (TMCS) anti-sticking layer.

For the chosen example of nanostructures, the bilayer stamp is very suitable. Indeed, the obtained nanodots dimensions are $\sim$ 28 nm of diameter, the periodicity of $\sim$ 78 nm, and the height of $\sim$ 60 nm. The figure 4 represents an AFM image of the hard-PDMS/PDMS stamp.

2.4. Soft UV-NIL process

2.4.1. *Optimization of pressure and decreasing of possible deformations for imprint step*

In order to realize the best pattern replication, the control of possible deformations of the replicated pattern is a very important point. Indeed, the PDMS stamp flexibility allows to obtain a conformal adhesion with the substrate at low pressure, during the imprint process. Nevertheless, structures deformations with a high aspect ratio can occur when the applied pressure increases due to the low Young's modulus. In the case where these deformations cannot be avoided, their control and reduction are keypoints and depend on the wished application. A study of the resolution was made by some groups. KarlSuss GmbH has studied the replication of nanoholes (340 nm of diameter) in AMONIL resist. They obtained a diameter of 340 nm ± 5%, and a period uniformity of 2 nm over a 6 inch area [32]. During the imprint step, the resist flow depends strongly on the applied pressure and determines the accuracy on the dimensions of the imprinted nanostructures. In order to reduce local distortions, the pressure of the imprint step must be minimized to obtain a good depth of

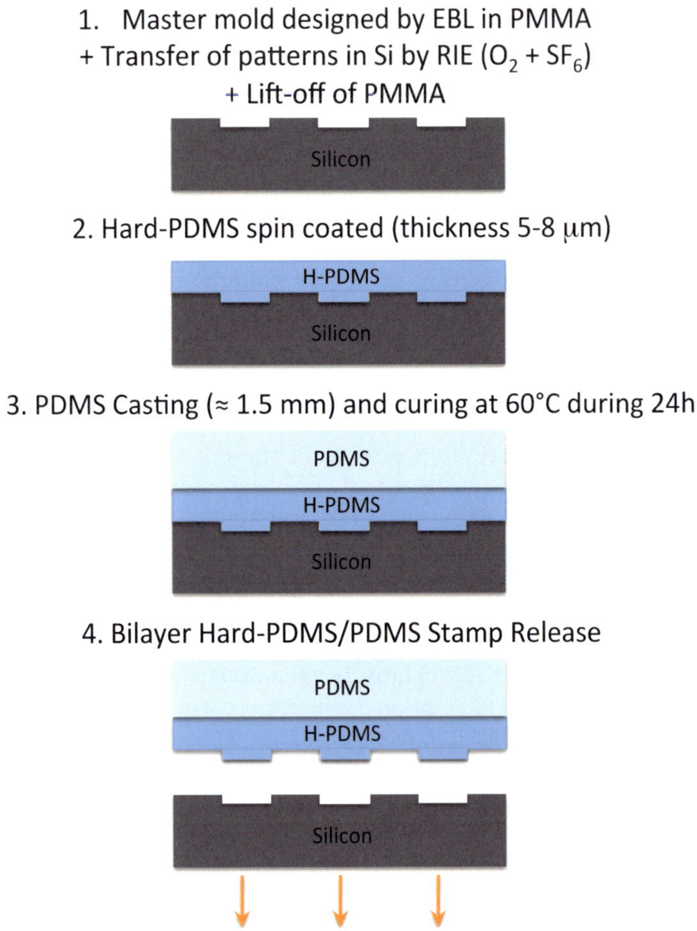

Figure 3. Principle scheme of the fabrication process of the hard-PDMS/PDMS stamp.

the resist in the stamp nanostructures. Cattoni *et al.* [33] demonstrated that the pressure could be reduced to 0.7 bar (thus that Shi et al. [19]) and combined with a UV exposure of 10 min (λ = 365 nm, dose of 2 J/cm^2), a high quality of nanostructures shape was obtained.

2.4.2. Optimization of the thickness of the resist residual layer

Another keypoint of the NIL process is the removal of the residual layer of the resist. In Thermal-NIL and standard UV-NIL, a rigid mold and a high pressure are used, and a thin residual layer of resist is mainly left between the mold protrusions and the substrate. It acts as a soft cushion layer that prevents direct impact of these fragile nanostructures and the substrate. The residual layer is typically withdrawn by RIE. The RIE step can strongly affect the initial shape and size of nanostructures. In addition, in the soft UV-NIL, which uses flexible stamps, the residual layer can be reduced by adapting the original resist thickness to the height (or depth following the desired pattern) of the stamp pattern. Several groups demonstrated the adaptation of the initial resist thickness to the height or depth of the

http://dx.doi.org/10.5772/56119

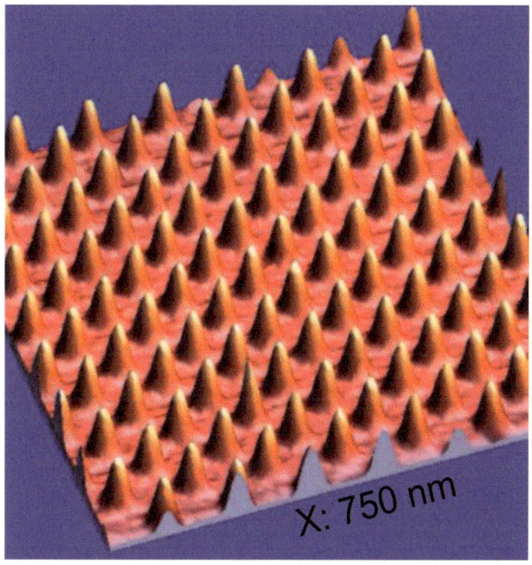

Figure 4. AFM image of the dots in the bilayer hard-PDMS/PDMS stamp (periodicity: ~ 78 nm, diameter (FWHM): ~ 28 nm, and height: ~ 60 nm).

stamp pattern in order to decrease the residual thickness of resist layer. One of these groups conducted a study on the AMONIL resist and they obtained a structure depth of 170 nm and a residual layer of 36 nm [32]. Thus, a very small thickness of the residual AMONIL layer was observed and allowed a good replication.

2.5. Soft UV-NIL for sub-30 nm plasmonic nanostructures fabrication

2.5.1. *Soft UV-NIL in AMONIL*

Various UV-sensitive resists as the NXR 2010 and the AMONIL are available. These two resists exhibit good performance for resolution and etching resistance. The AMONIL resist was used for its low cost compared to the NXR 2010 resist, and its excellent time of conservation. AMONIL resist is a mixture of organic and inorganic compounds having a surface energy of 39.5 mN/m. AMONIL MMS10 from AMO GmbH is used and spin coated on the top of a Ge/PMMA A2 bilayer (10 nm/100 nm thick, respectively), which allows the AMONIL lift-off after curing (see figure 5(a)). The Ge layer is used to improve the selectivity of the former one over the PMMA layer [20]. An AMONIL thickness of 70 nm is chosen in order to minimize the residual thickness of AMONIL. Then, the imprint process is performed in AMONIL with UV exposure at 365 nm wavelength for 10 min and a pressure of 0.7 bar. All these parameters were optimized for the fabrication of nanostructures, which use the bilayer hard-PDMS/PDMS stamp obtained from Si master mold. The figure 5(b) represents the imprint in AMONIL. The dimensions obtained for nanoholes imprinted in AMONIL are ~ 28 nm of diameter and ~ 78 nm of periodicity and these values are in good agreement with the dimensions of nanoholes of Si master mold.

2.5.2. *Plasmonic nanodisks fabrication*

Firstly, the residual AMONIL thickness in the ground of the nanoholes must be removed by a suitable RIE process. For the removal of the residual layer, the etch gases used for RIE

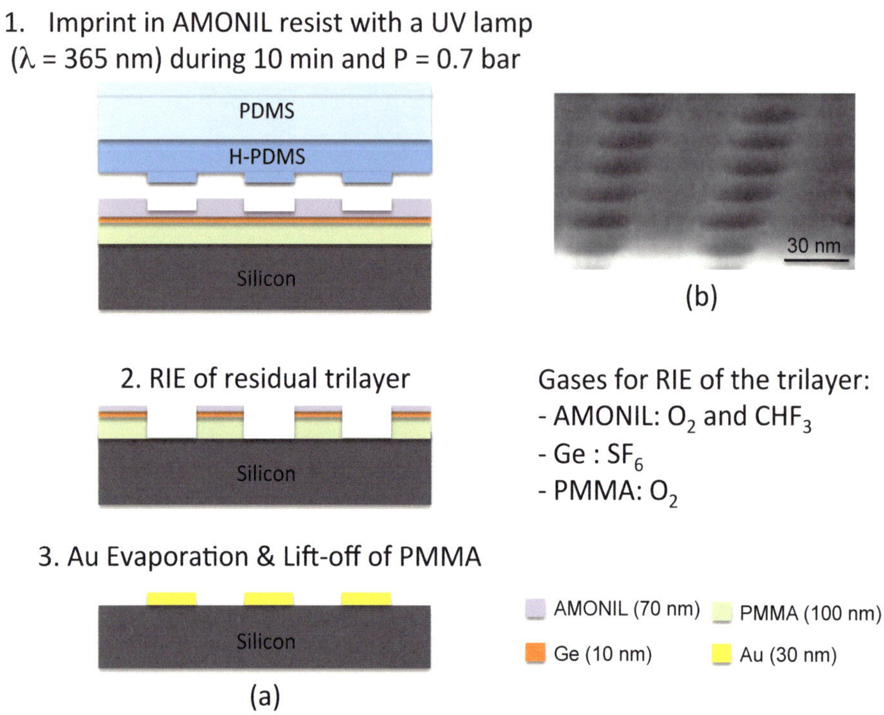

Figure 5. (a) Scheme of the trilayer soft UV-NIL process, and (b) SEM image of imprint in AMONIL with bilayer hard-PDMS/PDMS stamp.

are O_2 and CHF_3. Ge is removed by RIE using SF_6 [34]. For the removal of the PMMA A2, the gas used for RIE is O_2. A good selectivity between PMMA and AMONIL is obtained [15, 17]. The next step is to evaporate a gold thin layer (30 nm) in order to realize the plasmonic nanodisks. Previously, an adhesion layer (Cr) for gold is evaporated (3-5 nm). Then, a lift-off in acetone is used to remove the PMMA underlayer (+ AMONIL/Ge) in order to obtain the sub-30 nm plasmonic nanodisks. The figure 6 presents the results obtained with the bilayer hard-PDMS/PDMS stamp in AMONIL. We observe that the dimensions of plasmonic nanodisks are in good agreement with the dimensions obtained with the imprint in AMONIL. Then, the plasmonic nanostructures can be used as Localized Surface Plasmon Resonance biosensors [35].

3. Conclusion

In this chapter, we have demonstrated the fabrication of sub-30 nm plasmonic nanostructures with the soft UV-NIL technique. The soft UV-NIL is composed by three separate steps: the fabrication of the master mold, the replication of the flexible hard-PDMS/PDMS stamp from the Si master mold, and the imprinting process by using the bilayer hard-PDMS/PDMS stamp. All these steps are very important in order to obtain a very good quality of the final result, in terms of resolution and line edge roughness of the nanostructures. A master mold fabrication process based on EBL is presented in details. Then, the replication of the soft polymeric stamp, based on a composite hard-PDMS/PDMS bilayer, is presented. The ability

http://dx.doi.org/10.5772/56119

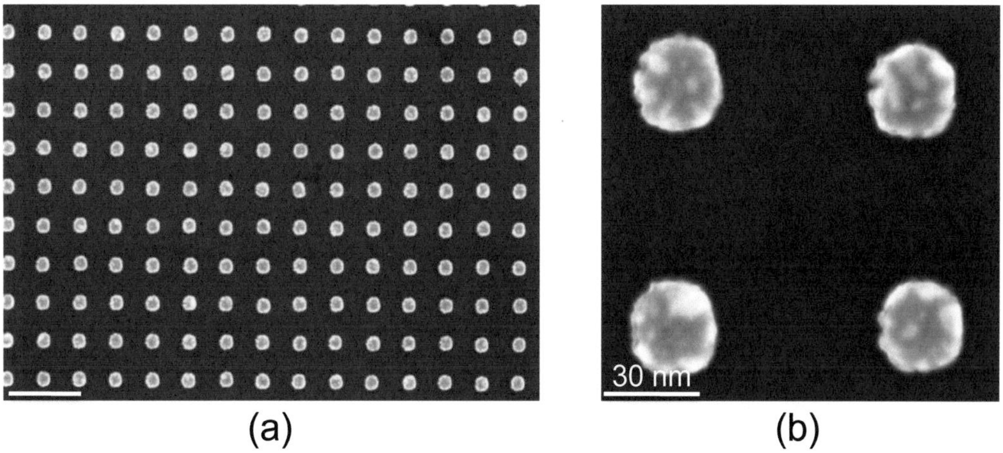

Figure 6. SEM images of plasmonic nanodisks: (a) a zone of some μm^2 (scale bar = 150 nm), (b) diameter $\sim$ 28 nm, and periodicity $\sim$ 78 nm.

of soft UV-NIL to replicate sub-30 nm nanostructures with high homogeneity at the whole pattern surface is demonstrated. To conclude, we present an example of a fabrication by soft UV-NIL on large area (1 cm^2) of plasmonic nanostructures with potential applications in biosensors and photonics. Finally, we believe that the soft UV-NIL technique will play quickly an important role as a powerful and versatile tool for the nanostructures fabrication.

Author details

Grégory Barbillon[1,2,*]

* Address all correspondence to: gregory.barbillon@laposte.net

1 Laboratory of Photonics and Nanostructures CNRS UPR 20, Marcoussis, France
2 Laboratory of Lasers Physics CNRS UMR 7538, University Paris 13, Villetaneuse, France

References

[1] Li M, Tan H, Wang J, Chou S.Y (2003) Large area direct nanoimprinting of SiO_2-TiO_2 gel gratings for optical applications. J. Vac. Sci. Technol. B 21: 660-663.

[2] Chegel V, Lucas B, Guo J, Lopatynskyi A, Lopatynska O, Poperenko L (2009) Detection of biomolecules using optoelectronic biosensor based on localized surface plasmon resonance. Nanoimprint lithography approach. Semiconductor Physics, Quantum Electronics & Optoelectronics 12 91-97.

[3] Jensen T.R, Duval M.L, Kelly K.L, Lazarides A.A, Schatz G.C, Van Duyne R.P (1999) Nanosphere Lithography : Effect of the External Dielectric Medium on the Surface Plasmon Resonance Spectrum of a Periodic Array of Silver Nanoparticles. J. Phys. Chem. B 103: 9846-9853.

[4] Barbillon G, Bijeon J.L, Lérondel G, Plain J, Royer P (2008) Detection of chemical molecules with integrated plasmonic glass nanotips. Surface Science 602: L119-L122.

[5] Faure A.C, Barbillon G, Ou M.G, Ledoux G, Tillement O, Roux S, Fabregue D, Descamps A, Bijeon J.L, Marquette C.A, Billotey C, Jamois C, Benyatou T, Perriat P (2008) Core/Shell nanoparticles for multiple biological detection with enhanced sensitivity and kinetics. Nanotechnology 19: 485103.

[6] Jiang X, Ji L, Chang A, Leung K.N (2003) Resolution improvement for a maskless microion beam reduction lithography system. J. Vac. Sci. Technol. B 21: 2724-2727.

[7] Vieu C, Carcenac F, Pépin A, Chen Y, Mejias M, Lebib A, Manin-Ferlazzo L, Couraud L, Launois H (2000) Electron beam lithography: resolution limits and applications. Applied Surface Science 164: 111-117.

[8] Barbillon G, Bijeon J.L, Plain J, Royer P (2009) Sensitive detection of biological species through localized surface plasmon resonance on gold nanodisks. Thin Solid Films 517: 2997-3000.

[9] Gates B.D, Xu Q, Stewart M, Ryan D, Wilson C.G, Whitesides G.M (2005) New Approaches to Nanofabrication: Molding, Printing, and Other Techniques. Chemical Reviews 105: 1171-1196.

[10] Krauss P.R, Chou S.Y (1997) Nano-compact disks with 400 Gbit/in^2 storage density fabricated using nanoimprint lithography and real with proximal probe. Appl. Phys. Lett. 71: 3174-3176.

[11] Chou S.Y, Krauss P.R, Renstorm P.J (1996) Imprint lithography with 25-Nanometer resolution. Science 272: 85-87.

[12] Austin M.D, Ge H, Wu W, Li M, Yu Z, Wasserman D, Lyon S.A, Chou S.Y (2004) Fabrication of 5 nm linewidth and 14 nm pitch features by nanoimprint lithography. Appl. Phys. Lett. 84: 5299-5301.

[13] Jung G.Y, Johnston-Halperin E, Wu W, Yu Z, Wang S.Y, Tong W.M, Li Z, Green J.E, Sheriff B.A, Boukai A, Bunimovich Y, Heath J.R, Stanley Williams R (2006) Circuit Fabrication at 17 nm Half-Pitch by Nanoimprint Lithography. Nano Lett. 6: 351-354.

[14] Austin M.D, Zhang W, Ge H, Wasserman D, Lyon S.A, Chou S.Y (2005) 6 nm half-pitch lines and 0.04 μm^2 static random access memory patterns by nanoimprint lithography. Nanotechnology 16: 1058.

[15] Hamouda F, Barbillon G, Held S, Agnus G, Gogol P, Maroutian T, Scheuring S, Bartenlian B (2009) Nanoholes by soft UV nanoimprint lithography applied to study of membrane proteins. Microelectron. Eng. 86: 583-585.

[16] Hamouda F, Barbillon G, Gaucher F, Bartenlian B (2010) Sub-200 nm gap electrodes by soft UV nanoimprint lithography using polydimethylsiloxane mold without external pressure. J. Vac. Sci. Technol. B 28: 82-85.

[17] Barbillon G, Hamouda F, Held S, Gogol P, Bartenlian B (2010) Gold nanoparticles by soft UV nanoimprint lithography coupled to a lift-off process for plasmonic sensing of antibodies. Microelectron. Eng. 87: 1001-1004.

[18] Hamouda F, Sahaf H, Held S, Barbillon G, Gogol P, Moyen E, Aassime A, Moreau J, Canva M, Lourtioz J.M, Hanbücken M, Bartenlian B (2011) Large area nanopatterning by combined anodic aluminum oxide and soft UV-NIL technologies for applications in biology. Microelectron. Eng. 88: 2444-2446.

[19] Shi J, Chen J, Decanini D, Chen Y, Haghiri-Gosnet A.M (2009) Fabrication of metallic nanocavities by soft UV nanoimprint lithography. Microelectron. Eng. 86: 596-599.

[20] Chen J, Shi J, Decanini D, Cambril E, Chen Y, Haghiri-Gosnet A.M (2009) Gold nanohole arrays for biochemical sensing fabricated by soft UV nanoimprint lithography. Microelectron. Eng. 86: 632-635.

[21] Pfeiffer K, Fink M, Ahrens G, Gruetzner G, Reuther F, Seekamp J, Zankovych S, Torres C.S, Maximov I, Beck M, Graczyk M, Montelius L, Schulz H, Scheer H.C, Steingrueber F (2002) Polymer stamps for nanoimprinting. Microelectron. Eng. 61-62: 393-398.

[22] Posognano D, D′Amone S, Gigli G, Cingolani R (2004) Rigid organic molds for nanoimprint lithography by replica of high glass transition temperature polymers. J. Vac. Sci. Technol. B 22: 1759-1763.

[23] Kang H, Lee J, Park J, Lee H.H (2006) An improved method of preparing composite poly(dimethylsiloxane) moulds. Nanotechnology 17: 197-200.

[24] Bender M, Plachteka U, Ran J, Fuchs A, Vratzov B, Kurz H, Glinsner T, Lindner F (2004) High resolution lithography with PDMS molds. J. Vac. Sci. Technol. B 22: 3229-3232.

[25] Hsia K.J, Huang Y, Menard E, Park J.U, Zhou W, Rogers J, Fulton J.M (2005) Collapse of stamps for soft lithography due to interfacial adhesion. Appl. Phys. Lett. 86: 154106.

[26] Schmid H, Biebuyck H, Michel B, Martin O.J.M (1998) Light-coupling masks for lensless, sub-wavelength optical lithography. Appl. Phys. Lett. 72: 2379.

[27] Choi D.G, Yu H.K, Yang S.M (2004) 2D nano/micro hybrid patterning using soft/block copolymer lithography. Mater. Sci. Eng. C 24: 213-216.

[28] Odom T.W, Love J.C, Wolfe D.B, Paul K.E, Whitesides G.M (2002) Improved pattern transfer in soft lithography using composite stamps. Langmuir 18: 5314-5320.

[29] Koo N, Bender M, Plachetka U, Fuchs A, Wahlbrink, T, Bolten J, Kurz H (2007) Improved mold fabrication for the definition of high quality nanopatterns by soft UV-nanoimprint lithography using diluted PDMS material. Microelectron. Eng. 84: 904-908.

[30] Schmid H, Michel B (2000) Siloxane Polymers for High-Resolution, High-Accuracy Soft Lithography. Macromolecules 33: 3042-3049.

[31] Plachteka U, Bender M, Fuchs A, Vratzov B, Glinsner T, Lindner F, Kurz H (2005), Wafer scale patterning by soft UV-Nanoimprint Lithography. Microelectron. Eng. 73/74: 167-171.

[32] Ji R, Hornung M, Verschuuren M.A, van de Laar R, van Eekelen J, Plachetka U, Moeller M, Moormann C (2010) UV enhanced substrate conformal imprint lithography (UV-SCIL) technique for photonic crystals patterning in LED manufacturing. Microelectron. Eng. 87: 963-967.

[33] Cattoni A, Cambril E, Decanini D, Faini G, Haghiri-Gosnet A.M (2010) Soft UV-NIL at 20 nm scale using flexible bi-layer stamp casted on HSQ master mold. Microelectron. Eng. 87: 1015-1018.

[34] Chen J, Shi J, Cattoni A, Decanini D, Liu Z, Chen Y, Haghiri-Gosnet A.M (2010) A versatile pattern inversion process based on thermal and soft UV nanoimprint lithography techniques. Microelectron. Eng. 87: 899-903.

[35] Barbillon G (2011) Soft UV Nanoimprint Lithography: A Tool to Design Plasmonic Nanobiosensors. In: Kostovski G, editor. Advances in Unconventional Lithography. Rijeka: InTech. pp. 3-14.

Section 4

Fabrication of 3D Nano-Structure

Chapter 9

The Fabrication of High Aspect Ratio Nanostructures on Quartz Substrate

Khairudin Mohamed and Maan M. Alkaisi

Additional information is available at the end of the chapter

http://dx.doi.org/10.5772/56315

1. Introduction

Recent developments in nano-scale devices have imposed many complex patterns with high aspect ratio nanostructures in its design. High aspect-ratio nanostructures have many applications such as X-ray diffractive optical elements [1,2], nano-electro-mechanical-system (NEMS), fuel cell electrodes [3] and nanoimprint molds [4]. However, fabricating the high aspect-ratio nanostructures is still a challenging problem. For silicon technology, Bosch process [5] which is based on alternating multiple steps of etching and sidewall passivation is normally employed to achieve a high aspect ratio nanostructure. Other alternative is the cryogenic process [6] which is cooling the silicon substrates to cryogenic temperature using liquid nitrogen in order to achieve vertical sidewall profiles during etching. These processes however are not suitable for quartz.

In addition, although a large number of articles on high aspect-ratio silicon structures have been reported, limited information is available for quartz etching process in achieving high aspect-ratio nanostructures. Fabricating high aspect-ratio nanostructures on glass or quartz would open several new possibilities in the MEMS/NEMS field and especially in BioMEMS.

In this chapter, we explained the pattern transfer process required on quartz substrate using CHF_3/Ar reactive ion etching (RIE) technique. The fabrication of feature sizes below 100 nm on quartz substrates will be demonstrated by understanding first the etching mechanism of quartz and then finding an optimised RIE process.

2. The fabrication method

Quartz is an insulating and hard substrate material in which patterning on top of its surface using electron beam lithography (EBL) is very challenging. Surface charging is the major issue for pattern writing on the insulating substrates using an EBL technique. The trapped and built-up charges by the e-beam exposure on the insulating substrates surface may deflect or distort the e-beam positioning, and eventually may cause undesired effects. A conductive layer is required to ground the trapped charges in minimising the surface charges. There are several ways to suppress the charging effects by draining the charges to the ground, such as thin metallic or carbon coating on top or underneath the resist layer [7].

This chapter presents a number of approaches attempted for grounding the trapped charges for suppressing the surface charging effects during the pattern definition on quartz substrate. This also applies in fabricating three-dimensional nanostructures on insulating substrates [8]. The deposition of thin metallic layer on top of imaging resist layer is a common practise but requires wet etching process using acidic solution to remove the metallic layer in post e-beam exposure process. Most lithographers will avoid this technique as this process is the source of particles contamination and other incomplete removal issues.

We developed and investigated two methods of charge suppression using a thin metallic coating on quartz substrate and a conductive polymer coating on top of the imaging resist layer. The processes involved in each method are illustrated in Figure 1(a) and (b).

2.1. Thin metallic coating on quartz substrate

In this technique as illustrate in Figure 1(a), a cleaned quartz substrate was firstly sputtered with a 5 nm thick Tungsten (W) as a charge dissipation layer using Edward Auto500 Magnetron Sputtering system prior to the poly-methyl-methacrylate (PMMA) bi-layer resist coating. The thin Tungsten layer can be stripped off easily using a short sulphur hexafluoride (SF_6) plasma etching at a later stage.

A positive PMMA bi-layer resist was spun coated onto the tungsten coated quartz substrate. Each layer of PMMA was spun coated at spinning speed of 4000 rpm for one minute. The first layer of 4% low molecular weight (LMW) PMMA was spun followed by hard baking at 185°C for 30 minutes and then allowed to cool down to room temperature. Then the second layer of 2.5% high molecular weight (HMW) PMMA was spun and hard baked at 185°C for 30 minutes. The LMW PMMA layer thickness was 80 nm and the HMW PMMA was 120 nm when measured using a Dektak surface profiler system. The total PMMA bi-layer thickness was approximately 200 nm.

Pattern definition using single pass line (SPL) method of e-beam exposure was carried out using Raith-150 EBL tool with voltage acceleration of 10 keV, an aperture size of 30 microns and e-beam dosage of 110 μC/cm^2. The e-beam exposed sample was then developed in a MIBK:IPA 1:3 developer at a temperature of 23 ^{o}C for 30 seconds. Short O_2 plasma is recommended for descumming the residual resist layers. The O_2 plasma exposure duration of 15 seconds at 100 W of O_2 plasma power should be appropriate to clear the residual layer of 10 nm to 20 nm for 100 nm feature, however, longer O_2 plasma exposure may enlarge the pattern size.

A 40 nm thick of nichrome (NiCr) 80/20 99.999% was then thermally deposited on the developed sample using a thermal metal evaporator system. The lift-off process was carried out to remove the unwanted resist and metal layers by soaking the sample in acetone for about three hours. A very short (about 10 seconds) RIE plasma process with SF_6 etchant gas was utilized to remove the exposed tungsten layer.

Finally an RIE process with CHF_3/Ar chemistry was used to etch the quartz substrate anisotropically. By using this technique, two metal layers (NiCr and W) were left on the top of the fabricated nanostructures. Another technique of fabricating nanostructures on quartz substrate is explained next.

2.2. Conductive polymer coating on top of resist

The nanofabrication process using this technique is illustrated in Figure 1(b). In this technique, a water soluble conductive polymer, poly (3,4 – Etylenedioxythiophene) / poly(styrenesulfonate) (PEDOT/PSS) was used as the charge dissipation layer [8].

A positive PMMA bi-layer resist was spun coated onto the cleaned quartz substrate as explained in previous section. PEDOT/PSS was then spun coated on top of PMMA bi-layer at spinning speed of 5000 rpm for one minute to achieve a 30 nm layer thickness.

The only issue with PMMA bi-layer resist was that its hydrophobic surface caused great difficulties for the watery solution to adhere to its surface. One well-known method of reducing the hydrophobicity or in other words, increasing the wettability of the PMMA bi-layer surface is by O_2 plasma surface treatment. Only very short O_2 plasma was needed to improve the wettability of the PMMA bi-layer surface.

The sample was then e-beam exposed using Raith-150 EBL system for pattern definition process as explained in previous section. Prior to the pattern development process, the unexposed PEDOT/PSS layer was removed by rinsing with deionised water (DIW) at room temperature for one minute.

In pattern definition process for feature size larger than 100 nm, the PEDOT/PSS layer was exposed with larger e-beam bombardment on a larger area which leads to the diffusion of PEDOT/PSS molecules into PMMA layer. This made the removal of PEDOT/PSS layer using DIW at room temperature became difficult. One option to resolve this issue is by increasing the DIW temperature to 45 ^{o}C and placed in ultrasonic bath for a couple of seconds.

The e-beam exposed samples were then developed in MIBK:IPA 1:3 at a temperature of 23 ^{o}C for 30 seconds and followed by metallization process. In additive pattern transfer employed, a 40 nm NiCr 80/20 99.999% layer was deposited as a pattern masking layer using thermal evaporator as explained in previous section. In order to realize the pattern, the unwanted NiCr layer was lifted off by soaking in acetone for about 3 hours.

At final stage, RIE process with CHF_3/Ar chemistry was used to etch the quartz substrate anisotropically. Using this technique, only a NiCr masking layer is left on top of the fabricated nanostructure. This metal layer can later be removed by the Chrome etch acidic solution.

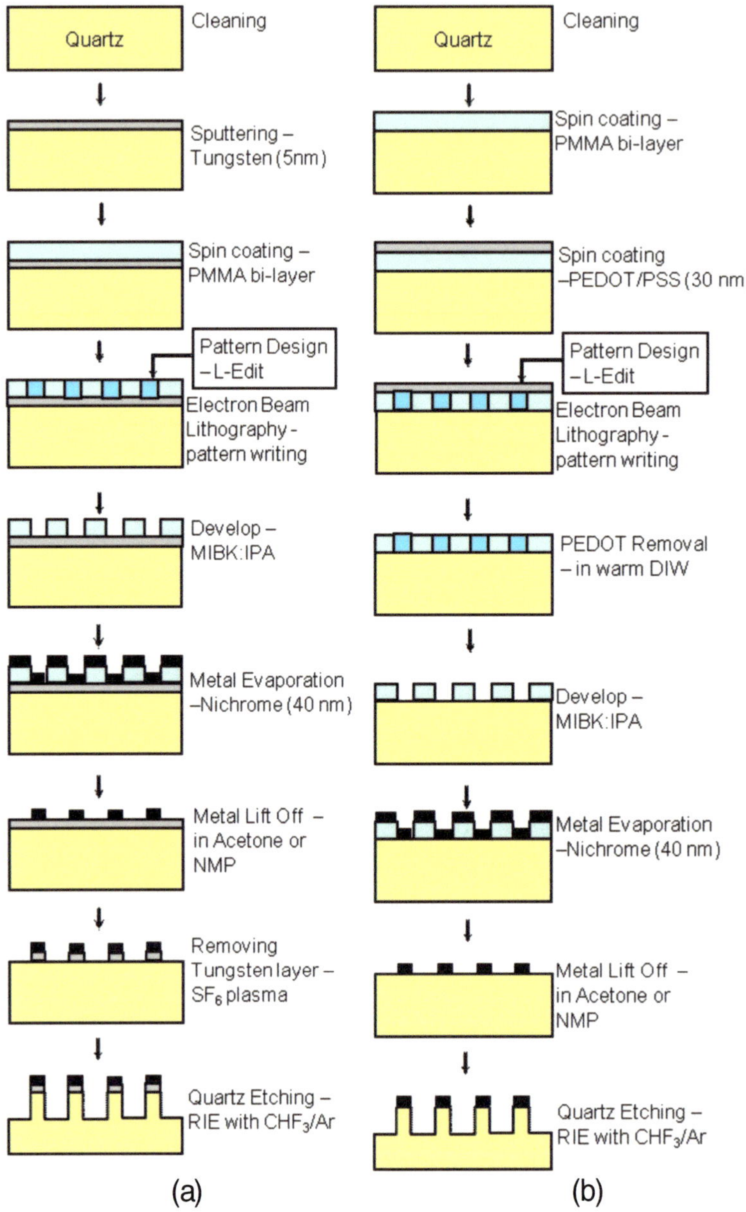

Figure 1. The fabrication process of nanostructures on quartz substrate using various charge suppression method (a) thin metallic coating (tungsten) on quartz substrate and (b) conductive polymer (PEDOT/PSS) coating on top of the resist.

3. The pattern transfer process

The RIE pattern transfer process employing CHF_3/Ar gas mixture was carried out using Oxford Plasmalab 80Plus etcher. A 100 nm diameter silicon wafer with a thick NiCr coated layer was clamped physically onto the power electrode as a sacrificial masking layer. The electrical power was supplied to the electrode at a radio frequency (RF) of 13.56 MHz to produce plasma discharge.

Trifluoromethane (CHF_3 or Freon 23, sometimes called Halocarbon 23) was used as the main processing gas owing to its moderate F/C ratio resulting in a moderate etching rate which is essential in controlling the etching profiles. For etching the narrow high aspect ratio nanostructures, there are limited ions and neutral transport present within the trench which limits the chemical reactions. Hence, large aspect ratio and dense nanostructures are etched more slowly than low aspect ratio nanostructures [9]. Moderate to low etching rate is appropriate for this work.

Oxide such as silicon dioxide (quartz) etching using CHF_3 etchant can only take place for RF bias values above 55 V at 1 mTorr etching pressure [10]. A higher RF bias value is required for quartz etching if operating at a higher etching pressure. A bias voltage of -330 V was achieved by this work showing good directionality of the plasma bombardments onto substrate surface.

An additive inert gas argon (Ar) was added to stabilize plasmas and to add inert ion bombardment of a surface, resulting in more directional or anisotropic etching [11] for a steep sidewall profile. The pattern transfer onto quartz substrates is obtained by CHF_3/Ar RIE with the optimised parameter as in Table 1.

RIE Parameter	Setting/measured
Gases	CHF_3/Ar
Flow rate	50/30 sccm
Pressure	20 mTorr
Temperature	295 K
RF Power	200 W
RF Power density	2.5 W/cm^2
Bias Voltage	-330 V
Etch rate	10 nm/min

Table 1. An optimized CHF_3/Ar RIE parameter for 2D pattern transfer onto quartz substrate

3.1. The etching chemistries

In fluorine–containing plasma for example, surface reaction, etching and polymerization can occur at the same time. The domination of certain reactions is dependent on the gas feed, the operating parameters and the chemical nature of the polymer/substrate and electrode material and geometry [12].

In a quartz etching process using CHF_3 etchant, the free fluorine radical, F^* and radicals such as CF_x are created by the plasma discharge and the etching chemistries can be described as follows;

$$e^{-} + CHF_3 \rightarrow CHF_2 + F^{*} + \text{radicals} + 2\,e^{-} \quad (1)$$

$$SiO_2 + xF^{*} \rightarrow SiF_x + O_2 \quad (2)$$

$$CF_x\text{radicals} + 2O_2 \rightarrow CO + CO_2 + COF_2 \quad (3)$$

F* is the reactive fluorine atom and SiO_2 is the quartz.

RIE plasma generates the reactive fluorine atom F*, radicals and ions from the supplied etchant gas CHF_3 (Eqn. 1). The bombardments of the heavy argon ions break the Si-O bonds and then the disassociated silicon ions react with F free radicals to form a SiF_x. The etching of quartz consumes the F atom to form SiF_x and oxygen radicals (Eqn. 2). The resultant CF_x radicals tend to deposit polymer film on all surfaces, but the oxygen liberated in the etching of quartz reacts with the CF_x radicals to form volatile CO, CO_2 and COF_2 (Eqn. 3)[13]. Thus, both F and C are consumed but the polymer deposition releases the F atoms that enrich the F content hence, increasing the F/C ratio.

When the F concentration in the plasma is high, the CF_x radicals are destroyed by recombination at various surfaces of the plasma reactor and probably lead to the re-creation of volatile CF_4. When the F concentration is low, surface production mechanism dominates the production of CF_2 which leads to the formation of a polymer layer at the surface (polymerazation) [14]. The vertical sidewalls are created by the polymerization of the CF_x passivation layer.

In anisotropic etching, vertical sidewalls are the location where chemical reaction less occur and less exposure to ions bombardments, hence polymerization will build up on these locations. This process is called sidewall passivation in which the undercut etchings are prevented and steep sidewall profiles are realized.

3.2. The etching mechanism

The etching mechanism can be described as illustrated in Figure 2. There are two steps involved in the CHF_3/Ar RIE plasma process. Firstly the bombardment of heavy inert ions of argon mechanically breaks the bond of the substrate elements (Si-O). The high energy and heavy argon ions are accelerated by RIE plasma and act as bullets in bombarding substrate and mask surfaces. Si-O bonds are very strong and could only be dissociated by high energy plasma bombardment using heavy inert ions such as argon.

Then ion-induced chemical reaction at horizontal surface created the etching reaction products. The chemistry reactions will only take place when the freed ions/atoms are available after the breakage of the Si-O bonds.

The combination of these actions resulted in highly selective anisotropic etching of the substrate material. The etching selectivity of Chromium or NiChrome (NiCr) mask to quartz

http://dx.doi.org/10.5772/56315

substrate is about 18:1 [15]. Hence, the sample requires a NiCr masking layer of at least 35 nm thick to mask the pattern structures to achieve a 600 nm structure height on quartz substrate.

The etching process of quartz is the combination of physical etching by ions bombardment to break the bonding and chemical etching where the freed ions/atoms react to each other to create by-products and sucked away by the vacuum system. As the etching pressure used in this work are very low (< 30 mTorr), physical bombardment reactions are expected to dominate and less chemical reactions are involved in the etching process.

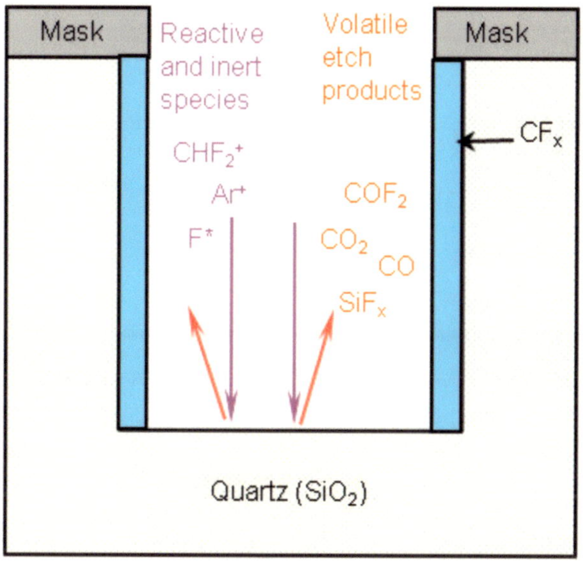

Figure 2. The schematic diagram of plasma etching mechanism, showing the fluorinated reactive gases and the by-products generated during the quartz etching process.

3.3. The etching cycle

Deep etching on quartz substrate requires a long etching time owing to very low etching rate of about 10 nm/min. Continuous heavy ions bombardments on the substrate surface for long period of time may resulted in increase of surface temperature. When the surfaces heat up, the F/C ratio could decrease, which would decrease the etching reactions as well. The etching rate could decrease to the point where the quartz etching may actually stops [16]. This can be avoided by etching in a short interval of time and repeat the etching cycle after the samples have cooled down.

In this research work, five minutes of cooling intervals on every 15 minutes of plasma etching process was employed to achieve a vertical sidewalls profile. This 'duty cycle' etching methods as illustrated in Figure 3 has shown beneficial to minimise heat generation effects and achieve high aspect ratio nanostructures with vertical sidewalls.

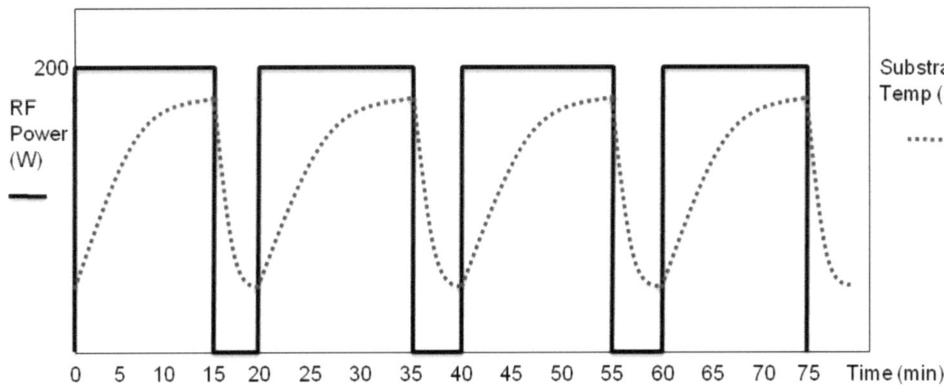

Figure 3. The profiles of the RIE RF power and the substrate temperatures in the 'duty cycle' etching method.

4. The fabrication results

Nanostructures with feature sizes of 20 nm, 40 nm, 50 nm, 60 nm, 70 nm and 90 nm have been attempted by this work. However, only the high aspect ratio nanostructures with feature sizes above 60 nm have been successfully fabricated on quartz substrates [17]. The subsequent figures show the SEM images of the results of the developed nanofabrication process in isometric angle.

Figure 4 shows an SEM image of the fabricated high aspect ratio structures on quartz substrate with dimensions of 60 nm lines width, 600 nm in height and 120 nm spacing. Hence, high aspect ratio nanostructures of up to 1:10 and sound vertical sidewall profiles have been demonstrated by this work. The second line from left is a bit shifted in the middle because of the presence of thin residual layer during metal deposition process and later shifted during the lift-off process.

Figure 5 shows another SEM image of high aspect-ratio nanostructures with dimensions of 90 nm lines width, 600 nm in height and 180 nm spacing. It has better sidewall, fine surface and flat top profiles as compared to 60 nm feature sizes nanostructure.

Figure 6 shows the SEM image of various high aspect-ratio nanostructures with feature sizes above 60 nm fabricated by this work. We have demonstrated various vertical sidewall nanostructures such as elbow corners, transistor-like profile and periodic lines. Almost all the nanostructures have vertical sidewall and flat top surface.

Figure 7 shows the line structures with width dimensions of 50 nm and 70 nm for comparison. The nanostructures of 50 nm in width failed to preserve the flat top surface which leads to uneven profile of top surface as compared to the 70 nm nanostructures.

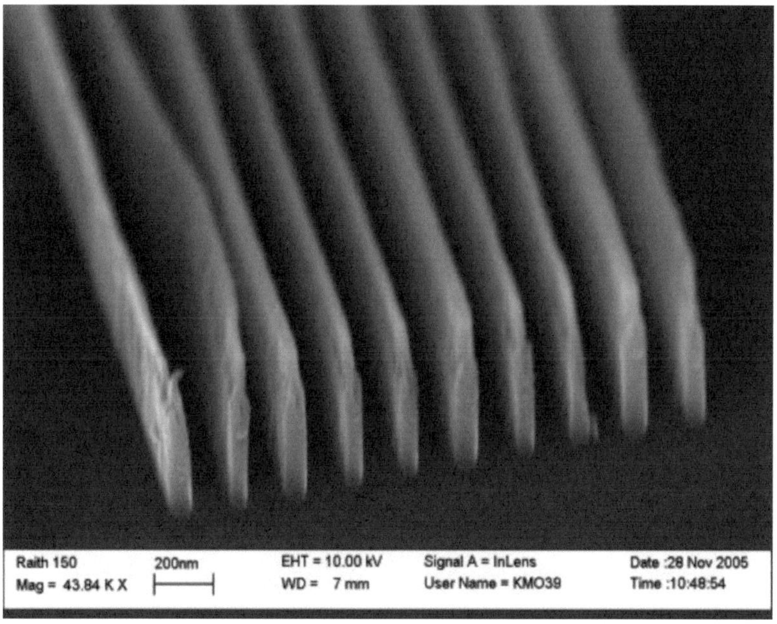

Figure 4. SEM images of the fabricated 2D high aspect-ratio nanostructures on quartz substrates showing the 60 nm lines structure with 600 nm height and 120 nm spacing.

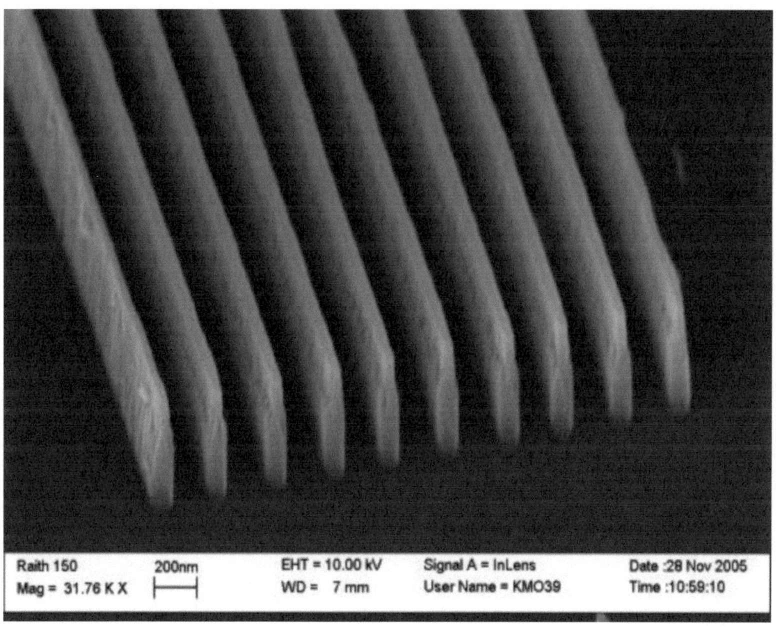

Figure 5. SEM images of the fabricated 2D high aspect-ratio nanostructures on quartz substrates showing the 90 nm lines structure with 600 nm height and 180 spacing

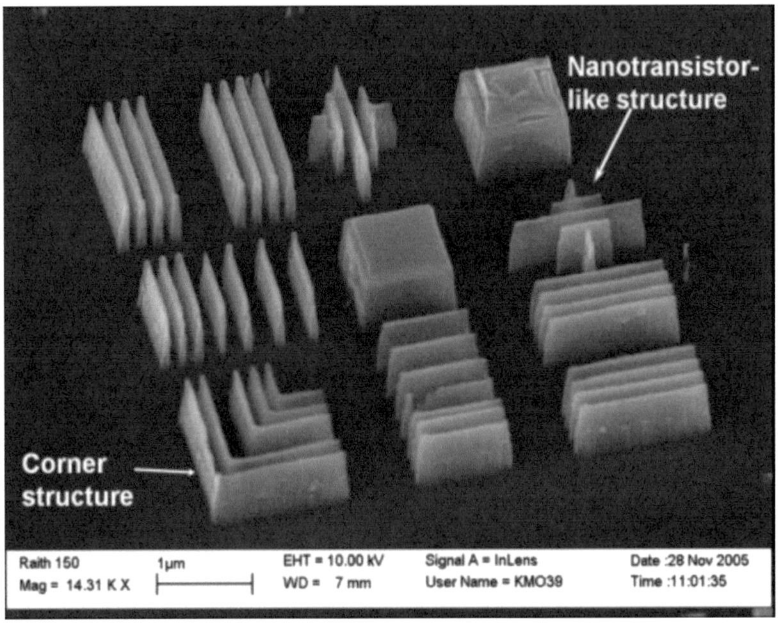

Figure 6. SEM image of the fabricated 2D high aspect-ratio nanostructures on quartz substrates with feature sizes of 60 nm width and 600 nm height lines, corner lines and transistor-like nanostructures.

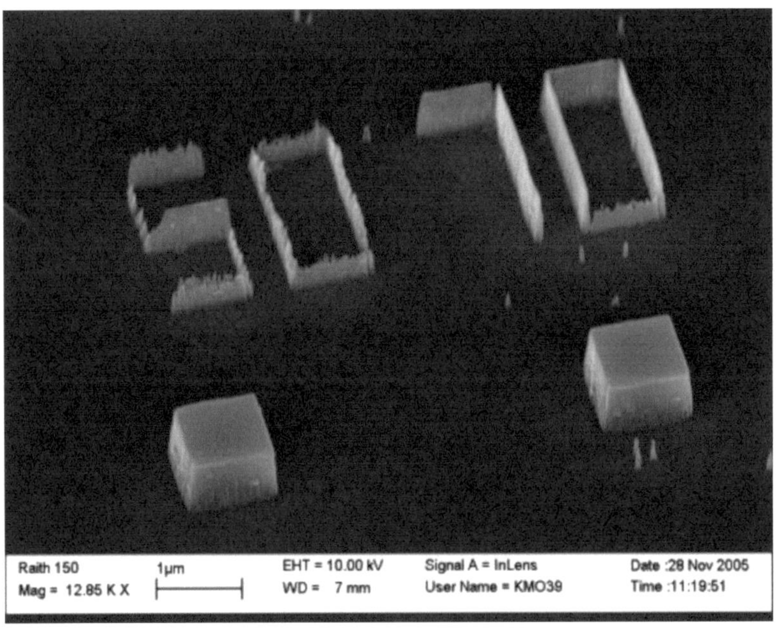

Figure 7. SEM image showing the comparison between the fabricated 50 nm and 70 nm nanostructures after the long etching process

http://dx.doi.org/10.5772/56315

5. Analysis and discussions

For feature sizes below 60 nm, the line structure failed to preserve the flat top surface as compared to the feature sizes above 60 nm. It could be to the insufficient of NiCr mask layer thickness in protecting or masking the top surface from ions bombardment. The NiCr masking layer was completely etched away earlier than the total etching time.

The uneven profile of NiCr masking layer could be due to blockage of the resist pattern during the metallization process. This could be explained schematically as illustrated in Figure 8, where it shows the diagram of metal deposition process for NiCr masking layer. During metal deposition process using thermal evaporator, depositing a metal layer through a narrow trench is difficult. Figure 8 also point out the location (at the entrance of the trench) of the shadowing effects and the difference between the measured coating thicknesses on top of the resist as compared to the actual NiCr mask thickness deposited on the substrate. This could be improved by using a thinner resist layer and a special developer that formulated for developing pattern with feature sizes less than 60 nm.

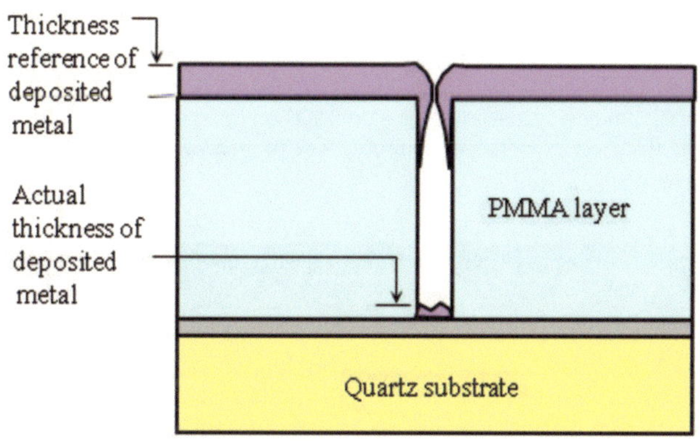

Figure 8. The schematic diagram describing the issue of metal deposition on feature sizes below 50 nm using PMMA bi-layer where shadowing effects causing insufficient metal deposition on fine feature sizes.

Figure 9 shows the SEM image of the deposited NiCr mask layer on quartz substrate where it shows 20 nm in width lines. The uneven brightness of the lines image could be due to uneven deposited thickness of NiCr mask layer. It has been observed that the NiCr layer thickness is less than 1/3 of the estimated thickness by metallization process due to shadowing effects.

Figure 10 shows the SEM image the 40 nm lines nanostructures after the etching process. The 40 nm lines feature failed to preserve the flat top surface due to the uneven and in sufficient NiCr mask layer thickness. The NiCr masking layer was etched away earlier than the total etching time.

Figure 11 shows the SEM image of 20 nm line feature with worse condition than 40 nm lines feature. The metal deposition process is more difficult for fine feature size which leads to lesser NiCr masking layer thickness. A new metal deposition method and new masking material could be attempted to resolve this issue.

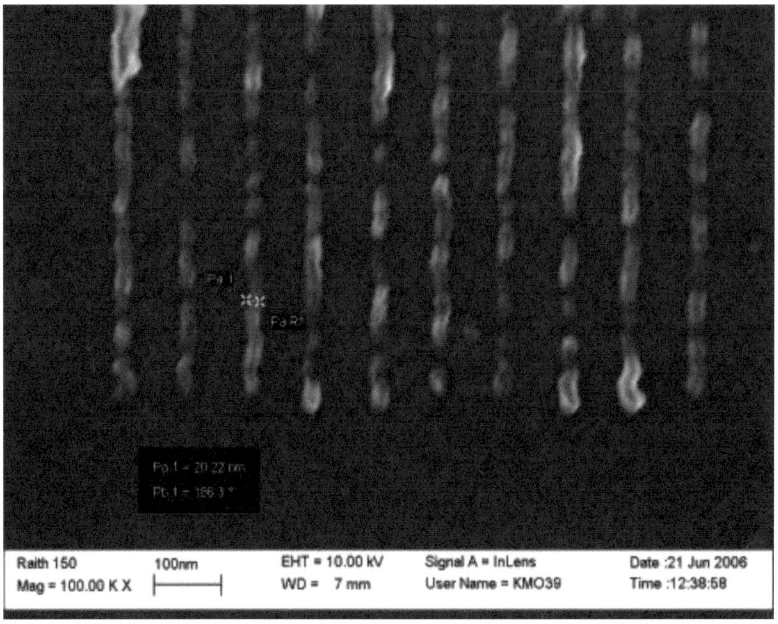

Figure 9. SEM image of 20 nm line pattern metallization (NiCr) on quartz substrate. The NiCr layer thickness is less than 1/3 of the estimated thickness by metallization process due to shadowing effects.

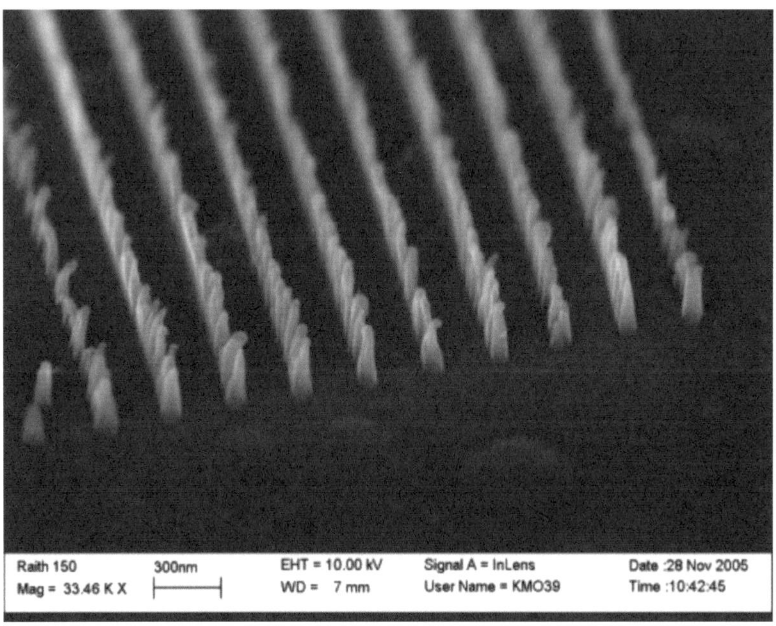

Figure 10. SEM images of the fabricated 2D nanostructure with feature sizes of 40 nm which failed to preserve the flat top profiles.

http://dx.doi.org/10.5772/56315

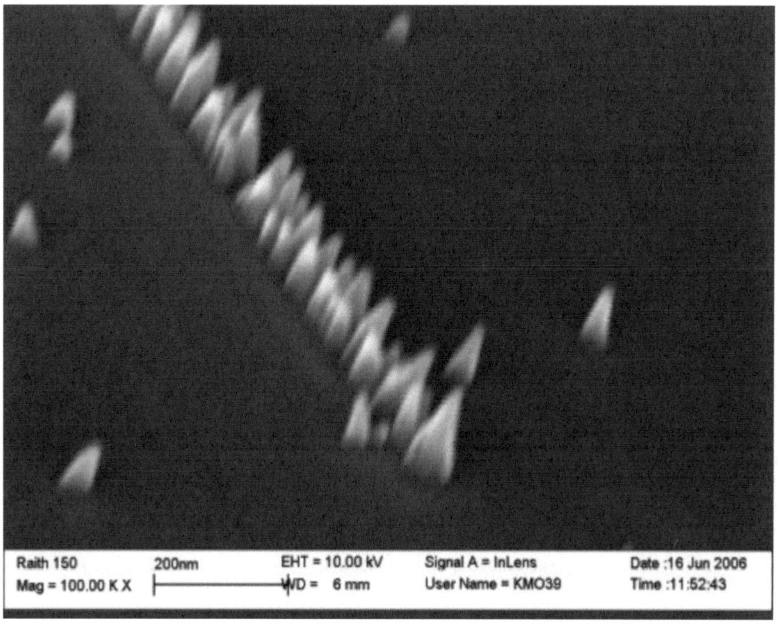

Figure 11. SEM images of the 20 nm lines structures which failed to preserve the flat top profiles after the etching process.

6. Conclusions

The fabrication of high aspect-ratio nanostructures on quartz substrates with feature sizes ranging from 60 nm to 100 nm has been demonstrated. The high aspect-ratio of up to 1:10 on 60 nm line nanostructures has been fabricated using the optimized CHF_3/Ar RIE plasma recipe. The fabrication of nanostructures with feature sizes below 60 nm was unsuccessful due to inadequate NiCr mask layer thickness. The five minutes cooling intervals on every 15 minutes of plasma etching process was found to achieve a 10 nm/min etch rate and steep vertical sidewalls profile.

Acknowledgements

Authors would like to acknowledge Universiti Sains Malaysia and Malaysian Ministry of Higher Education (MOHE) through Fundamental Research Grant Scheme (FRGS) 203/PMEKANIK/6071229 for funding this research project.

Author details

Khairudin Mohamed[1] and Maan M. Alkaisi[2]

1 Nanofabrication and Functional Materials Research Group, School of Mechanical Engineering, Engineering Campus, Universiti Sains Malaysia, Penang, Malaysia

2 MacDiarmid Institute for Advanced Materials and Nanotechnology, Department of Electrical and Computer Engineering, University of Canterbury, New Zealand

References

[1] Villa-comamala, J, Gorelick, S, Guzenko, V. A, Farm, E, Ritala, M, & David, C. Dense high aspect ratio hydrogen silsesquioxane nanostructures by 100 keV electron beam lithography", Nanotechnology, 21, 285305, (2010).

[2] Howells, M, Jacobsen, C, & Warwick, T. and van den Bos A Science of Microscopy (New York: Springer) chapter 13 (Principles and Applications of Zone Plate X-Ray Microscopes) (2007). , 835-926.

[3] Huang, M. -J, Yang, C. -R, Lee, R. -T, & Chiou, Y. -C. Fabrication of a fuel cell electrode with a high-aspect-ratio nanostructure array", Journal of Micromechanics and Microengineering, (2009). , 045003.

[4] Mohamed, K. PhD Thesis, University of Canterbury, New Zealand, (2009).

[5] Laemer, F. A. Schilp of Robert Bosch GmbH, "Method of anisotropically etching silicon, US Patent (1994). (5)

[6] Laemer, F, Franssila, S, Sainiemi, L, & Kolari, K. Deep Reactive Ion Etching" in Handbook of Silicon Based MEMS Materials and Technologies, (Eds. V. Lindroos, M. Tilli, A. Lehto and T. Motooka), Elseveir, (2010).

[7] Moreau, W. M. Semiconductor Lithography: Principles, Practices and Materials, Series Microdevices: Physics and Fabrication Technologies, Prenum Press, (1988).

[8] Mohamed, K, Alkaisi, M. M, & Blaikie, R. J. Surface charging suppression using PEDOT/PSS in the fabrication of three dimensional structures on quartz substrate", Microelectronic Engineering, 86, (2009). , 535-538.

[9] May, G. S, & Sze, S. M. Fundamentals of Semiconductor Fabrication, John Wiley and Sons, (2004).

[10] Oehrlein, G. S, Zhang, Y, Vender, D, & Haverlag, M. Fluorocarbon high density plasma: fluorocarbon film deposition and etching using CF4 and CHF3", Journal of Vacuum Science and Technology A: Vacuum, Surface, and Film, (1994). , 12, 323-332.

[11] Rossnagel, S. M, Cuomo, J. J, & Westwood, W. D. Handbook of Plasma Processing Technology: Fundamentals, Etching, Deposition, and Surface Interaction, William Andrew Inc., (1990).

[12] Chan, C. M, Ko, T. M, & Hiraoka, H. Polymer surface modification by plasmas and photons", Surface Science Reports, (1996). , 24, 1-54.

[13] Coulburn, J. W. Some fundamental aspects of plasma-assisted etching", in Handbook of Advanced Plasma Processing Techniques, R. J. Shul, S. J. Pearton, Eds. Springer, (2000). , 24.

[14] Booth, J. -P, & Cunge, G. CFx radical creation and destruction at surface in fluorocarbon plasmas", Journal of Plasma and Fusion Research, (1999). , 75, 821-829.

[15] Cheam, D. D, & Bergstrom, P. L. Optimization of focus ion beam patterning and reactive ion etching process of quartz imprint template for ultra violet nanoimprint lithography", in 26th Army Science Conference(ASC), JW Marriot Grande Lakes, Orlando, Florida, (2008).

[16] Joubert, O, Oehrlein, G. S, & Zhang, Y. Fluorocarbon high density plasma vs influence of aspect ratio on the etch rate of silicon dioxide in an electron cyclotron resonance plasma",Journal of Vacuum Science and Technology A: Vacuum, Surface, and Film, (1994). , 12, 658-664.

[17] Mohamed, K, & Alkaisi, M. M. Investigation of a nanofabrication process to achieve high aspect-ratio nanostructures on a quartz substrate", Nanotechnology, 24(1), 015302, (2013).

Chapter 10

Fabrication of 3D Micro- and Nano-Structures by Prism-Assisted UV and Holographic Lithography

Guomin Jiang, Kai Shen and Michael R. Wang

Additional information is available at the end of the chapter

http://dx.doi.org/10.5772/56417

1. Introduction

Lithography, the fundamental fabrication process of semiconductor devices, is playing a critical role nowadays in the fabrication of micro- and nano-structures especially for the realization of micro-electro-mechanical systems (MEMS), microfluidic devices, photonic crystals, photonic integrated circuits, micro-optics, and plasmonic optoelectronic devices. These devices have various practical applications including optical display, optical memory, optical interconnection for high speed computing systems, photonic planar lightwave circuits, medical fluidic filtering devices, drug delivery devices, solar energy devices, antireflection optical elements, and optical sensors.

Traditional photolithography, laser direct write maskless lithography, and gray-scale lithography are suitable for micro-structural patterning. E-beam lithography, EUV and X-ray lithography employing shorter wavelength beams can help improve pattern resolution for fabrication of finer scale nano-structures. Contact lithography including soft lithography and nano-imprint lithography offers capability of higher resolution nano-structure fabrication. However, most of these existing lithography tools are limited to the fabrication of two-dimensional (2D) micro- and nano-structures. Multilayer 2D patterning using the photo and e-beam lithography tools can yield 3D layered structures. Such fabricated structures due to multilayer lithography, would be very difficult to achieve high resolution layered thickness control and layer to layer alignment for finer resolution 3D micro-structures.

We present herein prism-assisted inclined ultraviolet (UV) lithography and holographic lithography (HL) for the fabrication of three-dimensional (3D) micro- and nano-structures in SU-8 photoresist. For inclined UV lithography, a prism is used as a refractor to deflect the incident UV light and expand the exposure beam angle range in the resist film. The sample internal surface reflection of the exposing UV light can facilitate the fabrication of symmetric structures. Prism with multidirectional side surfaces can be used to achieve one-step exposure fabrication of multidirectional slanted structures. For holographic lithography, a prism is used to form multi-directional interference beams that greatly simplify the beam splitting and redirecting in conventional HL and at the same time minimize the system vibration sensitivity. The prism-assisted HL is attractive for fast and large area realization of crystal structures, especially quasi-crystal structures. Some practical applications will be discussed.

2. Prism-assisted inclined UV lithography

Inclined UV lithography has recently been used for the fabrication of 3D microstructures (Beuret et al., 1994; Yoon et al., 2006; Han et al., 2004; Campo & Arzt, 2008; Campo & Greiner, 2007). It has demonstrated effective production of various 3D patterns for many practical applications. Two critical problems, namely limited exposure angle and complicated rotation process, have so far restricted the widespread use of the inclined UV lithography in the fabrication of 3D microstructures.

Many devices, such as optical pick-up heads (Huang et al., 2004), embedded waveguide mirrors (Dou et al., 2010), and sharped microneedles (Han et. al., 2007), require microstructures with large side surface angles measured from the normal direction of the resin surface. The widely used negative photoresist SU-8 (refractive index 1.67 at 365 nm) from MicroChem as an example when exposed directly in air, as shown in Fig. 1, the inclined exposure angle in the lithographic resin can in general be easily adjusted by changing the slanted stage angle. However, the refractive beam path bending associated to the large index difference between the air and photoresist has limited the exposure angles in the resin and thus the realization of large angle side surfaces.

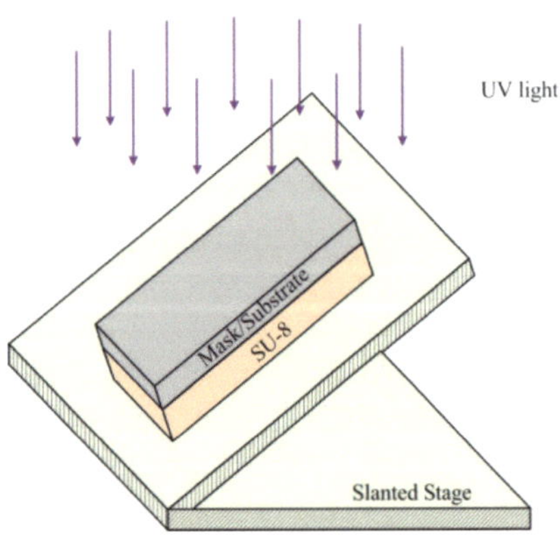

Figure 1. Schematic diagram of inclined UV lithography in the air.

In order to expand the exposure angle in the resin, one way is to immerse the lithographic resin in an index matching liquid (Huang et al., 2004; Han et. al., 2007; Ling & Lian, 2007), such as deionized (DI) water (n=1.33), heptane (n=1.39), or glycerol (n=1.6), which can effectively

http://dx.doi.org/10.5772/56417

minimize the light beam path bending, as shown in Fig. 2(a). Fig. 2(b) presents the simulation results of the exposure angle θ in the SU-8 versus slanted stage angle β in air, water, heptane and glycerol, respectively. The exposure angle can be easily increased beyond 45° with the use of an index matching liquid. However, this immersion method demands a sample settlement time in the index matching liquid to avoid uneven liquid surface and bubble formation. The presence of index matching liquid may affect the UV exposure properties of the resin, because of its influence of certain characteristics of the lithographic resin, such as water content.

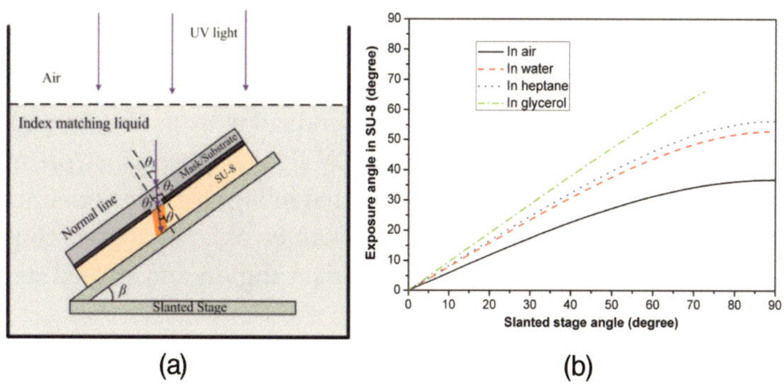

Figure 2. a) Schematic diagram of inclined UV lithography in an index matching liquid and (b) simulation of the exposure angle versus slanted stage angle in air, water, heptane, and glycerol.

Multi-directional inclined structures are normally applied in some more complex devices, such as microfilter and micromixer. Multi-step UV exposure with sample rotation has been proved effective. The slanted sample/mask holder of Fig. 3 can be used for such purpose. It would be more attractive to introduce a one-step fabrication technique for the realization of 3D structures with multi-directional inclined angles.

Figure 3. The sample/mask plate holder with rotation features for multi-directional inclined UV exposure.

2.1. Prism-assisted UV lithography for slanted structures with large exposure angles

As schematically presented in Fig. 4, the prism-assisted UV lithography can expand the exposure angle of slanted structures in the resin. A glass prism (refractive index 1.53 at 365 nm) acts as a refractor to deflect the direction of the incident UV light in the resin for the inclined UV lithographic exposure. This overcomes the exposure angle limitation due to the original large index difference between air and photoresist. The back-side UV exposure is applied here for accurate patterning benefiting from the intimate contact between the photomask/substrate and the resin. The prism side surface to bottom surface angle is α. Poly-dimethysiloxane (PDMS) serves as a good candidate for fast custom prototyping of the prism in house (Kang et al., 2006) with needed prism angle, benefiting from its low cost, molding flexibility, and easy angle control. The slanted stage provides an inclined angle β to the horizontal plane. The incident UV light at 365 nm wavelength is perpendicular to the horizontal plane. In the lithographic exposure process, the refractions of UV light happen at air/prism, prism/mask, and mask/SU-8 interfaces, as schematically illustrated in Fig. 4(b). The relationships between the incident and refractive angles on all interfaces can be obtained based on the Snell's law. The exposure angle θ in SU-8, as a function of the prism angle α and slanted stage angle β, can be written as (Jiang et al., 2012a)

$$\theta = \sin^{-1}\left\{\frac{n_{prism}}{n_{su-8}}\sin\left[\alpha - \sin^{-1}\left(\frac{n_{air}}{n_{prism}}\sin(\alpha-\beta)\right)\right]\right\},$$

where n denotes the refractive index of each medium.

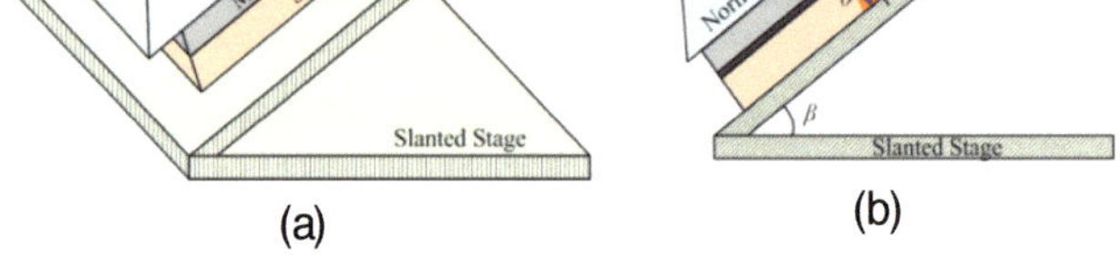

Figure 4. a) Schematic diagram of prism-assisted UV lithography for the expansion of the exposure angle in the resin. (b) Schematic of cross-sectional view of the UV light path bending.

The calculated relationships of exposure angle θ, the prism angle α, and the slanted stage angle β are shown in Fig. 5. The exposure angle in SU-8 can be easily expanded beyond 60° by using this method. Different combinations of α and β to realize 15° (green dash-dot line), 30° (black solid line), 45° (red dotted line), and 60° (blue dashed line) inclined surfaces are also highlighted in this figure. Obviously, we can expand and control the exposure angle in SU-8 by flexibly varying the combination of the prism angle and the slanted stage angle.

Using this technique, we have fabricated 3D slanted structures with different inclined angles of 15°, 30°, 45°, and 60° as illustrated in Fig. 6 (Jiang et al., 2012a). A 45° prism was used in the fabrication process, and the slanted stage angles were set to 0°, 27°, 54°, and 85°, respectively. Some functional and interesting inclined structures with 45° exposure angles (see Fig. 7) can be easily fabricated using this method.

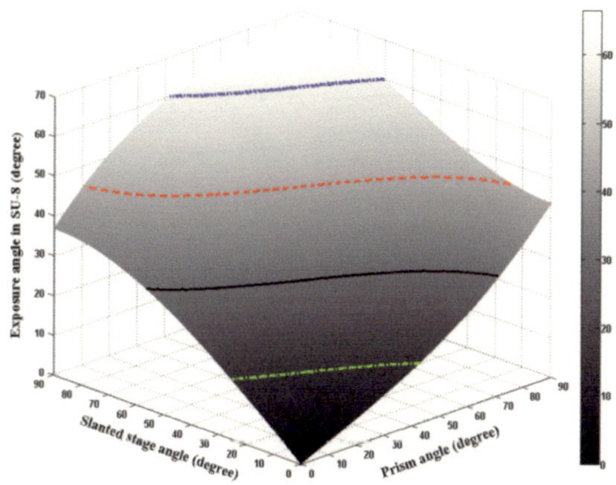

Figure 5. The exposure angle θin SU-8 as a function of the prism angle $\alpha$and slanted stage angle β.

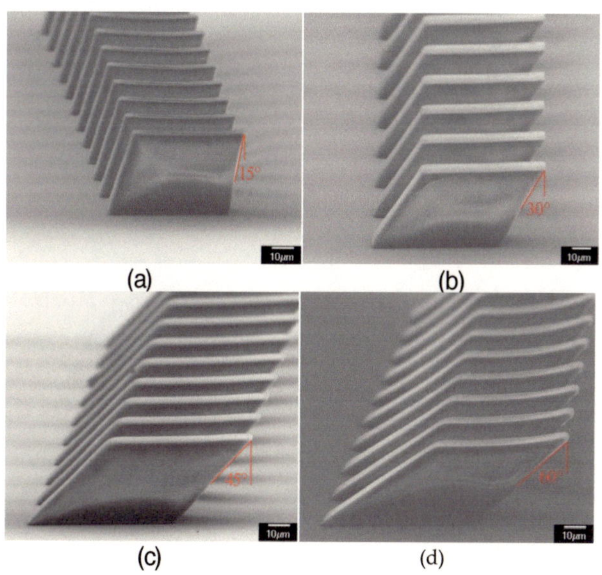

Figure 6. SEM images of the fabricated inclined structures with different exposure angles.

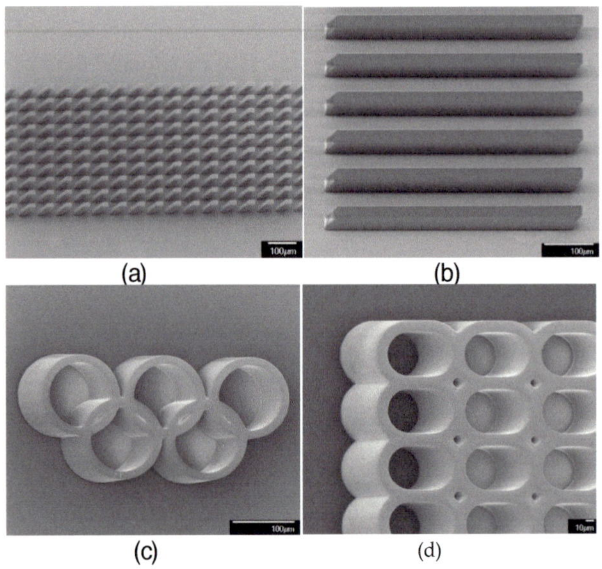

Figure 7. SEM images of the fabricated 3D inclined microstructures with 45o exposure angles.

The concept of utilizing the sample internal surface reflected exposing UV light to initiate cross-linking of the photoresist by increasing exposure times at a fixed UV light power (Campo & Arzt, 2008; Zhu et al., 2008) is given in Fig. 8. This is a simple yet efficient way to fabricate symmetric microstructures instead of twice exposures. The SEM images of the fabricated symmetric microstructures with 45° slanted surfaces on both sides are shown in Fig. 9 (Jiang et al., 2012b).

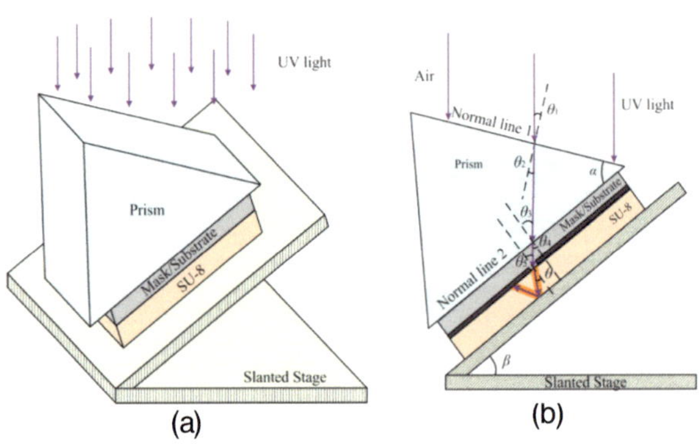

Figure 8. a) Schematic diagram of initiating cross-linking of the resin. (b) Schematic of cross-section view of the UV light path bending showing both incident and reflection exposures.

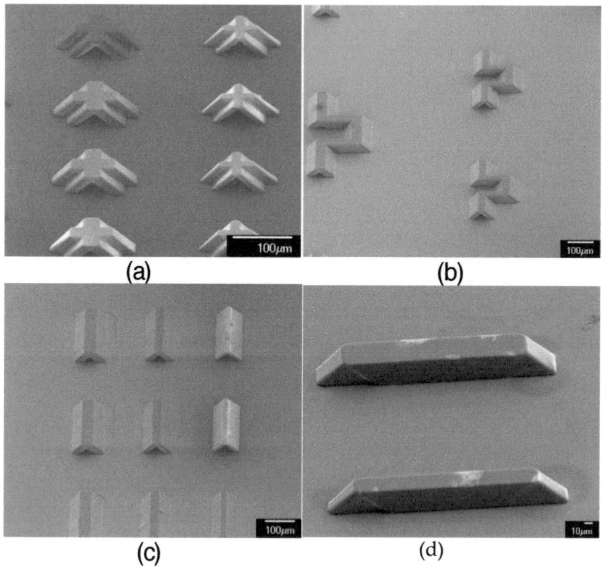

Figure 9. SEM images of the inclined structures with 45∘ exposure angles fabricated by internal reflected UV exposure beams.

2.2. Prism-assisted on-step UV lithography for multidirectional inclined structures

Besides the expansion of the exposure angle, the prism-assisted UV lithography is also attractive in one-step exposure fabrication of 3D structures with multi-directional inclined angles. The fabrication of V-cut structures with a right angle prism has been reported (Huang et al., 2008). A multi-surface optical prism of a polyhedron pyramid structure (a polyhedron pyramid prism) is introduced here to assist the fabrication of some particular microstructures as depicted in Fig. 10(a). Different side surfaces of the prism will simultaneously deflect UV light to different directions. Therefore, a set of slanted exposure beam columns will be formed by a set of refracted exposure beams from these side surfaces. It is thus possible to fabricate a complex multi-directional slanted structure in one step. Here, the slant stage angle β is set to be 0°. To illustrate this concept, we show the bending UV light paths from two side surfaces of the prism in Fig. 10(b).

A corner prism in Fig. 11 (a) was used in our experiment. It has a refractive index of 1.53 at 365 nm and a prism angle of 54.7° for all three side surfaces. The exposure angle θ in SU-8 caused by the refracted UV beams from each side-surface is 21°. When using a separated circular hole mask pattern, three slanted exposure beam columns are formed simultaneously by the refracted exposure beams from these three side surfaces. Thus, the exposure structure is an upside-down tripod structure. Fig. 11(b) and (c) present the fabricated upside-down tripod structures with different heights (Jiang et al., 2012a). Fig. 11(d) is the side view of the tripod structure when placing on its side which is dimensionally similar to that of Fig. 11(c).

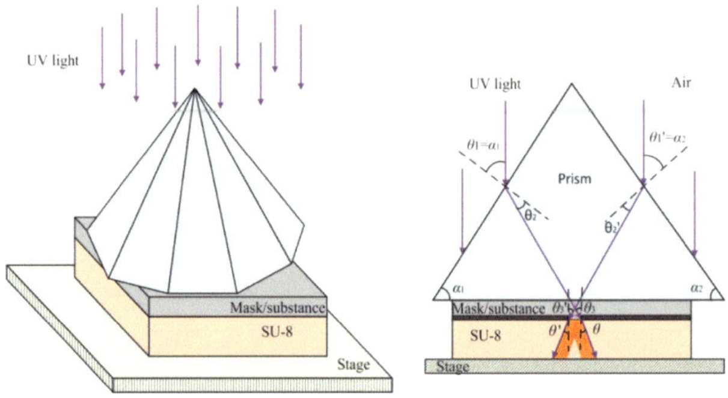

Figure 10. a) Schematic diagram of fabricating multi-directional structures using a polyhedron prism. (b) Schematic of a cross-sectional diagram showing the UV light paths.

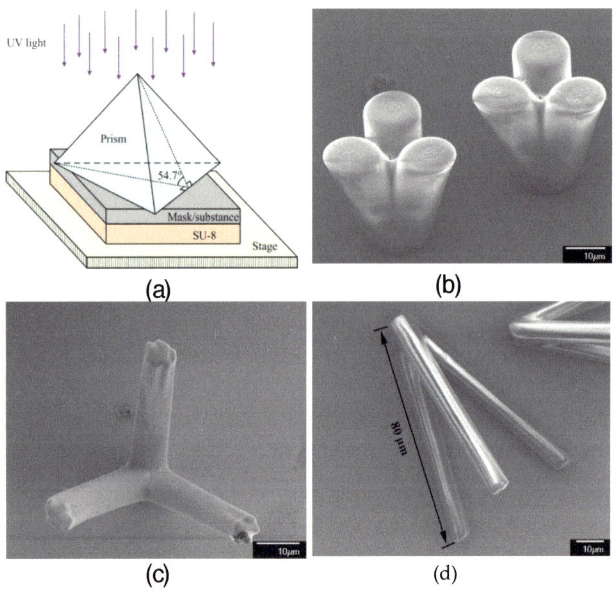

Figure 11. Fabrication of upside-down tripod structure by one-step exposure using a corner prism.

By arranging the basic upside-down tripod structures side by side, more complex 3D micro-structures can be obtained. We use a mask pattern of an equilateral triangular lattice of circular holes in the experiment. If one base side of the prism is parallel to one side of the triangular lattice (see Fig. 12(a)), the nearest three upside-down tripod structures will intersect as shown in Fig. 12(b). If the prism is rotated in-plane by 90°, one base side of the prism will be vertical to one side of the triangular lattice (see Fig. 12(c)). The nearest three upside-down tripod structures will not intersect with each other and a new structure as illustrated in Fig. 12(d) is obtained.

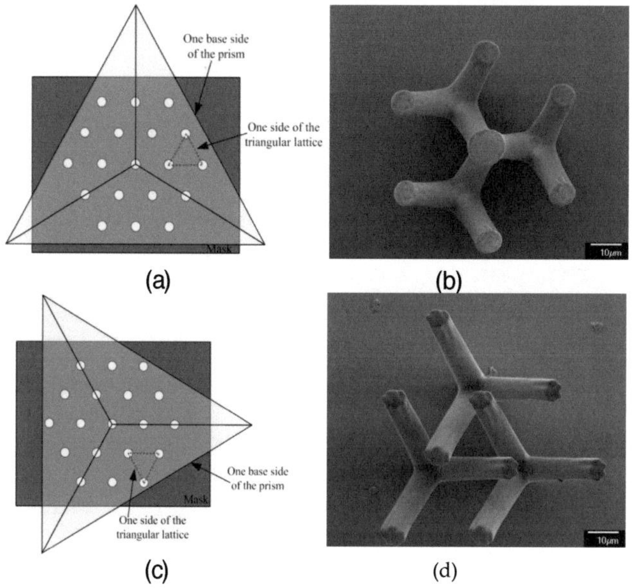

Figure 12. Two specific ways for the arrangement of the basic upside-down tripod structures.

With the setup in Fig. 12(a), the structure in Fig. 13(a) which owns a new layer with an equilateral triangular lattice of circular holes is formed when the height of the structure is 55 μm and the side length of the triangular lattice on the mask is 36 μm. If we change the setup as shown in Fig. 12(c), the 3D mesh structure highlighted in Fig. 13(b) is realized when the side length of triangular lattice on the mask is 28 μm and the height of the structure is 72 μm. This structure consists of three independent parts similar to that of Fig. 13(a). The three independent parts can be easily distinguished by following the color arrows indicated in Fig. 13(b), red arrows (the first part), blue arrows (the second part), and black arrows (the third part). Since this 3D mesh structure has three independent parts that are weaved closely together without intersecting, it may have a great strength and a greater flexibility. Therefore, this technique may be used to fabricate carbon fiber sheets with certain extensibility.

Fig. 13(c) and (d) presents a structure having one more layer than that of Fig. 12(a) which is achieved by adjusting the height of the structure to be 85 μm and the side length of the triangular lattice to be 28 μm. The periodicity of this 3D microstructure in the vertical direction can be realized by increasing the height of the structures or decreasing the side length of the triangular lattice on the mask. This method may be applied to the fabrication of 3D photonic crystals at optical wavelengths if its size and periodicity can be further reduced.

If we replace the corner prism in Fig. 11(a) with a cone prism as illustrated in Fig. 14(a), circular symmetric 3D structures can be obtained. In the experiment, the cone prism (refractive index 1.53 at 365 nm) has a 60° prism angle. Since all refracted UV light beams from cone surface intersect at the center of the base plane, by position a circle hole mask pattern at the base center of the prism, the exposure pattern on the photoresist will show a horn structure. The fabricated

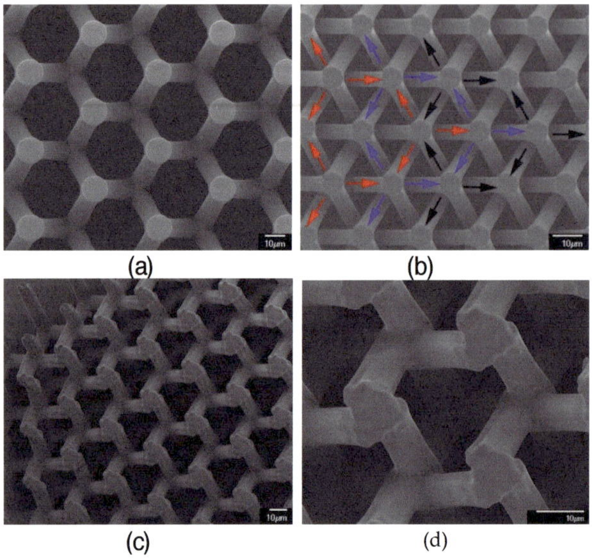

Figure 13. SEM images of the fabricated complex 3D microstructures using a corner prism.

horn structure is presented in Fig. 14(b) when using a mask with a 20 μm diameter circular hole pattern. The defects on the fabricated horn structure may be attributed to the misalignment between the center of circle pattern and the base center of prism and/or the imperfections of the cone prism.

If the circle hole mask pattern is not located at the center of the cone prism, the refracted UV light beams from cone surface to different directions will be asymmetric when passing through the circle hole mask pattern, so yielding asymmetric fan-shaped 3D structures. Fig. 14 (c) and (d) presents the asymmetric exposure patterns on the photoresist, when the circle holes mask pattern is 3 mm and 10 mm away from the base center of the prism, respectively. Because these holes are close to each other as a group but much farther away from the base center of the prism, these holes received similar asymmetric exposure and thus resulted in similar lithographic patterns. Clearly, different displacement from the base center of the prism will result in different asymmetric exposures.

When using a polyhedron pyramid prism for one-step UV lithographic fabrication of 3D multi-directional slanted structures, attention should be paid on the exposure dosage. There is certain beam energy lost as the UV light passes through the prism and interfaces. The exposure energy should be well controlled in the experiment. Fig. 15 shows the fabrication result of a similar structure as shown in Fig. 12(c) without enough exposure dosage. The nearest three upside-down tripod structures are intersecting, as we mentioned earlier. However, after photoresist developing, the inclined pillars cannot support the intersection resulting in the structural collapse. Over exposure may avoid the structural collapse but will introduce surface reflection exposure and structure size expansion that is also not desirable.

http://dx.doi.org/10.5772/56417

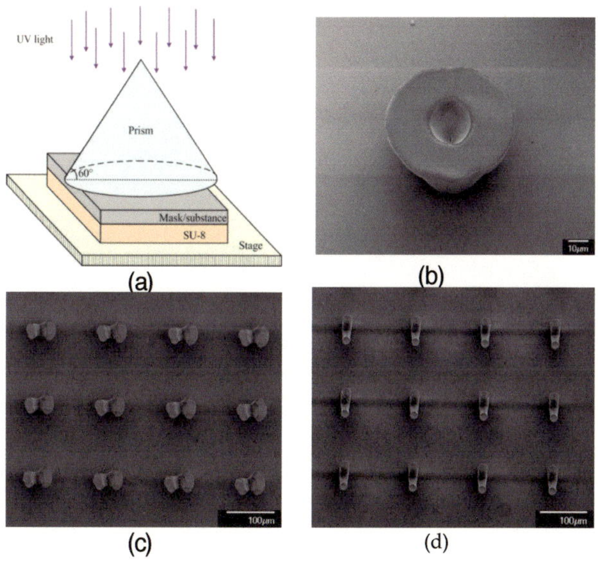

Figure 14. Fabrication of horn and fan-shaped structures by one-step exposure using a cone prism.

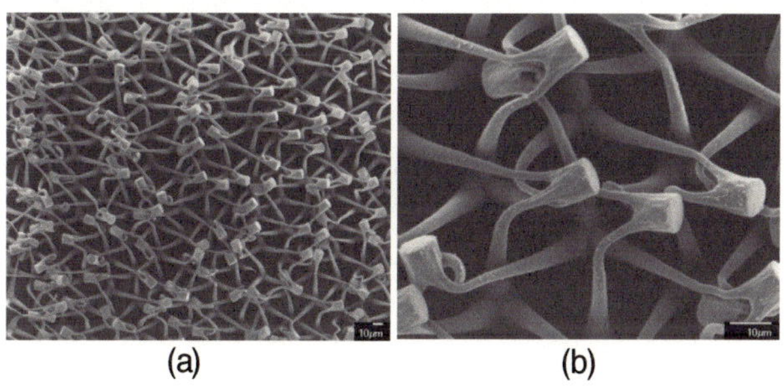

Figure 15. SEM images of the fabricated similar structure of Fig. 12(c) without enough exposure dosage with (b) as magnified view of a local portion of (a).

For multi-directional exposure, attention should also be paid on the effective exposure area of the prism, which also has a great influence on the formation of multi-directional slanted structures. Each side surface has its own projected exposure area (the projection of the side surface to the prism base), so the total effective exposure area means the common area of these projected exposure areas. Take a rotationally symmetric polyhedron pyramid prism as an example, the effective exposure area is determined by the slanted angle α of prism, the number of side surfaces n, and the circumradius r of the polygonal base. Fig. 16(a) shows the calculated results of effective exposure area (blue circles) and its percentage on the base area (red triangles) as a function of the number of side surfaces, at α = 60° and r= 25 mm. The effective exposure area and its percentage decrease rapidly as the increase of the number of the side surfaces. Fig. 16(b) (blue solid line with α = 60° and n= 3) shows that the effective exposure area can be enlarged by increasing the circumradius of the polygonal base when the number of the side surfaces is fixed. We also found that the percentage of the effective exposure area is independent of the circumradius of the base, as indicated in Fig. 16(b) (red dashed line). Therefore the number of side surfaces and the circumradius of the base need to be reasonably selected in structural fabrication design.

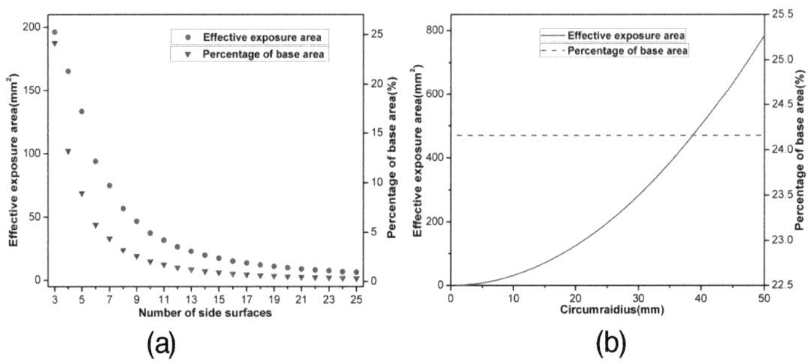

Figure 16. The effective exposure area and its percentage of base area (a) depend on the number of side surface and (b) depend on the circumradius of the base.

The prism-assisted inclined UV lithography is a flexible lithographic technique for the fabrication of 3D microstructures with different side surface angles. The one-step multi-directional exposure fabrication is attractive for fabricating complex 3D microstructures. The prism-assisted fabrication is also useful for the realization of periodic or quasi-periodic nanostructures through interference based holographic lithography discussed below.

3. Prism-assisted holographic lithography for nanostructures

The fabrication of periodic micro- and nano-structures can also be performed by a prism assisted holographic lithography technique. Holographic lithography has been used to fabricate photonic crystals (Yablonovitch, 1987; John, 1987) and metamaterial structures (Pendry et al., 1999; Smith et al., 2004; Soukoulis et al., 2007). Improvements in the HL for minimizing vibration sensitivity and simplifying fabrication process have been a constant and challenging issue. Besides holographic lithography (Berger et al., 1997; Shoji & Kawata, 2000; Campbell et al., 2000) that has demonstrated its suitability for fabrication of periodic micro- and nano-structures, there are other reported fabrication techniques like semiconductor lithography (Fleming & Lin, 1999) and chemical self-assembly (Zhou et al., 2000). The semi-conductor lithography is a very expensive and slow process. It is also difficult to fabricate large area 3D structures. The chemical self-assembly technique is only capable of fabricating face-centered cubic structure with frequent appearance of defects. Holographic lithography is so far a low-cost promising technique in generating multiple interference exposure beams on photoresist for the realization of photonic crystal structures with defect-free micro- or nano-structures over a large area.

http://dx.doi.org/10.5772/56417

Holographic lithography based on the optical setup design can be classified as multiple beams single-exposure (Campbell et al., 2000; Wang et al., 2003; Ullal et al., 2004; Mao et al., 2005, 2006, 2007), two beams multi-exposure (Orlic et al., 2011), and Lloyd's mirror system (Choi & Kim, 2006; Jesson et al., 2007). The multiple beams single-exposure has the advantage of both simple fabrication process (one exposure) and tunable period and structure (1D, 2D and 3D, even quasi-crystals). However, this traditional method requires two independent steps: one is splitting the laser into multi-beams by beam splitters and then align the multi-beams to one point for interference and exposure (Campbell et al., 2000), which introduce significant adjustment complexity. The differences in the optical path lengths and angles among these beams are difficult to eliminate even with a highly skilled optical scientist. Furthermore, it requires very stable optical setup due to the vibration sensitivity of the HL system.

Recently, Wang et al. (2003) demonstrated a top-cut triangular prism (TCTP) that can be used to split one incident beam into four beams and automatically overlay them in the bottom of prism to form a 3D interference pattern in the light sensitive recording materials. They have significantly improved the stability of the optical setup and simplify the alignment of the multi-beams. Herein, we report our improved prism-assisted HL fabrication of micro- and nano-structures.

3.1. Theory of multi-beams interference

The 2D or 3D periodic patterns can be formed by interference of multiple coherent laser beams. In general, the interference equation is given by

$$I=\sum_i |E_i|^2+\sum_{i,j} 2E_iE_j\cos\theta_{ij}\cos[(K_i-K_j)r+(\varphi_i-\varphi_j)]$$

where E_i, K_i and φ_i are the amplitude, wave vector and initial phase of the ith plane wave, and θ_{ij} is the angle of the polarizations between i,jth plane waves. For simplicity, we neglect the minor amplitude difference of different plane waves (assuming ideal equal amplitude beams), and treat all the initial phases as the same. We mainly discuss the wave vectors K_is on how they affect the final pattern formation and add the influence of polarization to the pattern contrast.

It is well known that two beams of coherence light can form 1D periodic pattern, with periodic bright and dark stripes as shown in Fig. 17.

As we increase the number of interference beams, we define $G_k=K_i-K_j$ as the vector from K_j to K_i representing the reciprocal vectors of the periodic lattice structures. Once we determine the independent vectors G_ks, we also determine the lattice structure. The two beams have two wave vectors K_1 and K_2, but only one independent $G_1=K_1-K_2$ (the vector $G_2=K_2-K_1=-G_1$ is not independent). Thus, the interference pattern is 1D stripe. If the number of beams is more than 2, and $(G_k=K_i-K_j)$s are all in the same plane but not parallel, we can generate 2D interference patterns. For example, with three symmetrical beams, the interference pattern is a hexagonal arrangement, as shown in Fig. 18(a). Four symmetrical beams would produce a

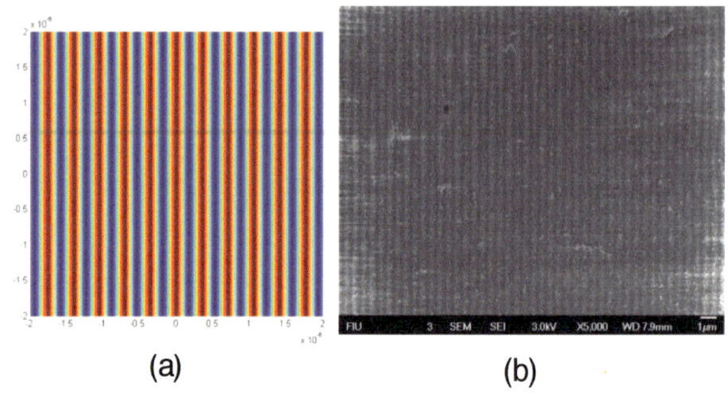

Figure 17. a) The simulated pattern of two beams interference in Matlab and (b) the observed exposed and developed pattern in photoresist under SEM.

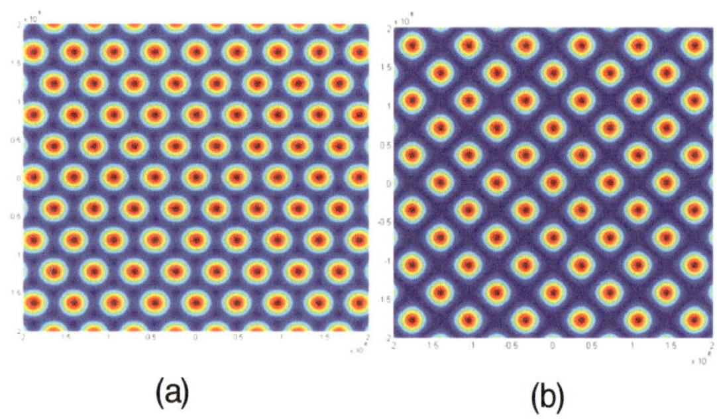

Figure 18. Simulated patterns come from the interference of three (a) and four (b) symmetrical beams.

square pattern shown in Fig. 18(b). These patterns can be realized in a photoresist after exposure and development.

These simple period patterns can be generated by laser or e-beam direct writing with relatively simple software control. The most attractive features of the HL is its capability in generating 3D and quasi-crystal patterns that are difficult by other pattern generating technique. Recently, Meisel *et al.*(2004) demonstrated the interference of umbrella-like four beams (this arrangement of beams will create 3 independent G_ks in 3D space) can form different crystal structures, such as simple cubic and face-centered cubic structure, simulated in Fig. 19.

Furthermore, these kinds of umbrella-like four beams can be easily realized by a top-cut triangular prism. Other groups have reported the use of prism-assist HL to create many different 3D structures (Wu et al., 2004; Xu et al., 2009; Park et al., 2011). Meanwhile, Wang *et al.* (2003) firstly present 5-fold symmetry quasi-crystal by 5 symmetrical beams in half space, simulated in Fig. 20(a). Other symmetrical multi-beams (such as 5, 8, 10…) can yield novel quasi-crystal patterns, shown in Fig. 20(b). These structures have advantages over normal photonic crystals, like higher symmetry and lower refractive index requirement for photonic bandgap forming.

http://dx.doi.org/10.5772/56417

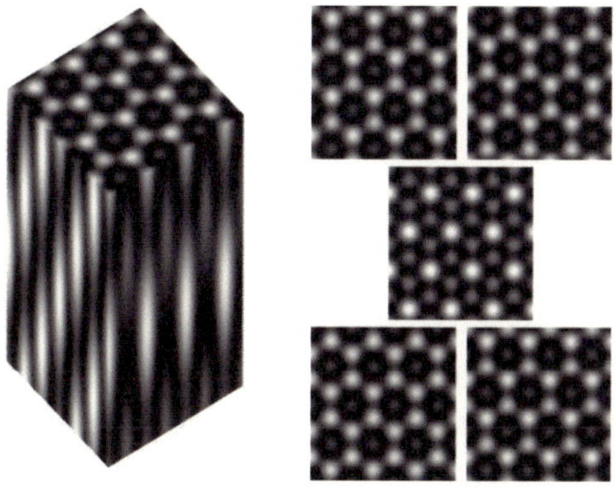

Figure 19. Simulated face-centered cubic structure (a) and the periodic changes of pattern on different X-Y planes (b).

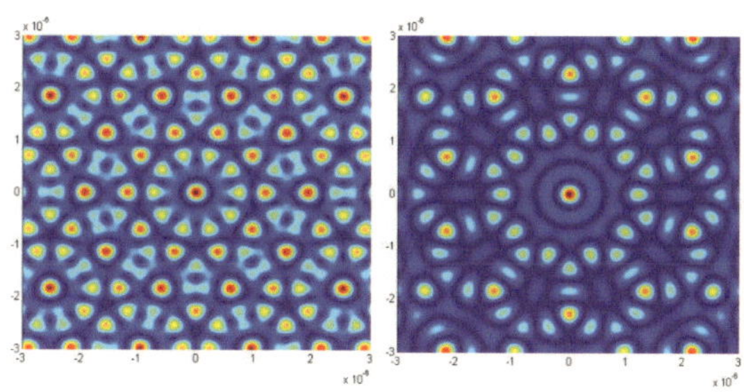

Figure 20. Simulated quasi-periodic pattern of five (a) and ten (b) symmetrical beams

The polarization states of these beams have significant effect to the interference patterns including the contrast and the pattern shape. Commonly, the intensity of cross terms $E_i E_j$ will be influenced by cos θ_{ij}, that will affect the contrast of final pattern. Su *et al.* (2003) investigated the relationship between the polarization and the contrast of the interference pattern, and detailed the optimum selection of polarizations for 3- and 4-beams. Furthermore, choosing special polarizations of different beams will significantly modify the final interference patterns, resulting in wood like 3D pattern (Park et al., 2011) or U-shaped pattern (Yang et al., 2008) for metameterial.

We introduce a novel HL method to create photonic quasi-crystals by fewer beams. Naturally, the same number of beams may be chosen as the fold symmetry, namely 5 beams for 5-fold symmetry, 8 for 8, 10 for 10 and so on (Wang et al., 2003). As the beam number is increased, the contrast of pattern will dramatically reduce due to the huge number of cross terms in the interference equation, making the fabrication very difficult or even impossible. Mao *et al.* (2006) demonstrated earlier that the design and fabrication of higher order photonic quasi-crystals with less number of laser beams. A quasi-periodic structure is regarded as a high dimensional periodic lattice projected onto a 3D space. There is an association between the high-dimensional space and the number of independent reciprocal vectors G_ks. If a (v+1)-dimensional space contains an N-fold symmetric periodic lattice, the smallest number v is given by the Euler totient function (Rabson et al. 1991)

$$v = \phi(N) = N\frac{(p_1 - 1)}{p_1}\frac{(p_2 - 1)}{p_2}\cdots$$

where p_i s are distinct prime factors of N. Here we found that N-fold symmetry quasi-crystal does not need the same number of beams. For example, an 8-fold symmetry with the prime factor of 2, we have $v + 1 = 8(1-1/2)+1 = 5$, that means we just need 5 beams to form an 8-fold photonic quasi-crystal. Furthermore, 10- and 12-fold symmetry quasi-crystals also only need 5 beams. The arrangement of interference beams and simulation of the interference pattern are illustrated in Fig. 21.

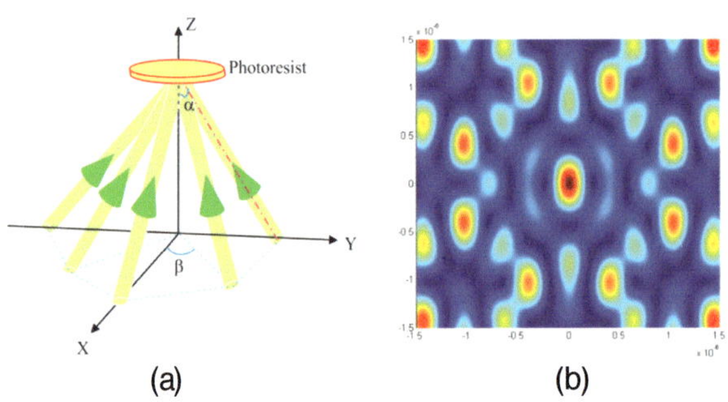

Figure 21. a) The arrangement of 5 beams in ¼ space for eight fold symmetrical interference pattern, where $\alpha = 26.3°$ and $\beta = 45°$. (b) Simulated interference pattern of the special five beams.

3.2. Fabrication of micro- and nano-structures by holographic lithography

3.2.1. Specially designed prism

Generally speaking, the geometry structure of the five coherent beams as shown in Fig. 21 can be configured by beam splitters and mirrors. But this type of conventional setup is subjected to critical alignment and sensitivity to slight positional deviation and minor phase shift, which lead to instability of final interference pattern. The specially configured prism of Fig. 22 can directly split an incident laser beam into five laser beams almost without relative phase shift and difference in optical path length, and overlap in the bottom of the prism. Figure 22(a) illustrates our designed prism with a 45° angle between the side surface and the bottom. For refractive index n=1.51, the emergent laser beams from bottom will illuminate the photoresist with an angle of 26.3°. The specially configured prism consists of two parts – cone and cylinder. The diameter of the bottom is 25.4 mm and the heights of the cone and cylinder are 12.7 mm and 5 mm, respectively.

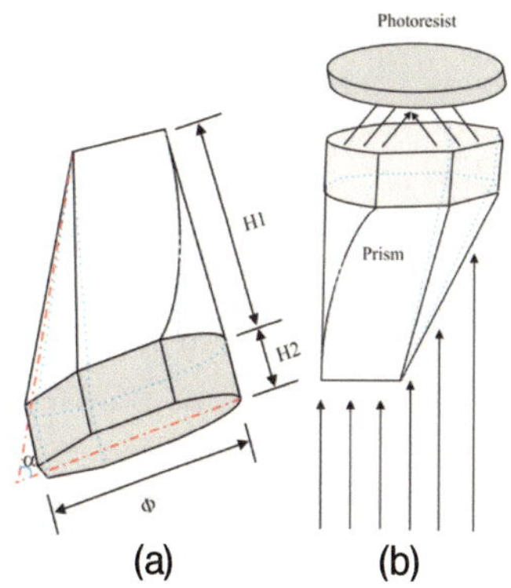

Figure 22. a) Specially configured prism for holographic eight-fold photonic quasi-crystal with only five continuous surfaces out of symmetrical eight side surfaces. Here Φ = 25.4 mm, H_1 = 12.7 mm, H_2 = 5 mm, and α= 45°. (b) Optical setup showing incident laser beam, the prism, and the photoresist.

A unique feature of this specially designed prism is that it has only five continuous surfaces out of symmetrical eight surfaces of the prism. Compared with eight beams from eight side surfaces to form eight-fold symmetric quasi-periodicity, this design of using only five continuous surfaces will dramatically improve the contrast of interference.

3.2.2. Prism-assisted fabrication of 8-fold quasi-crystals with 5 beams

In experiments, the photoresist solution was prepared by mixing 1.0 wt% of a photoinitiator, Cyclopentadienyl(fluorene)iron(II) hexafluorophosphate (Aldrich), in an SU-8 solution (Micro-Chem, SU-8 3010 diluted by SU-8 thinner (3:2)), that is sensitive to green light including the 532 nm wavelength. This photoresist was spin coated (4000 rpm for 30s) on glass plates and pre-baked at ~91°C for about 1 h. A 532 nm linearly polarized laser beam from a diode pumped solid state laser (Crystalaser) with output power of 60 mW was expanded by spatial filter and collimated by a collimation lens. The expanded laser beam was incident from the top side of the prism and recombined at the bottom of the prism, or the plane of photoresist. After about 10s of exposure, the sample was post-baked at ~93 °C for about 1 h to complete polymerization and developed in SU-8 developer (Micro-Chem) for 1 h, followed by rinse in isopropanol. The surface morphologies were investigated by scanning electron microscopy (SEM; JEOL JSM-7000F).

Fig. 23 exhibits four SEM images of holographic eight-fold quasi-crystals. For an appropriate exposure time, an obtained eight-fold quasi-crystal with clear features is shown in Fig. 23(a) and (b). Fig. 23(b) is a magnified image of the same structure of (a), showing more details. When over exposed, the dots will connect with each other as shown in Fig. 23(c) and (d), due to the degree of over exposures, but still revealing eight-fold symmetry. Therefore, eight-fold quasi-crystals can be fabricated by prism assist five beam single-exposure holographic lithography instead of usual eight beams. Additionally, if only use three continuous surfaces, we can get some novel patterns, such as bias short line hexagonal arrangement, as shown in Fig. 24.

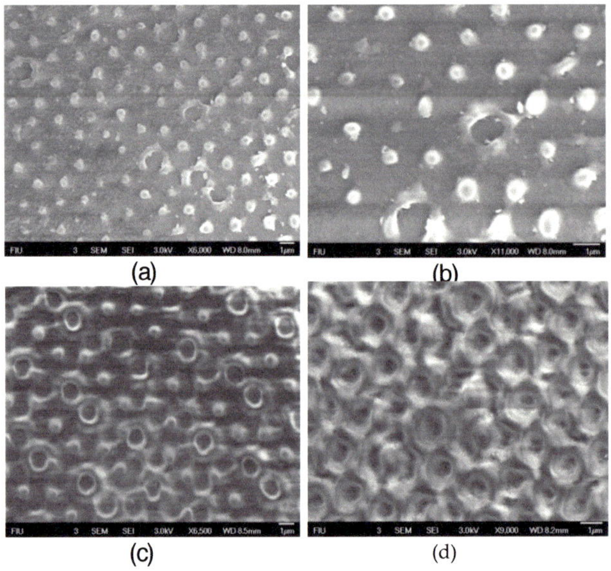

Figure 23. a) After an appropriate exposure time, we get clear picture of eight fold quasi-periodic pattern and (b) its local magnified picture with more details. (c) and (d) show the quasi-periodic patterns realized by slightly and highly over exposures, respectively.

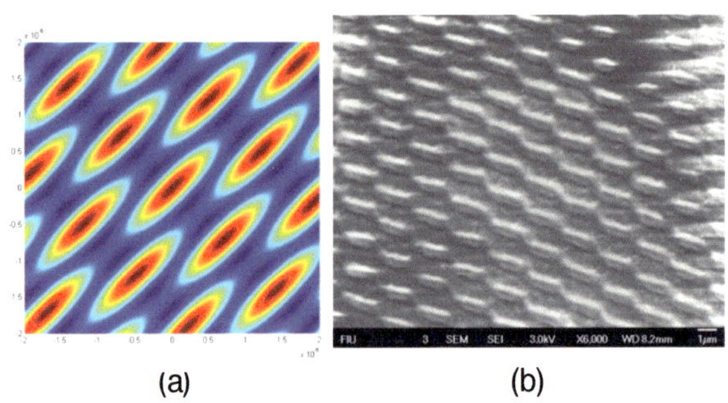

Figure 24. Simulation (a) and SEM picture (b) of three continuous side surfaces interference of this special prism.

4. Application of micro- and nano-structures

Many practical applications of 3D micro- and nano-structures have been reported, such as V-grooves for fiber holder (Ling & Lian, 2007), mesh structures for microfilter (Sato et al., 2004), inclined surfaces for the optical pick-up head (Huang et al., 2004), pyramid array for optical display (Yoon et al., 2006; Shieh et al., 2005), some complex microstructures for micromixer (Baek et al., 2011; Sato et al., 2006) and drug delivery (Han et al., 2007; Yoon et al., 2011), and periodic array for photonic crystal. In this section, we will demonstrate our exploitation of

http://dx.doi.org/10.5772/56417

micro- and nano-structures applications on 45° inclined mirrors and microfilter system, fabricated by the prism-assisted photolithography.

4.1. 45° inclined mirrors for card-to-backplane optical interconnect

Optical interconnects have been extensively researched for high-speed computing systems because of the speed limitations and drawbacks of electrical routing on boards (Doany et al., 2009). Polymer waveguides with 45° inclined mirrors are important components in optical interconnect (Lee et al., 2009; Wang et al., 2005; Wang et al., 2008; Glebov et al., 2005). We fabricated 45° inclined mirror structures in the master photoresist by prism-assisted UV lithography (Jiang et al., 2012c). Sample surface reflected UV light was utilized to eliminate undercut structures and to accomplish the inclined mirror surfaces on both ends of the straight waveguide segments by one-step UV exposure, as shown in Fig. 25. High quality microfluidic channels are then transferred to a PDMS mold. The vacuum assisted microfluidic soft lithography fabrication (Flores et al., 2008) using UV curable core waveguide resin yield the polymer waveguides with 45° inclined surfaces. Fig. 26 illustrates the SEM image of imprinted polymer waveguide before aluminum local coating. After aluminum local coating on the slanted surfaces followed by lower index cladding layer over coating it results in embedded polymer waveguides for optical interconnection. Surface normal optical coupling from the waveguide has been demonstrated supporting the card-to-backplane optical interconnects.

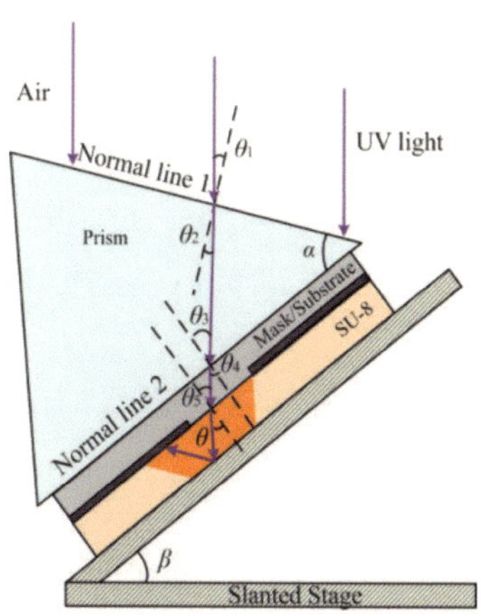

Figure 25. Schematic diagram of fabricating 45° inclined mirror surfaces on both ends of the straight waveguide.

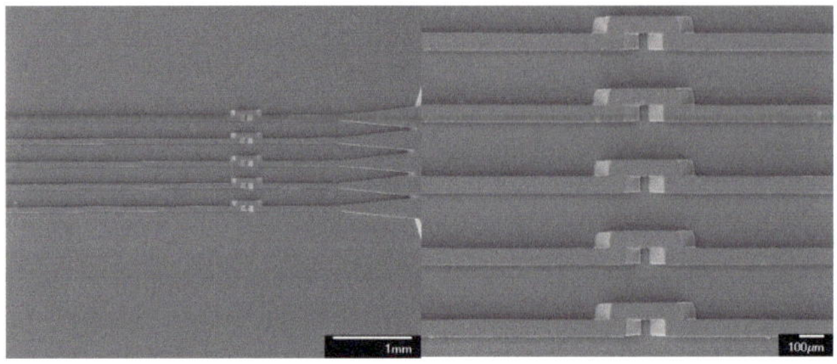

Figure 26. SEM images of imprinted polymer waveguide before aluminum local coating.

4.2. Mesh structures for microfilter system

A right angle prism (refractive index 1.53 at 365 nm) was ultilized for the one-step fabrication of micromesh structures in our experiment, as shown in Fig. 27. Fig. 28 (a) and (c) illustrates the fabricated micromesh structures with different mesh size of 20 μm and 10 μm respectively, which can be easily found from their enlarge parts in Fig. 28(b) and (d). The mesh size can be easily adjusted by changing the distance between rectangles on photomask. We can increase number of the mesh layers by enlarging the height of structure or decreasing the period of the rectangle array. After the fabrication of the micromesh structures, a PDMS mode with microfluidic channel on the bottom is placed on the top of the mesh structures for the micro-filter system. This microfluidic channel should own the same width and height as these mesh structures.

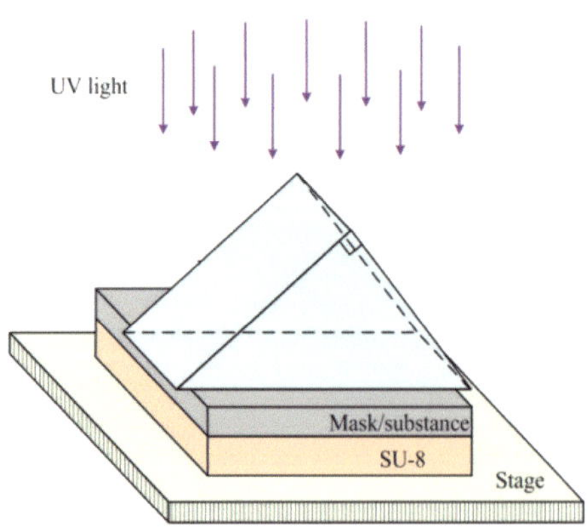

Figure 27. Schematic diagram of fabrication of mesh structures by a right angle prism.

http://dx.doi.org/10.5772/56417

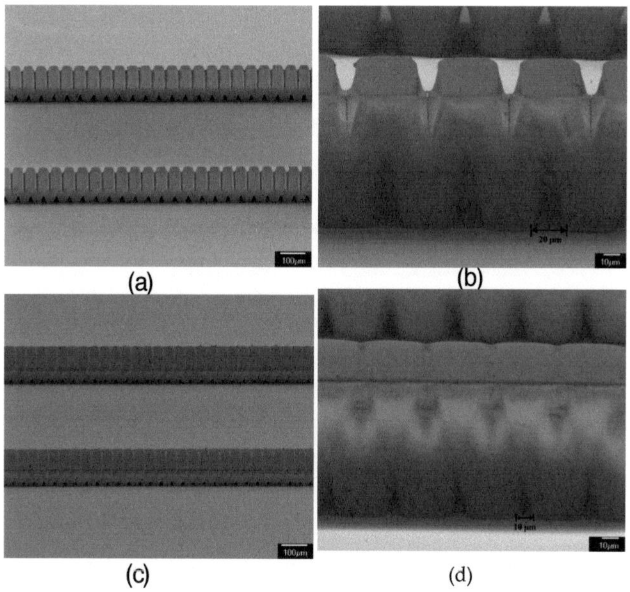

Figure 28. SEM images of micromesh structures with different mesh size.

5. Conclusion

In this chapter, the prism-assisted UV lithography and holographic lithography are introduced for the fabrication of 3D micro- and nano-structures. For prism-assisted inclined UV lithography, a prism is used as a refractor to deflect the incident UV light and control the exposure beams directions in the resist. Slanted structures with exposure angles ranging from 0° to 60° can be easily achieved. The sample internal surface reflection of the exposing UV light can be further utilized for the fabrication of symmetric structures. The fabrication of multi-directional slanted structures can be simplified by one-step UV exposure using a prism with multi-directional side surfaces. For holographic lithography, a prism is used to simply the process in conventional HL and at the same time improve the system stability. The prism-assisted HL is advantageous for realization of periodic sub-micro and nano structures over a large area. Some practical applications of our fabricated structures, such as 45° inclined mirrors and microfilter system, are also discussed. With setup simplicity, cost effectiveness and fabrication flexibility, the prism-assisted UV and holographic lithography techniques should aid in the fabrication of various 3D micro- and nano-structures.

Acknowledgements

The work is sponsored in part by the National Science Foundation. The authors thank Hui Lu, Sarfaraz Baig, and Bing Li for helpful discussions.

Author details

Guomin Jiang, Kai Shen and Michael R. Wang

University of Miami, USA

References

[1] Baek, S, & Song, S. A. (2011). A one-step photolithography method for fabrication of a staggered herringbone mixer using inclined UV lithography. J. Micromech. Microeng. 21 077001

[2] Berger, V, Gauthier-lafaye, O, & Costard, E. (1997). Photonic band gaps and holography. J. Appl. Phys. , 82, 60-64.

[3] Beuret, C, Racine, G. A, Gobet, J, Luthier, R, & De Rooij, N. F. (1994). Microfabrication of 3D multidirectional inclined structures by UV lithography and electroplating. Proc. IEEE Int. Conf. on Micro Electro Mechanical Syst. (MEMS 1994) , 81-85.

[4] Campbell, M, Sharp, D. N, Harrison, M. T, Denning, R. G, & Turberfield, A. J. (2000). Fabrication of photonic crystals for the visible spectrum of holographic lithography. Nature, 404, 53-56.

[5] Campo, A. d, & Arzt, E. (2008). Fabrication approaches for generating complex micro- and nanopatterns on polymeric surfaces. Chem. Rev. , 108-911.

[6] Campo, A. d, & Greiner, C. (2007). SU-8 a photoresist for high-aspect-ratio and 3D submicron lithography. J. Micromech. Microeng. , 17, R81-R95.

[7] Choi, C. –H, & Kim, C. J. (2006). Fabrication of dense array of tall nanostructures over a large sample area with sidewall profile and tip sharpness control. Nanotechnol. , 17, 5326-5333.

[8] Doany, F. E, Schow, C. L, Baks, C. W, Kuchta, D. M, Pepeljugoski, P, Schares, L, Budd, R, Libsch, F, Dangel, R, Horst, F, Offrein, B. J, & Kash, J. A. (2009). 160 Gb/s Bidir- Lithography ectional Polymer-Waveguide Board-Level Optical Interconnects Using CMOS-Based Transceivers. IEEE Trans. Adv. Package , 32, 345-359.

http://dx.doi.org/10.5772/56417

[9] Doua, X, Wangb, X, Huanga, H, Lina, X, Dinga, D, Pana, D. Z, & Chen, R. T. (2010). Polymeric waveguides with embedded micromirrors formed by Metallic Hard Mold. Opt. Express , 18, 378-385.

[10] Fleming, J. G, & Lin, S. -Y. (1999). Three-dimensional photonic crystal with a stop band from 1.35 to 1.95 μm. Opt. Lett. , 24, 49-51.

[11] Flores, A, Song, S, Baig, S, & Wang, M. R. (2008). Vacuum-assisted microfluidic technique for fabrication of guided wave devices. IEEE Photon. Technol. Lett. , 20, 10 1246-1248.

[12] Glebov, A. L, Roman, J, Lee, M. G, & Yokouchi, K. (2005). Optical interconnect modules with fully integrated reflector mirrors. IEEE Photon. Technol. Lett. , 17, 13 1540-1542.

[13] Han, M, Hyun, D. –H, Park, H. –H, Lee, S. S, Kim, C. H, & Kim, C. G. (2007). A novel fabrication process for out-of-plane microneedle sheets of biocompatible polymer. J. Micromech. Microeng. , 17, 1184-1191.

[14] Han, M, Lee, W, Lee, S, -K, & Lee, S. S. (2004). 3D microfabrication with inclined/rotated UV lithography. Sensors Actuators A , 111, 14-20.

[15] Huang, Y, -J, Chang, T. -L, Chou, H. –P, & Lin, C. -H. (2008). A novel fabrication Method for forming inclined groove-based microstructures using optical elements. Japanese Journal of Applied Physics , 47, 5287-5290.

[16] Hung, K. Y, Hu, H. T, & Tseng, F. G. (2004). Application of 3D glycerol-compensated inclined-exposure technology to an integrated optical pick-up head. J. Micromech. Microeng. , 14, 975-983.

[17] Jesson, D. E, Pavlov, K. M, Morgan, M. J, & Usher, B. F. (2007). Imaging surface topography using Lloyd's mirror in photoemission electron microscopy. Phys. Rev. Lett. 99, 016103

[18] Jiang, G, Baig, S, & Wang, M. R. (2012). 3D microstructures fabricated by prism assisted inclined UV lithography. Proc. SPIE 8249 82490L

[19] Jiang, G, Baig, S, & Wang, M. R. (2012). Prism-Assisted Inclined UV Lithography for 3D Microstructures Fabrication. J. Micromech. Microeng. 22 085022

[20] Jiang, G, Baig, S, & Wang, M. R. (2012). Soft Lithography Fabricated Polymer Waveguides with 45° Inclined Mirrors for Card-to-Backplane Optical Interconnects. Proc. SPIE, 8267, Jan. 24,

[21] John, S. (1987). Strong localization of photons in certain disordered dielectric superlattices. Phys. Rev. Lett. , 58, 2486-2489.

[22] Kang, W. -J, Rabe, E, Kopetz, S, & Neyer, A. (2006). Novel exposure methods based on reflection and refraction effects in the field of SU-8 lithography. J. Micromech. Microeng. , 16, 821-831.

[23] Lee, W, Hwang, S. H, Lim, J. W, & Rho, B. S. (2009). Polymeric Waveguide Film with Embedded Mirror for Multilayer Optical Circuits. Photon. Technol. Lett. , 21, 12-14.

[24] Ling, Z, & Lian, K. (2007). SU-8 3D microoptic components fabricated by inclined UV lithography in water. Microsyst. Technol. , 13, 245-251.

[25] Mao, W. D, Liang, G. Q, Pu, Y. Y, Wang, H. Z, & Zeng, Z. H. (2007). Complicated three-dimensional photonic crystals fabricated by holographic lithography Appl. Phys. Lett. 91, 261911

[26] Mao, W. D, Liang, G. Q, Zou, H, & Wang, H. Z. (2006). Controllable fabrication of two-dimensional compound photonic crystals by single-exposure holographic lithography. Opt. Lett. , 31, 1708-1710.

[27] Mao, W. D, Liang, G. Q, Zou, H, Wang, H. Z, Zhang, R, & Zeng, Z. H. (2006). Design and fabrication of two-dimensional holographic photonic quasi crystals with high-order symmetries. J. Opt. Soc. Am. B , 23, 2046-2050.

[28] Mao, W. D, Wathuthanthri, I, & Choi, C. -H. (2011). Tunable two-mirror interference lithography system for wafer-scale nanopatterning. Opt. Lett. , 36, 3176-3178.

[29] Mao, W. D, Zhong, Y. C, Dong, J. W, & Wang, H. Z. (2005). Crystallography of two dimensional photonic lattices formed by holography of three noncoplanar beams. J. Opt. Soc. Am. B , 22, 1085-1091.

[30] Meisel, D. C, Wegener, M, & Busch, K. (2004). Three-dimensional photonic crystals by holographic lithography using the umbrella configuration: Symmetries and complete photonic band gaps. Phys. Rev. B 70, 165104

[31] Orlic, S, Müller, C, & Schlösser, A. (2011). All-optical fabrication of three-dimensional photonic crystals in photopolymers by multiplex-exposure holographic recording. Appl. Phys. Lett. 99, 131105

[32] Park, S, Miyake, M, Yang, S, & Braun, P. (2011). Cu2O Inverse Woodpile Photonic Crystals by Prism Holographic Lithography and Electrodeposition. Adv. Mater. , 24, 2749-2752.

[33] Pendry, J. B, Holden, A. J, Robbins, D. J, & Stewart, W. J. (1999). Magnetism from conductors and enhanced nonlinear phenomena. IEEE Trans. Microwave Theory Tech. , 47, 2075-2084.

[34] Rabson, D. A, Mermin, N. D, Rokhsar, D. S, & Wright, D. C. (1991). The space groups of axial crystals and quasicrystals. Rev. Mod. Phys. , 63, 699-733.

[35] Sato, H, Kakinuma, T, Go, J. S, & Shoji, S. (2004). In-channel 3-D micromesh structures using maskless multi-angle exposures and their microfilter application. Sensors Actuators A , 111, 87-92.

[36] Sato, H, Yagyu, D, Ito, S, & Shoji, S. (2006). Improved inclined multi-lithography using water as exposure medium and its 3D mixing microchannel application. Sensors Actuators A , 128, 183-190.

[37] Shieh, H. -P. D, Huang, Y. -P, & Chien K. -W. (2005). Micro-optics for Liquid Crystal Displays Applications. IEEE/OSA journal of display technology , 1, 62-76.

[38] Shoji, S, & Kawata, S. (2000). Photofabrication of three-dimensional photonic crystals by multibeam laser interference into a photopolymerizable resin. Appl. Phys. Lett. 11 76, 2668-2670.

[39] Smith, D. R, Pendry, J. B, & Wiltshire, M. C. K. (2004). Metamaterials and negative refractive index. Science , 305, 788-792.

[40] Soukoulis, C. M, Linden, S, & Wegener, M. (2007). Physics: negative refractive index at optical wavelengths. Science , 315, 47-49.

[41] Su, H. M, Zhong, Y. C, Wang, X, Zheng, X. G, Xu, J. F, & Wang, H. Z. (2003). Effects of polarization on laser holography for microstructure fabrication. Phys. Rev. E 67, 056619

[42] Ullal, C. K, Maldovan, M, Thomas, E. L, Chen, G, Han, Y. J, & Yang, S. (2004). Photonic crystals through holographic lithography: Simple cubic, diamond-like, and gyroid-like structures. Appl. Phys. Lett. , 84, 5434-5436.

[43] Wang, G. P, Tan, C, Yi, Y, & Shan, H. (2003). Holography for one-step fabrication of three-dimensional metallodielectric photonic crystals with a single continuous wavelength laser beam. J. Mod. Opt. , 50, 2155-2161.

[44] Wang, L, Wang, X, Jiang, W, Choi, J, Bi, H, & Chen, R. T. (2005). 45° polymer-based total internal reflection coupling mirrors for fully embedded intraboard guided wave optical interconnects. Appl. Phys., 87, 141110

[45] Wang, X, Jiang, W, Wang, L, Bi, H, & Chen, R. T. (2008). Fully embedded board-level optical interconnects from waveguide fabrication to devices integration. J. Lightw. Technol. , 26, 243-250.

[46] Wang, X, Ng, C. Y, Tam, W. Y, Chan, C. T, & Sheng, P. (2003). Large-area two-dimensional mesoscale quasi-crystals. Adv. Mater. , 15, 1526-1528.

[47] Wang, X, Xu, J. F, Su, H. M, Zeng, Z. H, Chen, Y. L, Wang, H. Z, Pang, Y. K, & Tam, W. Y. (2003). Three-dimensional photonic crystals fabricated by visible light holographic lithography. Appl. Phys. Lett. , 82, 2212-2214.

[48] Wu, L, Zhong, Y, Chan, C. T, Wong, K. S, & Wang, G. P. (2005). Fabrication of large area two- and three-dimensional polymer photonic crystals using single refracting prism holographic lithography. Appl. Phys. Lett. 86, 241102

[49] Xu, D, Chen, K. P, Harb, A, Rodriguez, D, Lozano, K, & Lin, Y. K. (2009). Phase tunable holographic fabrication for three-dimensional photonic crystal templates by using a single optical element. Appl. Phys. Lett. 94, 231116

[50] Yablonovitch, E. (1987). Inhibited Spontaneous Emission in Solid-State Physics and Electronics. Phys. Rev. Lett. , 58, 2059-2062.

[51] Yang, Y, Li, Q, & Wang, G. P. (2008). Design and fabrication of diverse metamaterial structures by holographic lithography. Optics Express 16, 11275-11280.

[52] Yoon, Y. –K, Park, J. -H, & Allen, M. G. (2006) Multidirectional UV lithography for complex 3-D MEMS structures. IEEE J. Microelectromech. Syst. 15, 1121-1130.

[53] Yoon, Y. –K, Park, J. -H, Lee, J. -W, Prausnitz, M. R, & Allen, M. G. (2011). A thermal microjet system with tapered micronozzles fabricated by inclined UV lithography for transdermal drug delivery. J. Micromech. Microeng. 21 025014

[54] Zhou, J, Zhou, Y, Ng, S. L, Zhang, H. X, Que, W. X, Lam, Y. L, Chan, Y. C, & Kam, C. H. (2000). Three-dimensional photonic band gap structure of a polymer-metal com18 posite. Appl. Phys. Lett. 76, 3337

[55] Zhu, Z, Zhou, Z. -F, Huang, Q. -A, & Li, W. -H. (2008). Modeling simulation and experimental verification of inclined UV lithography for SU-8 negative thick photoresists. J. Micromech. Microeng. 18 125017